NEW PROGRESS ON GRAPHENE RESEARCH

Edited by **Jian Ru Gong**

New Progress on Graphene Research
http://dx.doi.org/10.5772/3358
Edited by Jian Ru Gong

Contributors

Alexander Feher, Eugen Syrkin, Sergey Feodosyev, Igor Gospodarev, Kirill Kravchenko, Fei Zhuge, Miroslav Pardy, Tong Guo-Ping, Victor Zalipaev, Michael Forrester, Dariush Jahani, Tao Tu, Wenge Zheng, Bin Shen, Wentao Zhai, Mineo Hiramatsu

Published by InTech

Janeza Trdine 9, 51000 Rijeka, Croatia

Publishing Process Manager Dejan Grgur
Technical Editor InTech DTP team
Cover InTech Design team

Additional hard copies can be obtained from orders@intechopen.com

New Progress on Graphene Research, Edited by Jian Ru Gong
p. cm.
ISBN 978-953-51-1091-0

For EU product safety concerns:
IN TECH d.o.o., Prolaz Marije Krucifikse Kozulić 3, 51000 Rijeka, Croatia,
info@intechopen.com or visit our website at intechopen.com.

Meet the editor

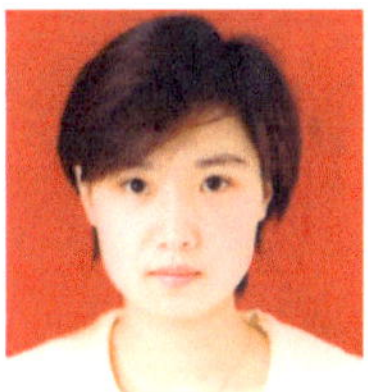

Jian Ru Gong is a Professor at National Center for Nanoscience and Technology, China. She received her PhD in Physical Chemistry from Institute of Chemistry, Chinese Academy of Sciences under the guidance of Prof. Chunli Bai (2005), and was a postdoctoral fellow with Prof. Charles M. Lieber at Harvard University (2005-2008). Her current research interest focuses on design and preparation of novel nanoscale materials and devices, and their applications for green renewable energy and biomedicine.

Contents

Preface

Graphene is a one-atom-thick and two-dimensional repetitive hexagonal lattice sp^2-hybridized carbon layer. The extended honeycomb network of graphene is the basic building block of other important allotropes of carbon. 2D graphene can be wrapped to form 0D fullerenes, rolled to form 1D carbon nanotubes, and stacked to form 3D graphite. Depending on its unique structure, graphene yields many excellent electrical, thermal, and mechanical properties. It has been interesting to both theoreticians and experimentalists in various fields, such as materials, chemistry, physics, electronics, and biomedicine, and great progress have been made in this rapid developing arena.

The aim of publishing this book is to present the recent new achievements about graphene research on a variety of topics. And the book is divided into two parts: Part I, from theoretical aspect, Graphene tunneling (Chapter 1), Localized states of Fabry-Perot type in graphene nanoribbons (Chapter 2), Electronic properties of deformed graphene nanoribbons (Chapter 3), The Čererenkov effect in graphene-like structures (Chapter 4), and Electronic and vibrational properties of adsorbed and embedded carbon nanofilms with defects (Chapter 5) are elaborated; Part II, from experimental aspect, Quantum transport in graphene quantum dots (Chapter 6), Advances in resistive switching memories based on graphene oxide (Chapter 7), Surface functionalization of graphene with polymers for enhanced properties (Chapter 8), and Carbon nanowalls: synthesis and applications (Chapter 9) are introduced. Also, in-depth discussions ranging from comprehensive understanding to challenges and perspectives are included for the respective topic. Each chapter is relatively independent of others, and the Table of Contents we hope will help readers quickly find topics of interest without necessarily having to go through the whole book.

Last, I appreciate the outstanding contributions from scientists with excellent academic records, who are at the top of their fields on the cutting edge of technology, to the book. Research related to graphene updates every day, so it is impossible to embody all the progress in this collection, and hopefully it could be of any help to people who are interested in this field.

Prof. Jian Ru Gong
National Center for Nanoscience and Technology, Beijing
P. R. China

Section 1

Theoretical Aspect

Chapter 1

Electronic Tunneling in Graphene

Dariush Jahani

Additional information is available at the end of the chapter

http://dx.doi.org/10.5772/51980

1. Introduction

In this chapter the transmission of massless and massive Dirac fermions across two-dimensional p-n and n-p-n junctions of graphene which are high enough so that they correspond to 2D potential steps and square barriers, respectively is investigated. It is shown that tunneling without exponential damping occurs when an relativistic particle is incident on a very high barrier. Such an effect has been described by Oskar Klein in 1929 [1] (for an historical review on klein paradox see [2]). He showed that in the limit of a high enough electrostatic potential barrier, it becomes transparent and both reflection and transmission probability remains smaller than one [3]. However, some later authors claimed that the reflection amplitude at the step barrier exceeds unity [4,5], implying that transmission probability takes the negative values.

Throughout this chapter, these negative transmission and higher-than-unity reflection probability is refereed to as the Klein paradox and not to the transparency of the barrier in the limit $V_0 \to \infty$ (V_0 is hight of the barrier). However, by considering the massless electrons tunneling through a potential step which can correspond to a p-n junction of graphene, as the main aim in the first section, it is be clear that the transmission and reflection probability both are positive and the Klein paradox is not then a paradox at all. Thus, one really doesn't need to associate the particle-antiparticle pair creation, which is commonly regarded as an explanation of particle tunneling in the Klein energy interval, to Klein paradox. In fact it will be revealed that the Klein paradox arises because of not considering a π phase change of the transmitted wave function of momentum-space which occurs when the energy of the incident electron is smaller than the height of the electrostatic potential step. In the other words, one arrives at negative values for transmission probability merely because of confusing the direction of group velocity with the propagation direction of particle's wave function or equivalently- from a two-dimensional point of view- the propagation angle with the angle that momentum vector under the electrostatic potential step makes with the normal incidence. Then our attentions turn to the tunneling of massless electrons into a barrier with the hight V_0 and width D. It will be found that the probability for an electron (approaching perpendicularly) to penetrate the barrier is equal to one, independent of V_0 and D. Although this result is very interesting from the point of view of fundamental research, its presence in graphene is unwanted when it comes to applications of graphene to nano-electronics because the pinch-off of the field effect transistors may be very ineffective. One way to overcome these difficulties is by generating a gap in the graphene spectrum. From the point of view of Dirac fermions this is equivalent to the appearing of a mass term in relativistic equation which describes the low-energy excitations of graphene, i.e. 2D the massive Dirac equation:

$$H = -iv_F\boldsymbol{\sigma}.\boldsymbol{\nabla} \pm \Delta\sigma^z \tag{1}$$

where Δ is equal to the half of the induced gap in graphene spectrum and it's positive (negative) sign corresponds to the K (K') point. Then the exact expression for T in gapped graphene is evaluated. Although the presence of massless electrons which is an interesting aspect of graphene is ignored, it"l be seen that how it can save us from doing the calculation once more with zero mass on both sides of the barrier, but non-zero mass inside the barrier. This might be a better model for two pieces of graphene connected by a semiconductor barrier (see fig. 6). Another result that show up is that the expression for T in the former case shows a dependence of transmission on the sign of refractive index, n, while in the latter case it will be revealed that T is independent from the sign of n.

From the above discussion and motivated by mass production of graphene, using 2D massive Dirac-like equation, in the next sections, the scattering of Dirac fermions from a special potential step of height V_0 which electrons under it acquire a finite mass, due to the presence of a gap of 2Δ in graphene spectrum is investigated [2], resulting in changing of it's spectrum from the usual linear dispersion to a hyperbolic dispersion and then show that for an electron with energy $E < V_0$ incident on such a potential step, the transmission probability turns out to be smaller than one in normal incident, whereas in the case of $\Delta \to 0$, this quantity is found to be unity. In graphene, a p-n junction could correspond to such a potential step if it is sharp enough [6-7].

Here it should be noted that for building up such a potential step, finite gaps are needed to be induced in spatial regions in graphene. One of the methods for inducing these gaps in energy spectra of graphene is to grow it on top of a hexagonal boron nitride with the B-N distance very close to C-C distance of graphene [8,9,10]. One other method is to pattern graphene nanoribbons.[11,12]. In this method graphene planes are patterned such that in several areas of the graphene flake narrow nanoribbons may exist. Here, considering the slabs with SiO_2-BN interfaces, on top of which a graphene flake is deposit, it is then possible to build up some regions in graphene where the energy spectrum reveals a finite gap, meaning that charge carriers there behave as massive Dirac fermions while there can be still regions where massless Dirac fermions are present. Considering this possibility, therefore, the tunneling of electrons of energy E through this type of potential step and also an electrostatic barrier of hight V_0 which allows quasi-particles to acquire a finite mass in a region of the width D where the dispersion relation of graphene exhibits a parabolic dispersion is investigated. The potential barrier considered here is such that the width of the region of finite mass and the width of the electrostatics barrier is similar. It will be observed that this kind of barrier is not completely transparent for normal incidence contrary to the case of tunneling of massless Dirac fermions in gapless graphene which leads to the total transparency of the barrier

[13,14]. As mentioned it is a real problem for application of graphene into nano-electronics, since for nano-electronics applications of graphene a mass gap in itŠs energy spectrum is needed just like a conventional semiconductor. We also see that, considering the appropriate wave functions in region of electrostatic barrier reveals that transmission is independent of whether the refractive index is negative or positive[15-17]. There is exactly a mistake on this point in the well-known paper "The electronic properties of graphene" [18].

In the end, throughout a numerical approach the consequences that the extra π-shift might have on the transmission probability and conductance in graphene is discussed [19].

2. Quantum tunneling

According to classical physics, a particle of energy E less than the height V_0 of a potential barrier could not penetrate it because the region inside the barrier is classically forbidden, whereas the wave function associated with a free particle must be continuous at the barrier and will show an exponential decay inside it. The wave function must also be continuous on the far side of the barrier, so there is a finite probability that the particle will pass through the barrier(Fig. 1). One important example based on quantum tunnelling is α-radioactivity which was proposed by Gamow [20-22] who found the well-known Gamow formula. The story of this discovery is told by Rosenfeld [23] who was one of the leading nuclear physicist of the twentieth century.

In the following, before proceeding to the case of massless electrons tunneling in graphene, we concern ourselves to evaluation of transmission probability of an electron incident upon a potential barrier with height much higher than the electron's energy.

2.1. Tunneling of an electron with energy lower than the electrostatic potential

For calculating the transmission probability of an electron incident from the left on a potential barrier of hight V_0 which is more than the value of energy as indicated in the Figure 1 we consider the following potential:

$$V(x) = \begin{cases} 0 & x < 0 \\ V_0 & 0 < x < w \\ 0 & x > w \end{cases} \tag{2}$$

For regions I, the solution of Schrodinger's equation will be a combination of incident and reflected plane waves while in region II, depending on the energy, the solution will be either a plane wave or a decaying exponential form.

$$\psi_I = e^{ikx} + re^{-ikx} \tag{3}$$

$$\psi_{II} = ae^{iqx} + be^{-iqx} \tag{4}$$

$$\psi_{III} = te^{ikx} \tag{5}$$

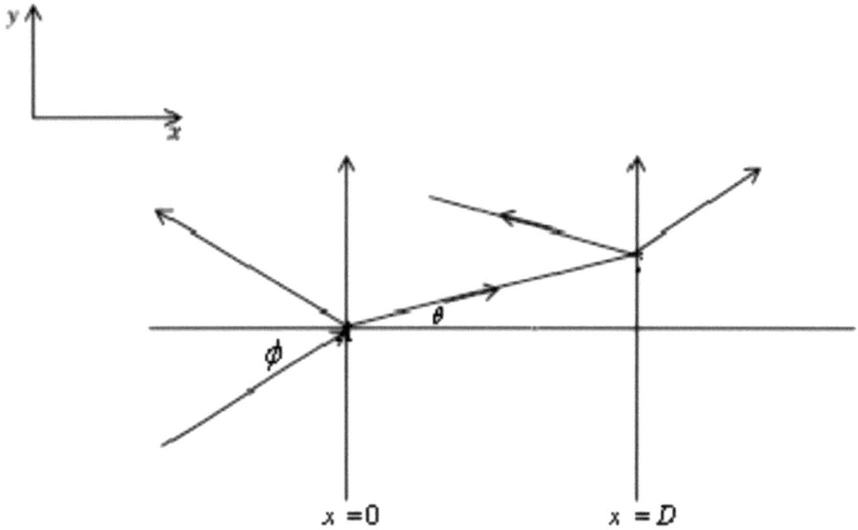

Figure 1. Schematic representation of tunneling in a 2D barrier.

where a, b, r, t are probability coefficients that must be determined from applying the boundary conditions. k and q are the momentum vectors in the regions I an II, respectively:

$$k = \sqrt{\frac{2mE}{\hbar^2}}, \tag{6}$$

$$q = \sqrt{\frac{2m(E - V_0)}{\hbar^2}}. \tag{7}$$

We know that the wave functions and also their first spatial derivatives must be continuous across the boundaries. Imposing these conditions yields:

$$\begin{cases} 1 + r = a + b \\ ik(1 - r) = iq(a - b) \\ ae^{iqD} + be^{-iqD} = te^{ikD} \\ iq(ae^{iqD} - be^{-iqD})a = ikte^{ikD} \end{cases} \tag{8}$$

The transmission amplitude, t is easily obtained:

$$t = \frac{4e^{-ikD}kq}{(q + k)^2 e^{-ikD} - (q - k)^2 e^{ikD}}, \tag{9}$$

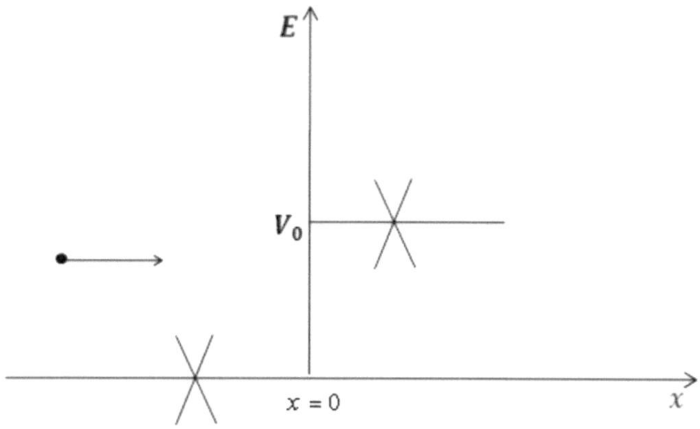

Figure 2. A p-n junction of graphene in which massless electrons incident upon an electrostatic region with no energy gap so that electrons in tunneling process have an effective mass equal to zero.

which from it the transmission probability T can be evaluated as:

$$T = |t|^2 = \frac{16k^2q^2}{(q+k)^2e^{-ikD} - (q-k)^2e^{ikD}}. \tag{10}$$

For energies lower than V_0, the wave decays exponentially as it passes through the barrier, since in this case q is imaginary. Also note that the perfect transmission happens at $qD = n\pi$ (n an integer). This resonance in transmission occurs physically because of instructive and destructive matching of the transmitted and reflected waves in the potential region. Now that we have got a insight on the quantum tunneling phenomena in non-relativistic limit, the next step is to extent our attentions to the relativistic case.

3. Massless electrons tunneling into potential step

Here, first a p-n junction of graphene which could be realized with a backgate and could correspond to a potential step of hight V_0 on which an massless electron of energy E is incident (see Fig 2) is considered. Two region, therefore, can be considered. The region for which $x < 0$ corresponding to a kinetic energy of E and the region corresponding to a kinetic energy of $E - V_0$. In order to obtain the transmission and reflection amplitudes, we first need to write down the following equation:

$$H = v_F \boldsymbol{\sigma}.\boldsymbol{p} + V(\mathbf{r}), \tag{11}$$

where

$$V(\mathbf{r}) = \begin{cases} V_0 & x > 0 \\ 0 & x < 0 \end{cases} \tag{12}$$

The above Dirac equation for $x > 0$ has the exact solutions which are the same as the free particle solutions except that the energy E can be different from the free particle case by the addition of the constant potential V_0. Thus, in the region II, the energy of the Dirac fermions is given by:

$$E = v_F\sqrt{q_x^2 + k_y^2} + V_0, \tag{13}$$

where **q** is the momentum in the region of electrostaic potential. The wave functions in the two regions can be written as:

$$\psi_I = \frac{1}{\sqrt{2}}\begin{pmatrix} 1 \\ \lambda e^{i\phi} \end{pmatrix} e^{i(k_x x + k_y y)} + \frac{r}{\sqrt{2}}\begin{pmatrix} 1 \\ \lambda e^{i(\pi-\phi)} \end{pmatrix} e^{i(-k_x x + k_y y)}, \tag{14}$$

and

$$\psi_{II} = \frac{t}{\sqrt{2}}\begin{pmatrix} 1 \\ \lambda' e^{i(\theta+\pi)} \end{pmatrix} e^{i(q_x x + k_y y)}, \tag{15}$$

where r and t are reflected and transmitted amplitudes, respectively, $\lambda' = sgn(E - V_0)$ is the band index of the wave function corresponding to the second region ($x > 0$) and $\phi = \arctan(\frac{k_y}{k_x})$ is the angle of propagation of the incident electron wave and $\theta = \arctan(\frac{k_y}{q_x})$ with

$$q_x = \pm\sqrt{[\frac{(V_0 - E)^2}{v_F^2}] - k_y^2}, \tag{16}$$

is the angle of the propagation of the transmitted electron wave[1] and not, as it should be, the angle that momentum vector **q** makes with the x-axis. The reason will be clear later.

The following set of equations are obtained, if one applies the continuity condition of the wave functions at the interface $x = 0$:

$$1 + r = t \tag{17}$$

$$\lambda e^{i\phi} - r\lambda e^{-i\phi} = \lambda' t e^{i\theta}, \tag{18}$$

which gives the transmission amplitude, t, as follows:

$$t = \frac{2\lambda\cos\phi}{\lambda' e^{i\theta} + \lambda e^{-i\phi}}. \tag{19}$$

Multiplying t by it's complex conjugate yields:

$$tt^* = \frac{2\cos^2\phi}{1 + \lambda\lambda' cos(\phi + \theta)}. \tag{20}$$

[1] By this definition θ falls in the range $-\frac{\pi}{2} < \theta < -\frac{\pi}{2}$.

Here it should be noted that the transmission probability, T, as we see later, is not simply given by tt^* unlike to the refraction probability, R, which is always equal to rr^*:

$$R = rr^* = \frac{1 - \lambda\lambda' cos(\phi - \theta)}{1 + \lambda\lambda' cos(\phi + \theta)}. \tag{21}$$

The reader can easily check that using the relation:

$$R + T = 1. \tag{22}$$

Physically the reason that T is not given by tt^* is because in the conservation law:

$$\nabla.\mathbf{j} + \frac{\partial}{\partial t}|\psi|^2, \tag{23}$$

which gives for the probability current

$$j = v_F \psi^\dagger \sigma \psi, \tag{24}$$

it is the probability current, $\mathbf{j}(x,y)$, that matters, which is not simply given by probability density $|\psi|^2$. The probability current also contains the velocity which means that if velocity changes between the incoming wave and the transmitted wave, T is not, therefore, given by $|t|^2$, however there is the ratio of the two velocities entering. Here, in order to find the transmission, since the system is translational invariant along the y-direction, we get

$$\nabla.\mathbf{j}(x,y) = 0, \tag{25}$$

which implies that:

$$j_x(x) = constant. \tag{26}$$

Hence one can write the following relation:

$$j_x^i + j_x^r = j_x^t, \tag{27}$$

where j_x^i, j_x^r and j_x^i denote the incident, reflected and transmission currents, respectively. From this equation it is obvious that:

$$1 = |r|^2 + |t|^2 \frac{\lambda\lambda' \cos\theta}{\sin\phi} \tag{28}$$

One can then obtain the transmission probability from the relation (R+T=1) as:

$$T = \frac{2\lambda\lambda' \cos\theta \cos\phi}{1 + \lambda\lambda' cos(\phi + \theta)}. \tag{29}$$

This equation shows that for an electron of energy $E > V_0$, the probability is positive and also less than unity, whereas for an electron of energy $E < V_0$, as in this case we have $\lambda = 1$ and $\lambda' = sgn(E - V_0) = -1$, we find that the probability is negative and therefore the reflection probability, R, exceeds unity as it is clear from (21). In fact the assumption of particle-antiparticle (in this case electron-hole) pair production at the interface was considered as an explanation of these higher-than-unity reflection probability and negative transmission and has been so often interpreted as the meaning of the Klein paradox. In particular, throughout this chapter, these features are refereed to as the Klein paradox.

Another odd result will be revealed, if we consider the normal incident of electrons upon the interface of the potential step. Assuming an electron propagating with propagation angle $\phi = 0$ on the potential step, we see that both R and T, in this case, become infinite which does not make sense at all because it would imply the existence of a hypothetical current source corresponding to the electron-hole pair creation at interface of the step. In other words no known physical mechanism can be associated to this results.

As it will be clear in what follows the negative T and higher than one reflection probability that equations (29) and (21) imply, arises from the wrong considered direction of the momentum vector, $\mathbf{q}$, of the wave function in the region II. In fact, in the case of $E < V_0$, momentum and group velocity v_g which is evaluated as:

$$\mathbf{v}_g = \frac{\partial E}{\partial q_x} = \frac{q_x}{E - V_0}, \tag{30}$$

have opposite directions because we assumed that the transmitted electron moves from left to right and therefore v_g must be positive implying that q_x has to assign it's negative value, meaning that the direction of momentum in the region II differs by 180 degree from the direction of which the wave packed propagates. In the other words in the case of $E < V_0$, the phase of the transmitted wave function in momentum-space undergoes a π change in transmitting from the region I to region II. Thus, the appropriate wave functions in the momentum space, ψ_{II}, is:

$$\psi_{II} = \frac{t}{\sqrt{2}} \begin{pmatrix} 1 \\ \lambda' e^{i(\theta+\pi)} \end{pmatrix}, \tag{31}$$

which from them T and R are given by:

$$T = -\frac{2\lambda\lambda' \cos\theta \cos\phi}{1 + \lambda\lambda' \cos(\phi + \theta)}. \tag{32}$$

$$R == \frac{1 + \lambda\lambda' \cos(\phi - \theta)}{1 - \lambda\lambda' \cos(\phi + \theta)}. \tag{33}$$

These expressions now reveal that both transmission and reflection probability are positive and less than unity. It also shows that if electron arrives perpendicularly upon the step, the probability to go through it is one which is is related to the well-known "absence of backscattering" [24] and is a consequence of the chirality of the massless Dirac electrons [25]. Notice that in the limit $V_0 >> E$, since in this case $q_x \to \infty$ and therefore $\theta \to 0$, transmission and reflection probability are:

$$T(\phi) = \frac{2\cos\phi}{1 + \cos\phi}, \tag{34}$$

and

$$R(\phi) = \frac{1 - \cos\phi}{1 + \cos\phi}. \tag{35}$$

As it is clear in the case of normal incident the p-n junction become totally transparent, i.e. $T(0) = 1$.

4. Ultra-relativistic tunneling into a potential barrier

In this section the scattering of massless electrons of energy E by a n-p-n junction of graphene which can correspond to a square barrier if it is sharp enough I address as depicted in figure 3. By writing the wave functions in the three regions as:

$$\psi_I = \frac{1}{\sqrt{2}} \begin{pmatrix} 1 \\ \lambda e^{i\phi} \end{pmatrix} e^{i(k_x x + k_y y)} + \frac{r}{\sqrt{2}} \begin{pmatrix} 1 \\ \lambda e^{i(\pi - \phi)} \end{pmatrix} e^{i(-k_x x + k_y y)}, \tag{36}$$

$$\psi_{II} = \frac{a}{\sqrt{2}} \begin{pmatrix} 1 \\ \lambda' e^{i\theta_t} \end{pmatrix} e^{i(q_x x + k_y y)} + \frac{b}{\sqrt{2}} \begin{pmatrix} 1 \\ \lambda' e^{\imath(\pi - \theta_t)} \end{pmatrix} e^{i(q_x x + k_y y)}, \tag{37}$$

$$\psi_{III} = \frac{t}{\sqrt{2}} \begin{pmatrix} 1 \\ \lambda e^{i\phi} \end{pmatrix} e^{i(k_x x + k_y y)}, \tag{38}$$

we'll be able to calculate T only by imposing the continuous condition of wave function at the boundaries and not it's derivative. Note that, in the case of $E < V_0$, $\theta_t = \theta + \pi$ is the angle of momentum vector $\mathbf{q}$, measured from the x-axis while θ is the angle of propagation of the wave packed and, therefore, shows the angle that group velocity, v_g, makes with the x-axis[2].

[2] Notice that if one consider the case $E > V_0$, one then see that $\theta_t = \theta$, implying that momentum and group velocity are parallel.

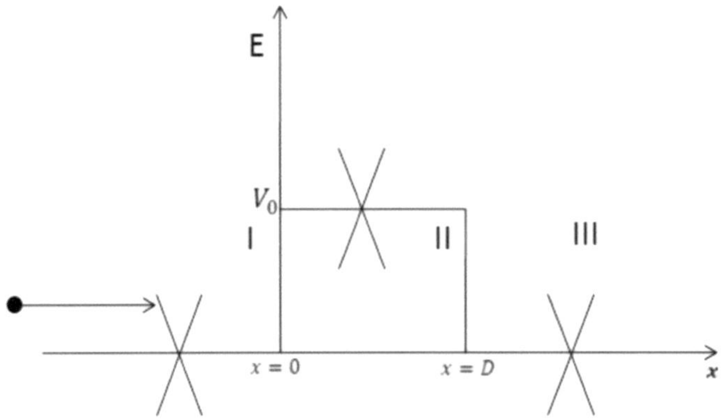

Figure 3. an one dimensional schematic view of a n-p-n junction of gapless graphene. In all three zones the energy bands are linear in momentum and therefore we have massless electrons passing through the barrier.

By applying the continuity conditions of the wave functions at the two discontinuities of the barrier ($x = 0$ and $x = D$), the following set of equations is obtained:

$$1 + r = a + b \tag{39}$$

$$\lambda e^{i\phi} - \lambda r e^{-i\phi} = \lambda' a e^{i\theta_t} - \lambda' b e^{-i\theta_t} \tag{40}$$

$$a e^{iq_x D} + b e^{-iq_x D} = t e^{ik_x D} \tag{41}$$

$$\lambda' a e^{i\theta_t + iq_x D} - \lambda' b e^{-i\theta_t - iq_x D} = \lambda t e^{i\phi + ik_x D}. \tag{42}$$

Here, as previous sections, the transmission amplitude in the first region (incoming wave) is set to 1. For solving the above system of equations with respect to transmission amplitude, t, we first determine a from (41) which turns out to be:

$$a = t e^{-iq_x D + ik_x D} - b e^{-2iq_x D}, \tag{43}$$

and then substituting it in equation (42), b can be evaluated as:

$$b = \frac{t e^{iq_x D + ik_x D}(\lambda' e^{i\theta_t} - \lambda e^{i\phi})}{2\lambda' \cos\theta_t} \tag{44}$$

Now equation (40) by the use of relation (39) could be rewritten as follows:

$$2\lambda \cos\phi = a(\lambda' e^{i\theta_t} + \lambda e^{-i\phi}) - b(\lambda' e^{-i\theta_t} - \lambda e^{-i\phi}). \tag{45}$$

Thus, by plugging a and b into this equation, after some algebraical manipulation t can be determined as:

$$t = -e^{-ik_x D} \frac{4\lambda\lambda' \cos\phi \cos\theta_t}{e^{iq_x D}[2 - 2\lambda\lambda' \cos(\phi - \theta_t)] - e^{-iq_x D}[2 + 2\lambda\lambda' \cos(\phi + \theta_t)]} \tag{46}$$

Up to now, we have only obtained the transmission amplitude and not transmission probability. One can multiply t, by itŠs complex conjugation and get the exact expression for the transmission probability of massless electrons as:

$$T(\phi) = \frac{\cos^2\phi \cos^2\theta_t}{(\cos\phi \cos\theta_t \cos(q_x D))^2 + \sin^2(q_x D)(1 - \lambda\lambda' \sin\phi \sin\theta_t)^2} \tag{47}$$

It is evident that $T(\phi) = T(-\phi)$ and for values of $q_x D$ satisfying the relation $q_x D = n\pi$, with n an integer, the barrier becomes totally transparent, as in this case we have $T(\phi) = 1$. Another interesting result will be obtained when we consider the scattering of an electron incident on the barrier with propagation angle $\phi = 0$ ($\phi \to 0$ leading to $\theta_t \to 0$ and π for the case of $E > V_0$ and $E < V_0$, respectively) which imply that, no matter what the value of $q_x D$ is, the barrier becomes completely transparent, i.e. T(0) = 1. However for applications of graphene in nano-electronic devices such as a graphene-based transistors this transparency of the barrier is unwanted, since the transistor can not be pinched off in this case, however, in the next section by evaluating the transmission probability of a n-p-n junction of graphene which quasi-particles can acquire a finite mass there, it will be clear that transmission is smaller than one and therefore suitable for applications purposes. Turning our attention back to expression (47), it is clear that if one considers the cases $E > V_0$ and $E < V_0$ with the same magnitude for x-component of momentum vector **q**, corresponding to same values for $|V_0 - E|$, would arrive at the same results for transmission probability, irrespective of whether the energy of incident electron is higher or smaller than the hight of the barrier[3]. This is a very interesting result because it shows that transmission is independent of the sign of refractive index n of graphene, since for the case of $E < V_0$ group velocity and the momentum vector in the region II have opposite directions and graphene, therefore, meets the negative refractive index. There is a mistake exactly on this point in [18]. In this paper the angle that momentum vector **q** makes with the x-axis have been confused with the propagation angle θ. In fact the negative sign of q_x have not been considered there and therefore expression for T which is written there as

$$T(\phi) = \frac{\cos^2\phi \cos^2\theta}{(\cos\phi \cos\theta \cos(q_x D))^2 + \sin^2(q_x D)(1 - \lambda\lambda' \sin\phi \sin\theta)^2}, \tag{48}$$

results in different values for probability when $|E - V_0|$ is the same for both cases of $E > V_0$ and $E < V_0$. In other words, the π phase change of the transmitted wave function

[3] Because if we assume that energy of incident electron is smaller than height of the barrier, the band index λ' assigns it's negative value, meaning that the transmission angle θ_t is $\theta_t = \theta + \pi$ and therefore we get $\sin\theta_t = -\sin\theta$.

in momentum-space in the latter case is not counted in. It is worth noticing that both expressions for normal incident lead to same result $T(0) = 1$.

For a very high potential barrier ($V_0 \to \infty$), we have $\theta \to 0$, π, and, therefore, we arrive at the following result for T:

$$T(\phi) = \frac{\cos^2\phi}{\cos^2\phi\cos^2(q_x D) + \sin^2(q_x D)} = \frac{\cos^2\phi}{1-\cos^2(q_x D)\sin^2\phi}, \tag{49}$$

which reveals that for perpendicular incidence the barrier is again totally transparent.

5. Tunnelling of massive electrons into a p-n junction

In the two previous sections the tunneling of massless Dirac fermions across p-n and n-p-n junctions was covered. In this section the massive electrons tunneling into a two dimensional potential step (n-p junction) of a gapped graphene which shows a hyperbolic energy spectrum unlike to the linear dispersion relation of a gapless graphene is discussed (see Fig. 4). The low energy excitations, therefore, are governed by the two dimensional massive Dirac equation. Thus, in order to calculate the transmission probability, we first need to obtain the eigenfunctions of the following Dirac equation which describes the massive Dirac fermions in gapped graphene so that we'll be able to write down the wave functions in different regions:

$$H = v_F \boldsymbol{\sigma}.\boldsymbol{p} \, + \Delta\sigma^z, \tag{50}$$

where 2Δ is the induced gap in graphene spectrum and $\boldsymbol{\sigma} = (\sigma^x, \sigma^y)$ with

$$\sigma^x = \begin{pmatrix} 0 & 1 \\ 1 & 0 \end{pmatrix}, \quad \sigma^y = \begin{pmatrix} 0 & -i \\ i & 0 \end{pmatrix}, \quad \sigma^z = \begin{pmatrix} 1 & 0 \\ 0 & -1 \end{pmatrix}, \tag{51}$$

the i=x,x,z, Pauli matrix. Now for obtaining the eigenfunctions one may rewrite the Hamiltonian as:

$$H = \begin{pmatrix} \Delta & v_F|\mathbf{p}|e^{-i\varphi_\mathbf{p}} \\ v_F|\mathbf{p}|e^{i\varphi_\mathbf{p}} & \Delta \end{pmatrix}, \tag{52}$$

where

$$\varphi_\mathbf{p} = \arctan(p_y/p_x). \tag{53}$$

As one can easily see the corresponding eigenvalues are given by:

$$E = \lambda\sqrt{\Delta^2 + v_F^2\mathbf{P}^2}, \tag{54}$$

http://dx.doi.org/10.5772/51980

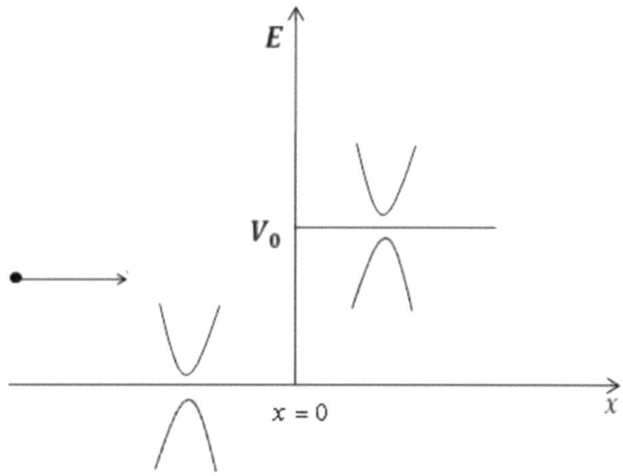

Figure 4. Massive Dirac electron tunneling into a step potential of graphene. As it is clear an opening gap in graphene spectrum makes electrons to acquire an effective mass of $\Delta/2v_F^2$ in both regions

where $\lambda = \pm$ correspond to the positive and negative energy states, respectively. Now in order to obtain the eigenfunctions, one can make the following ansatz:

$$\psi_{\lambda,\mathbf{k}} = \frac{1}{\sqrt{2}} \begin{pmatrix} u_\lambda \\ v_\lambda \end{pmatrix} e^{i(k_x x + k_y y)}, \tag{55}$$

where we've used units such that $\hbar = 1$. Plugging the above spinors into the corresponding eigenvalue equation then gives:

$$u_\lambda = \sqrt{1 + \frac{\lambda\Delta}{\sqrt{\Delta^2 + v_F^2 k^2}}}, \quad v_\lambda = \lambda \sqrt{1 - \frac{\lambda\Delta}{\sqrt{\Delta^2 + v_F^2 k^2}}} e^{i\varphi_k}. \tag{56}$$

The wave functions, therefore are given by:

$$\psi_{\lambda,\mathbf{k}} = \frac{1}{\sqrt{2}} \begin{pmatrix} \sqrt{1 + \frac{\lambda\Delta}{\sqrt{\Delta^2 + v_F^2 k^2}}} \\ \lambda \sqrt{1 - \frac{\lambda\Delta}{\sqrt{\Delta^2 + v_F^2 k^2}}} e^{i\varphi_\mathbf{k}} \end{pmatrix} e^{i(k_x x + k_y y)}. \tag{57}$$

It is clear that in the limit $\Delta \to 0$, one arrives at the same eigenfunctions

$$\psi_{\lambda,\mathbf{k}} = \frac{1}{\sqrt{2}} \begin{pmatrix} 1 \\ \lambda e^{i\varphi_\mathbf{k}} \end{pmatrix} e^{i(k_x x + k_y y)}, \tag{58}$$

as those of massless Dirac fermions in graphene.

Now that we have found the corresponding eigenfunctions of Hamiltonian (4.52), assuming an electron incident upon a step of height V_0, we can write the single valley Hamiltonian as:

$$H = v_F \boldsymbol{\sigma}.\boldsymbol{p} + \Delta\sigma^z + V(\mathbf{r}), \tag{59}$$

where $V(\mathbf{r}) = 0$ for region I ($x < 0$) and for the region II ($x > 0$), massive Dirac fermions feel a electrostatic potential of hight V_0 with the kinetic energy $E - V_0$. The wave functions in the two regions then are:

$$\psi_I = \frac{1}{\sqrt{2}} \begin{pmatrix} \alpha \\ \gamma\lambda e^{i\phi} \end{pmatrix} e^{i(k_x x + k_y y)} + \frac{r}{\sqrt{2}} \begin{pmatrix} \alpha \\ \gamma\lambda e^{i(\pi-\phi)} \end{pmatrix} e^{i(-k_x x + k_y y)} \tag{60}$$

and

$$\psi_{II} = \frac{t}{\sqrt{2}} \begin{pmatrix} \beta \\ \lambda'\eta e^{i\theta_t} \end{pmatrix} e^{i(q_x x + k_y y)}, \tag{61}$$

where in order to make things more simple, the following abbreviations is introduced:

$$\alpha = \sqrt{1 + \frac{\lambda\Delta}{\sqrt{\Delta^2 + v_F^2(k_x^2 + k_y^2)}}}, \quad \gamma = \sqrt{1 - \frac{\lambda\Delta}{\sqrt{\Delta^2 + v_F^2(k_x^2 + k_y^2)}}}, \tag{62}$$

$$\beta = \sqrt{1 + \frac{\lambda'\Delta}{\sqrt{\Delta^2 + v_F^2(q_x^2 + k_y^2)}}}, \quad \eta = \sqrt{1 - \frac{\lambda'\Delta}{\sqrt{\Delta^2 + v_F^2(q_x^2 + k_y^2)}}}. \tag{63}$$

Imposing the continuity conditions of ψ_I and ψ_{II} at the interface leads to the following system of equations:

$$\alpha + \alpha r = \beta t, \tag{64}$$

$$\lambda\gamma e^{i\phi} - \lambda\gamma r e^{-i\phi} = \lambda'\eta t e^{i\theta_t}, \tag{65}$$

which solving them with respect to r and t gives

$$r = \frac{\lambda e^{i\phi} - \lambda' \frac{\alpha\eta}{\beta\gamma} e^{i\theta_t}}{\lambda' \frac{\alpha\eta}{\beta\gamma} e^{i\theta_t} + \lambda e^{-i\phi}}, \tag{66}$$

and

$$t = \frac{2\lambda \cos\phi}{\frac{\eta}{\gamma}\lambda' e^{i\theta_t} + \frac{\beta}{\alpha}\lambda e^{-i\phi}}. \tag{67}$$

From (1.66) it is straightforward to show that R is:

$$R = \frac{N_r - 2\lambda\lambda' S_r \cos(\phi - \theta_t)}{N_r + 2\lambda\lambda' S_r \cos(\phi + \theta_t)}, \tag{68}$$

where

$$\begin{aligned} N_r &= \frac{\beta^2\gamma^2 + \alpha^2\eta^2}{\beta^2\gamma^2} \\ &= 2\frac{E|V_0 - E| - \lambda\lambda'\Delta^2}{E|V_0 - E| - \lambda\lambda'\Delta^2 - \lambda|V_0 - E|\Delta + \lambda' E\Delta} \\ &= 2\frac{E|V_0 - E| - \lambda\lambda'\Delta^2}{(|V_0 - E| + \lambda'\Delta)(E - \lambda\Delta)} \end{aligned} \tag{69}$$

and

$$\begin{aligned} S_r &= \frac{\alpha\eta}{\beta\gamma} \\ &= \frac{E|V_0 - E| - \lambda\lambda'\Delta^2 + \lambda' E\Delta - \lambda|V_0 - E|\Delta}{E|V_0 - E| - \lambda\lambda'\Delta^2 - \lambda' E\Delta + \lambda|V_0 - E|\Delta} \\ &= \frac{(|V_0 - E| + \lambda'\Delta)(E - \lambda\Delta)}{(|V_0 - E| - \lambda'\Delta)(E + \lambda\Delta)} \end{aligned} \tag{70}$$

In the limit $\Delta \to 0$ we get the same reflection as that of massless case. In the limit of no electrostatic potential we arrive at the logical result $R = 0$. This is important because we see later that for a special potential step in this limit R is not zero. Now one remaining problem is to calculate the transmission probability. So, considering equation (67) and:

$$j_x^{in} = \lambda\alpha\gamma\cos\phi, \quad j_x^r = -\lambda\alpha\gamma\cos\phi, \quad j_x^t = \lambda'\eta\beta\cos\theta_t \tag{71}$$

T is found to be:

$$\begin{aligned} T &= |t|^2 \frac{\lambda\lambda'\eta\beta\cos\theta_t}{\alpha\gamma\cos\phi} \\ &= \frac{4\lambda\lambda' S_t \cos\phi\cos\theta_t}{N_t + 2S_t\lambda\lambda'\cos(\phi + \theta_t)}, \end{aligned} \tag{72}$$

where the following abbreviations is defined:

$$S_t = \frac{\eta\beta}{\alpha\gamma} = \left[\frac{v_F^2 q^2}{\Delta^2 + v_F^2 q^2}\frac{\Delta^2 + v_F^2 k^2}{v_F^2 k^2}\right]^{\frac{1}{2}} = \frac{q}{k}\frac{E}{|V_0 - E|}, \tag{73}$$

and

$$N_t = \frac{\eta^2\alpha^2 + \beta^2\gamma^2}{\alpha^2\gamma^2} = 2\frac{E(E|V_0 - E| - \lambda\lambda'\Delta^2)}{v_F^2 k^2 |V_0 - E|}. \tag{74}$$

At this point one can obtain $T(0)$ as follows:

$$T(0) = 2\frac{v_F^2|k_x||q_x|}{E|V_0 - E| - \lambda\lambda'\Delta^2 + v_F^2|k_x||q_x|}. \tag{75}$$

Note that S_t and N_t are positive. It is clear that in the case of $V_0 \to 0$ and $V_0 \to \infty$ T is one. Also note that in the limit of $\Delta \to 0$, as:

$$E|V_0 - E| = v_F^2|k_x||q_x|, \tag{76}$$

we see that probability is unity in agreement with result obtained for massless case. Another interesting result that expression for T shows is that probability is not independent of the band index contrary to the a gapless step that leaded to no independency to band index, λ and λ'.

6. The barrier case

Opening nano-electronic opportunities for graphene requires a mass gap in it's energy spectrum just like a conventional semiconductor. In fact the lack of a bandgap on graphene, can limit graphene's uses in electronics because if there is no gaps in graphene spectrum one can't turn off a graphene-made transistor . In this section, motivated by mass production of graphene, we obtain the exact expression for transmission probability of massive Dirac fermions through a two dimensional potential barrier which can correspond to a n-p-n junction of graphene, and show that contrary to the case of massless Dirac fermions which results in complete transparency of the potential barrier for normal incidence, the probability transmission, T, in this case, apart from some resonance conditions that lead to the total transparency of the barrier, is smaller than one. An interesting result is that in the case of q_x satisfy the relation $q_x D = n\pi$, where n is an integer, we again see that tunneling is easier for a barrier than a potential step, i.e the resonance tunneling is occurred.

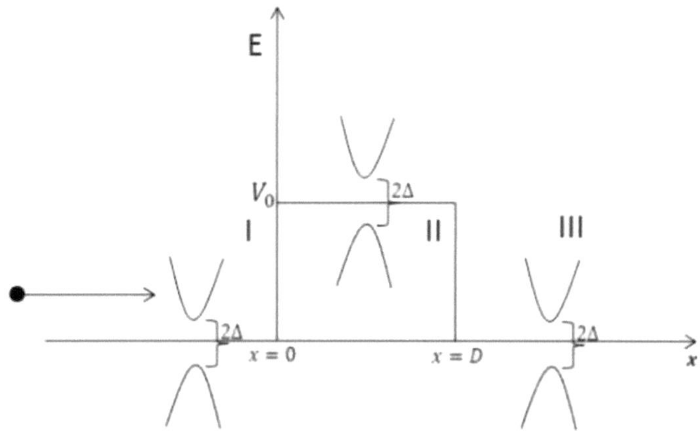

Figure 5. An massive electron of energy E incident on a potential barrier of hight V_0 and thickness of about 50 nm. The opening gap in the all three zones are of the same value and therefore the tunneling phenomenon occurs in a symmetric barrier.

As depicted in the figure 5 there are three regions. The first is for $x < 0$ where the potential is equal to zero. The second region is for $0 < x < D$ where there is a electrostatic potential of hight V_0 and finally, the third region is defined for $x > 0$ and as well as the first region we have $V_0 = 0$. At this point, using equations of previous sections, we are able to write the wave functions in these three different regions in terms of incident and reflected waves. The wave function in region I is then given by:

$$\psi_I = \frac{1}{\sqrt{2}} \begin{pmatrix} \alpha \\ \lambda\gamma e^{i\phi} \end{pmatrix} e^{i(k_x x + k_y y)} + \frac{r}{\sqrt{2}} \begin{pmatrix} \alpha \\ \lambda\gamma e^{i(\pi-\phi)} \end{pmatrix} e^{i(-k_x x + k_y y)}. \quad (77)$$

In the second region we have:

$$\psi_{II} = \frac{a}{\sqrt{2}} \begin{pmatrix} \beta \\ \lambda' \eta e^{i\theta_t} \end{pmatrix} e^{i(q_x x + k_y y)} + \frac{b}{\sqrt{2}} \begin{pmatrix} \beta \\ \lambda' \eta e^{i(\pi-\theta_t)} \end{pmatrix} e^{i(-q_x x + k_y y)}. \quad (78)$$

In the third region we have only a transmitted wave and therefore the wave function in this region is:

$$\psi_{III} = \frac{t}{\sqrt{2}} \begin{pmatrix} \alpha \\ \lambda\gamma e^{i\phi} \end{pmatrix} e^{i(k_x x + k_y y)} \quad (79)$$

With the continuity of the spinors at the discontinuities, we arrive at the following set of equations:

$$\alpha + \alpha r = \beta a + \beta b \quad (80)$$

$$\lambda\gamma e^{i\phi} - \lambda\gamma r e^{-i\phi} = \eta\lambda' a e^{i\theta_t} - \eta\lambda' b e^{-i\theta_t} \quad (81)$$

$$\beta a e^{iq_x D} + \beta b e^{-iq_x D} = \alpha t e^{ik_x D} \tag{82}$$

$$\eta\lambda' a e^{i\theta_t + iq_x D} - \eta\lambda' b e^{-i\theta_t - iq_x D} = \gamma\lambda t e^{i\phi + ik_x D} \tag{83}$$

Here in order to obtain the transmission T we first solve the above set of equations with respect to transmission amplitude t. So we first need to calculate the coefficients r,a, and b. From (82), a can be written as follows:

$$a = \frac{\alpha}{\beta} t e^{-iq_x D + ik_x D} - b e^{-2iq_x D}, \tag{84}$$

which writing it with respect to transmission amplitude requires to plug b which one can obtain it using the equation (83) as:

$$b = \frac{t e^{iq_x D + ik_x D}(\lambda' \frac{\alpha\eta}{\beta} e^{i\theta_t} - \lambda\gamma e^{i\phi})}{2\lambda'\eta\cos\theta_t}, \tag{85}$$

into the corresponding equation for a. Rewriting (81) by the use of relation $\alpha + \alpha r = \beta a + \beta b$ as:

$$2\lambda\cos\phi = a(\lambda' \frac{\eta}{\gamma} e^{i\theta_t} + \lambda \frac{\beta}{\alpha} e^{-i\phi}) - b(\lambda' \frac{\eta}{\gamma} e^{-i\theta_t} - \lambda \frac{\beta}{\alpha} e^{-i\phi}), \tag{86}$$

and then using the equations (85) and (86), the expression for transmission amplitude yields:

$$t = \frac{-4e^{-ik_x D}\lambda\lambda'\cos\phi\cos\theta}{[e^{iq_x D}(N - 2\lambda\lambda'\cos(\phi-\theta)) - e^{-iq_x D}(N + 2\lambda\lambda'\cos(\phi+\theta))]}, \tag{87}$$

where

$$N = \frac{\eta\alpha}{\beta\gamma} + \frac{\beta\gamma}{\eta\alpha}. \tag{88}$$

It is straightforward to show that:

$$N = 2\frac{E|V_0 - E| - \lambda\lambda'\Delta^2}{v_F^2 kq}, \tag{89}$$

where

$$E = \sqrt{\Delta^2 + v_F^2(k_x^2 + k_y^2)} \tag{90}$$

$$|V_0 - E| = \sqrt{\Delta^2 + v_F^2(q_x^2 + k_y^2)} \tag{91}$$

$$k = \sqrt{k_x^2 + k_y^2} \tag{92}$$

$$q = \sqrt{q_x^2 + k_y^2}. \tag{93}$$

Finally by multiplying t by it's complex conjugation, one can obtain the exact expression for the probability transmission of massive electrons, T, as:

$$T(\phi) = \frac{\cos^2\phi\cos^2\theta}{(\cos\phi\cos\theta\cos(q_x D))^2 + \sin^2(q_x D)(\frac{N}{2} - \lambda\lambda'\sin\phi\sin\theta)^2}. \tag{94}$$

It is clear that in the Klein energy interval ($0 < E < V_0$), λ and λ' has opposite signs so that the term $N/2$ in the above expression is bigger than one and, therefore, we see that unlike to the case of massless Dirac fermions which results in complete transparency of the potential barrier for normal incidence, the transmission T for massive quasi-particles in gapped graphene is smaller than one something that is of interest in a graphene transistor. It is obvious that substituting Δ with $-\Delta$ does not change the T, and hence the result for the both Dirac points is the same, as it should be.

Now considering an electron incident on the barrier with propagation angle $\phi = 0$, we know that θ_t becomes 0 (π), depending on the positive (negative) sign of λ'. So in the normal incidence probability reads:

$$T(0) = \frac{2}{2 + (N-2)\sin^2(q_x D)} \tag{95}$$

Now if the following condition is satisfied:

$$q_x D = n\frac{\pi}{2}, \tag{96}$$

the equation for probability results in:

$$T(0) = \frac{2}{N} = \frac{v_F^2|k_x||q_x|}{E|V_0 - E| - \lambda\lambda'\Delta^2} \tag{97}$$

At this point it is so clear that the transmission depends on the sign of $\lambda\lambda' = \pm$. In the other words, this equation for the same values of $|V_0 - E|$, depending on whether E is higher or smaller than V_0, results in different values for T. The result that have not been revealed before. In the limit $|V_0| >> |E|$, the exact expression obtained for transmission would be simplified to:

$$T(\phi) \simeq \frac{\cos^2\phi}{1 - \sin^2\phi\cos^2(q_x D)} \tag{98}$$

which reveals that in this limit, $T(0)$ is again smaller than one while in the case of $q_x D$ satisfying the condition $q_x D = n\pi$, with n an integer, we still have complete transparency. Furthermore from equations (90) to (93) it is clear that in the limit $\Delta \to 0$, we get $N/2 = 1$ and, therefore, one arrives at the same expressions for $T(\phi)$ corresponding to the case of massless Dirac fermions i.e. equations (48) and (49). Notice that there is transmission resonances just like other barriers studied earlier. It is important to know that resonances occur when a p-n interface is in series with an n-p interface, forming a p-n-p or n-p-n junction.

7. Transmission into spatial regions of finite mass

In this section the transmission of massless electrons into some regions where the corresponding energy dispersion relation is not linear any more and exhibits a finite gap of Δ is discussed. Thus, the mass of electrons there can be obtained from the relation $mv_F^2 = \Delta$. Starting by looking at a two demential square potential step and after obtaining the probability of penetration of step by electrons, transmission of massless electrons into a region of finite mass is investigated and then see how it turns out to be applicable in a transistor composed of two pieces of graphene connected by a conventional semiconductor or linked by a nanotube.

7.1. Tunnelling through a composed p-n junction

In this section the scattering of an electron of energy E from a potential step of hight V_0 which allows massless electrons to acquire a finite mass in the region of the electrostatic potential is investigated(see Fig. 6). The electrostatic potential under the region of finite mass is:

$$V(\mathbf{r}) = \begin{cases} 0 & \mathrm{x} < 0 \\ V_0 & 0 < \mathrm{x} < D \\ 0 & \mathrm{x} > D \end{cases} \tag{99}$$

Assuming an electron of energy E, propagating from the left, the wave functions then in the two zones can be written as:

$$\psi_I = \frac{1}{\sqrt{2}} \begin{pmatrix} 1 \\ \lambda e^{i\phi} \end{pmatrix} e^{i(k_x x + k_y y)} + \frac{r}{\sqrt{2}} \begin{pmatrix} 1 \\ \lambda e^{i(\pi - \phi)} \end{pmatrix} e^{i(-k_x x + k_y y)} \tag{100}$$

$$\psi_{II} = \frac{t}{\sqrt{2}} \begin{pmatrix} \beta \\ \lambda' \eta e^{i\theta_t} \end{pmatrix} e^{i(q_x x + k_y y)} \tag{101}$$

where

$$\beta = \sqrt{1 + \frac{\lambda' \Delta}{\sqrt{\Delta^2 + v_F^2(q_x^2 + k_y^2)}}}, \quad \eta = \sqrt{1 - \frac{\lambda' \Delta}{\sqrt{\Delta^2 + v_F^2(q_x^2 + k_y^2)}}}, \tag{102}$$

and r and t are reflected and transmitted amplitudes, respectively. Applying the continuity conditions of the wave functions at $x = 0$ yields:

$$1 + r = \beta t \tag{103}$$

$$\lambda e^{i\phi} - r\lambda e^{-i\phi} = \lambda' \eta t e^{i\theta_t} \tag{104}$$

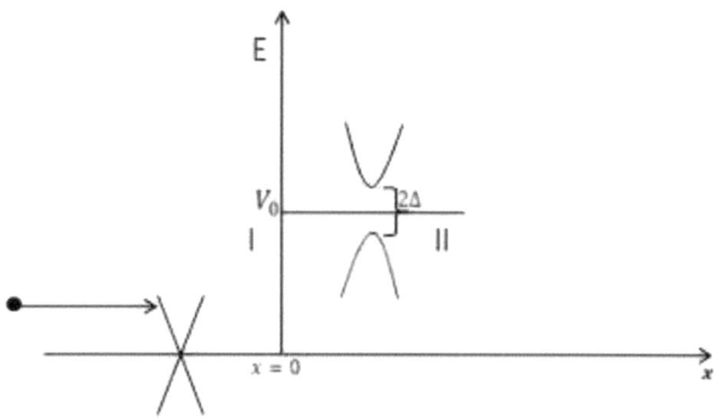

Figure 6. A special potential step of height V_0 and width D which massless electrons of energy E under it acquire a finite mass.

Solving the above equations gives us the following expression for $|t|^2$ and R:

$$|t|^2 = \frac{2\cos^2\phi}{1 + \lambda\lambda'\eta\beta cos(\phi + \theta_t)}, \tag{105}$$

and

$$R = rr^* = \frac{1 - \lambda\lambda'\eta\beta cos(\phi - \theta_t)}{1 + \lambda\lambda'\eta\beta cos(\phi + \theta_t)} \tag{106}$$

where

$$\eta\beta = \left[\frac{v_F^2(q_x^2 + k_y^2)}{v_F^2(q_x^2 + k_y^2) + \Delta^2}\right]^{\frac{1}{2}} = \frac{v_F q}{|V_0 - E|} \tag{107}$$

For obtaining the transmission probability we need to evaluate the x-component of probability current in two regions. Using equation (24) we get:

$$j_x^{in} = \lambda\cos\phi \tag{108}$$

$$j_x^r = -\lambda\cos\phi|r|^2 \tag{109}$$

$$j_x^t = \lambda'\eta\beta\cos\theta_t|t|^2. \tag{110}$$

Here notice that, using the probability conservation law and the fact that our problem is time independent and invariant along the y-direction, j_x, then has the same values in the two regions. So by the use of relation (27) the following equation come outs:

$$1 - |r|^2 = \frac{\lambda\lambda'\eta\beta\cos\theta_t}{\cos\phi}|t|^2, \tag{111}$$

which once again shows that the probability, T, is not given by $|t|^2$ and instead is:

$$T = \frac{\lambda\lambda'\eta\beta\cos\theta_t}{\cos\phi}|t|^2. \tag{112}$$

The probability, therefore, is given by:

$$T(\phi) = \frac{2\lambda\lambda'\eta\beta\cos\theta_t\cos\phi}{1+\lambda\lambda'\eta\beta cos(\phi+\theta_t)}. \tag{113}$$

This result shows that the relation $T(\phi) = T(-\phi)$. Thus, the induced gap in graphene spectrum has nothing to do with relation this relation. We now turn our attention to the case in which an electron is incident perpendicularly upon the step. The probability for this electron to penetrate the step is:

$$\begin{aligned} T(0) &= \frac{2\eta\beta}{1+\eta\beta} \\ &= \frac{2v_F|q_x|}{|V_0 - E| + v_F|q_x|}, \end{aligned} \tag{114}$$

which shows there is no way for the electron to pass into the step with probability equal to one. However if we consider a potential step which is high enough so that we'll be able to write

$$|V_0 - E| = \sqrt{v_F^2 q_x^2 + \Delta^2} \approx v_F|q_x|, \tag{115}$$

we see the step becomes transparent. So by increasing the potential's hight, more electrons can pass through the step. Notice that probability is independent of $\lambda\lambda'$ unlike to the result (72) [19]. Also note that in the limit $\Delta \to 0$, q_x we can write:

$$v_F|q_x| = |V_0 - E| \tag{116}$$

which immediately gives $T(0)$ as:

$$T(0) = 1, \tag{117}$$

Also note that since for normal incidence we have $E = v_F k_x$, from the equation (114) it is evident that in the case of no electrostatic potential ($V_0 = 0$) we get:

$$T = \frac{2q_x}{k_x + q_x} \quad , \quad R = \frac{k_x - q_x}{k_x + q_x}, \tag{118}$$

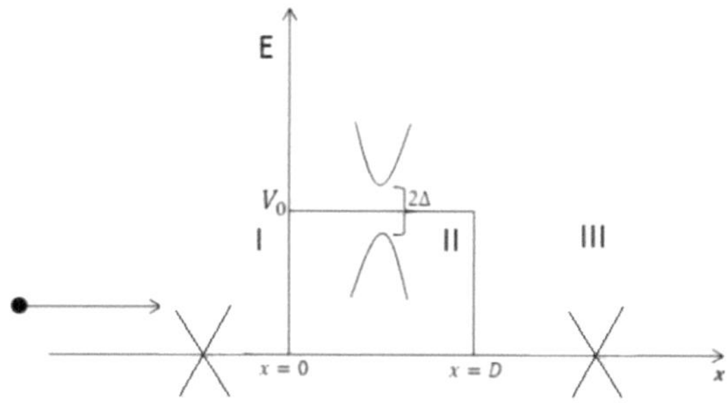

Figure 7. An massless electron of energy E incident (from the left) on a potential barrier of height V_0 and width D, which acquires a finite mass under the electrostatic potential, due to the presence of a gap of 2Δ in the region II. The effective mass of electron in this region is then $m = \Delta/v_F^2$

which shows that probability always remains smaller than one, as there is no way for k and q to be equal[4]. Turning our attention back to equation (113), we see that in the limit $\Delta \to 0$ one arrives at the following solution for T:

$$T = \frac{2\lambda\lambda' \cos\theta_t \cos\phi}{1 + \lambda\lambda' cos(\phi + \theta)}, \tag{119}$$

which is just the transmission of massless Dirac fermions through a p-n junction in gapless graphene. This expression now reveals in the limit $V_0 >> E \approx \Delta$ it can be simplified to the following equation

$$T = \frac{2\cos\phi}{1 + cos\phi} \quad , \quad R = \frac{1 - \cos\phi}{1 + cos\phi} \tag{120}$$

which show that for normal incidence the transmission and reflection probability are unity and zero, respectively.

Here, before proceeding to some numerical calculations in order to depict consequences that the π phase change might have on the probability, I attract the reader's attention to this fact that, the phase change of the wave function in momentum space is equivalent to the rotation of momentum vector, **q** by 180 degree, meaning that the direction of momentum and group velocity is antiparallel which itself lead to negative refraction in graphene reported by Cheianov [26,27]. As it clear for imaginary values of q_x an evanescent wave is created in the zone I and a total reflection is observed.

Now, before ending, in order to emphasize on the importance of the π-phase change mentioned earlier some numerical calculations depicting the transmission probability is shown in Fig. 8 which reveal a perceptible difference between result obtained based on considering the $\pi - shift$ and those obtained if one ignores it. As it is clear for an electron of energy $E = 85meV$, barrier thickness of $100nm$ and height of $V_0 = 200meV$ the probability gets smaller values if the extra phase is not considered. This means that considering the

[4] There is no need to say that when there is no electrostatic potential q_x is positive

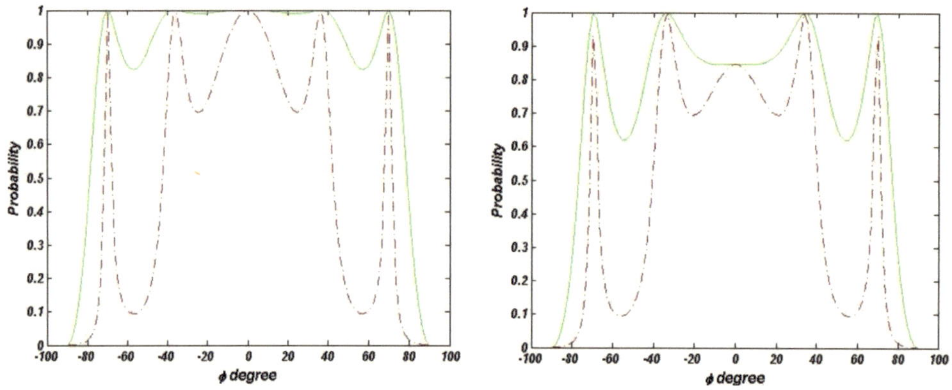

Figure 8. left: Transmission probability as a functions of incident angle for an electron of energy $E = 85meV$, $D = 100nm$ and $V_0 = 200meV$. Right: Transmission in gapped graphene for gap value of $20meV$ as a functions of incident angle for an electron of energy $E = 85meV$, $D = 100nm$ and $V_0 = 200meV$.

Büttiker formula [28] for conductivity lower conductance is predicted in absence of the extra phase. As it is clear the chance for an electron to penetrate the barrier increases if one chooses the appropriate wave function in the barrier.

The potential application of the theory of extra π phase consideration introduced in the previous sections [19] is that we can have higher conductivity in graphene-based electronic devices and also the results of this work is important in combinations of graphene flakes attached with different energy bands in order to get different kind of n-p-n junctions for different uses. Notice that for nanoelectronic application of graphene the existence of a mass gap in graphene's spectrum is essential because it leads to smaller than one transmission which is of most important for devices such as transistors and therefore the results derived in this work concerning gapped graphene could be applicable in nanoelectronic applications of graphene.

In the end of this chapter I would like to remind that one important result that obtained is that Klein paradox is not a paradox at all. More precisely, it was demonstrated theoretically that the reflection and transmission coefficients of a step barrier are both positive and less than unity, and that the hypothesis of particle-antiparticle pair production at the potential step is not necessary as the experimental evidences confirm this conclusion [29].

Author details

Dariush Jahani

Young Researchers Club, Kermanshah Branch, Islamic Azad University, Kermanshah, Iran

8. References

[1] O. Klein. Z. Phys. 53. 157 (1929).

[2] A. Calogeracos and N. Dombey, Contemp. Phys. 40, 313 (1999).

[3] D. Dragoman, ArXiv quant ph/0701083.

[4] A. Hansen and F. ravndal, Phys. scr. 23, 1036-1402 (1981)

[5] P. Krekora, Q. su, and R. Grobe, Phys. Rev. Lett. 92,040406-4 (2004).

[6] B. Huard et al., Phys. Rev. Lett. 98, 236803 (2007).

[7] N. Stander et al., Phys. Rev. Lett. 102, 026807 (2009).

[8] Gianluca Giovannetti, Petr A. Khomyakov, Geert Brocks, Paul J. Kelly, and Jeroen van den Brink, Phys. Rev. B 76, 73103 (2007).

[9] J. Zupan, Phys. Rev. B 6, 2477 (1972).

[10] J. Viana Gomes and N. M. R. Peres, J. Phys.: Condens. Matter 20, 325221 (2008).

[11] Young-Woo Son, Marvin L. Cohen, and Steven G. Louie, Phys. Rev. Lett. 97, 216803 (2006).

[12] Qimin Yan, Bing Huang, Jie Yu, Fawei Zheng, Ji Zang, Jian Wu, Bing-Lin Gu, Feng Liu, and Wenhui Duan, NanoLetters 6, 1469 (2007).

[13] Katsnelson, M. I., K. S. Novoselov, and A. K. Geim, Nature Physics 2, 620 (2006).

[14] Katsnelson, M. I., and K. S. Novoselov, Sol. Stat. Comm. 143, 3 (2007).

[15] M. Allesch, E. Schwegler, F. Gygi, G. Galli, J. Chem. Phys. 120, 5192 (2004).

[16] C. Millot, A. J. Stone, Mol. Phys. 77, 439 (1992).

[17] E. M. Mas et al., J. Chem. Phys. 113, 6687 (2000).

[18] A. Castro-Neto et al., Rev. Mod. Phys. 81, 109 (2009).

[19] Setare M R and Jahani D 2010 J. Phys.: Condens. Matter 22 245503

[20] G. Gamow, Quantum theory of atomic nucleus , Z. f. Phys. 51, 204 (1928).

[21] G. Gamow, Quantum theory of nuclear disintegration , Nature 122. 805 (1928).

[22] G. Gamow, Constitution of Atomic Nuclei and Radioactivity (Oxford University Press, London (1931).

[23] Leon Rosenfeld in Cosmology, Fusion and Other Matters, Edited by F. Reines (Colorado Associated University Press, (1972).

[24] A. A. Balandin, S. Ghosh, W. Bao, I. Calizo, D. Teweldebrhan, F. Miao, C. N. Lau, Nano Lett. (2008) .

[25] A. Shytov et al. , Phys Rev Lett, 10, 101 (2008).

[26] V.V. Cheianov, V. Falko, B.L. Altshuler, Science 315, 1252 (2007).

[27] V.V. Cheianov, V. Falko, Phys. Rev. B 74, 041403 (2006).

[28] Datta S 1995 Electronic Transport in Mesoscopic Systems (Cambridge: Cambridge University Press)

[29] J.R. Williams, L. DiCarlo, C.M. Marcus, "Quantum Hall effect in a gate-controlled p-n junction of graphene", Science 317, 638-641 (2007).

Chapter 2

Localised States of Fabry-Perot Type in Graphene Nano-Ribbons

V. V. Zalipaev, D. M. Forrester, C. M. Linton and F. V. Kusmartsev

Additional information is available at the end of the chapter

http://dx.doi.org/10.5772/52267

1. Introduction

Graphene has been spoken of as a "'wonder material"' and described as paradigm shifting in the field of condensed matter physics [1]. The exceptional behavior of single layer graphene is down to its charge carriers being massless, relativistic particles. The anomalous behavior of graphene and its low energy excitation spectrum, implies the emergence of novel electronic characteristics. For example, in graphene-superconductor-graphene junctions specular Andreev reflections occur [1] and in graphene p-n junctions a Veselago lens for electrons has been outlined [2]. It is clear that by incorporating graphene into new and old designs that new physics and applications almost always emerges. Here we investigate Fabry-Perot like localized states in graphene mono and bi-layer graphene. As one will no doubt appreciate, there are many overlaps in the analysis of graphene with the studies of electron transport and light propagation. When we examine the ballistic regime we see that the scattering of electrons by potential barriers is also described in terms of transmission, reflection and refraction profiles; in analogy to any wave phenomenon. Except that there is no counterpart in normal materials to the exceptional quality at which these occur, with electrons capable of tunneling through a potential barrier of height larger than its energy with a probability of one - Klein tunneling. So, normally incident electrons in graphene are perfectly transmitted in analogy to the Klein paradox of relativistic quantum mechanics. A tunable graphene barrier is described in [3] where a local back-gate and a top-gate controlled the carrier density in the bulk of the graphene sheet. The graphene flake was covered in poly-methyl-methacrylate (PMMA) and the top-gate induced the potential barrier. In this work they describe junction configurations associated with the carrier types (p, for holes and n for electrons) and found sharp steps in resistance as the boundaries between n-n-n and n-p-n or p-n-p configurations were crossed. Ballistic transport was examined in the limits of sharp and smooth potential steps. The PMMA is a transparent thermoplastic that has also been used to great effect in proving that graphene retains its 2D properties when embedded in a polymer heterostructure [4]. The polymers can be made to be sensitive to a specific stimulus that leads to a change in the conductance of the underlying graphene [4] and it is entirely likely that graphene based devices of the future will be hybrids including polymers that can control the carrier charge density. In [5] an experiment was performed to create a n-p-n junction to examine the ballistic regime. Oscillations in the conductance showed up as interferences between the two p-n interfaces and a Fabry-Perot resonator in graphene was created. When there was no magnetic field applied, two consecutive reflections on the p-n interfaces occurred with opposite angles, whereas for a small magnetic field the electronic trajectories bent. Above about 0.3 Tesla the trajectories bent sufficiently to lead to the occurrence of two consecutive reflections with the same incident angle and a π-shift in the phase of the electron. Thus, a half period shift in the interference fringes was witnessed and evidence of perfect tunneling at normal incidence accrued.

Quantum interference effects are one of the most pronounced displays of the power of wave quantum mechanics. As an example, the wave nature of light is usually clearly demonstrated with the Fabry-Perot interferometers. Similar interferometers may be used in quantum mechanics to demonstrate the wave nature of electrons and other quantum mechanical particles. For electrons they were first demonstrated in graphene hetero-junctions formed by the application of a top gate voltage [6]. These were simple devices consisting mainly of the resonant cavity, and with transport channels attached. These devices exhibited quantum interference in the regular resistance oscillations that arose when the gate voltage changed.

Within the conventional Fabry-Perot model [7, 8], the resistance peaks correspond to minima in the overall transmission coefficient. The peak separation can be approximated by the condition $2k_F L = 2\pi n$. The charge accumulates a phase shift of 2π after completing a single lap (the round-trip) $2L$ in the resonant cavity, where k_F is the Fermi wave vector of the charges, and L is the length of the Fabry-Perot cavity. This is the Fabry-Perot-like resonance condition: the fundamental resonance occurs when half the wavelength of the electron mode fits inside the p-n-p junction representing the Fabry-Perot cavity.

The simplest electron cavity, but still very effective, for the Fabry-Perot resonator may be formed by two parallel metallic wire-like contacts deposited on graphene [9]. There in a simple two terminal graphene structure there are clearly resolved Fabry-Perot oscillations. These have been observed in sub-100 nm devices. With a decrease of the size of the graphene region in these devices, the characteristics of the electron transport changes. Then the channel-dominated diffusive regime is transferred into the contact-dominated ballistic regime. This normally indicates that when the size of the cavity is about 100 nm or less the Fabry-Perot interference may be clearly resolved. The similar Fabry-Perot interferometer for Dirac electrons has been recently developed from carbon nanotubes [10].

Earlier work on the resistance oscillations as a function of the applied gate voltage led to their observation in the p-n-p junctions [6, 11]. It was first reported by Young and Kim [6], but the more pronounced observations of the Fabry-Perot oscillations have been made in the Ref. [11]. There high-quality n-p-n junctions with suspended top gates have been fabricated. They indeed display clear Fabry-Perot resistance oscillations within a small cavity formed by the p-n interfaces.

The oscillations arise due to an interference of an electron ballistic transport in the p-n-p junction, i.e. from Fabry-Perot interference of the electron and hole wave functions comprised between the two p-n interfaces. Thus, the holes or electrons in the top-gated region are multiply reflected between the two interfaces, interfering to give rise to standing waves, similar to those observed in carbon nanotubes [12] or standard graphene devices [13]. Modulations in the charge density distribution change the Fermi wavelength of the charge carriers, which in turn is altering the interference patterns and giving rise to the resistance oscillations.

In the present work we consider a simplest model of the Fabry-Perot interferometer, which is in fact the p-n-p or n-p-n junction formed by a one dimensional potential. We develop an exact quasi-classical theory of such a system and study the associated Fabry-Perot interference in the electron or hole transport.

Although graphene is commonly referred to as the "'carbon flatland"' there has been a feeling of discontent amongst some that the Mermin-Wagner theorem appeared to be contradicted. However, recent work shows that the buckling of the lattice can give rise to a stable 3D structure that is consistent with this theorem [14]. In what follows we present the general methodology for analysis of graphene nanoribbons using semiclassical techniques that maintain the assumption of a flat lattice. However, it should be mentioned that the effects found from these techniques are powerful in aiding our understanding of potential barriers and are an essential tool for the developing area of graphene barrier engineering. The natural state of graphene to accommodate defects or charged impurities is important for applications. The p-n interfaces described above may be capable of guiding plasmons and to create the electrical analogues of optical devices to produce controllable indices of refraction [15].

In Part *I* of this chapter we investigate the use of powerful semiclassical methods to analyze the relativistic electron and hole tunneling in graphene through a smooth potential barrier. We make comparison to the rectangular barrier. In both cases the barrier is generated as a result of an electrostatic potential in the ballistic regime. The transfer matrix method is employed in complement to the adiabatic WKB approximation for the Dirac system. Crucial to this method of approximation for the smooth barrier problem, when there is a skew electron incidence, is careful consideration of four turning points. These are denoted by x_i, $i = 1, 2, 3, 4$ and lie in the domain of the barrier. The incident electron energy in this scattering problem belongs to the middle part of the segment $[0, U_0]$, where U_0 is the height of the barrier, and essentially the incident parameter p_y should be large enough to allow normal and quasi-normal incidence.

Therefore, between the first two turning points, x_1 and x_2, and also between the next two, x_3 and x_4 there is no coalescence. Two columns of total internal reflection occur which have solutions that grow and decay exponentially. Looking away from the close vicinity of the asymptotically small boundary layers of x_i, there exists five domains with WKB type solutions (See Fig. 2): three with oscillatory behavior and two exhibiting asymptotics that are exponentially growing and decaying. Combining these five solutions is done through applying matched asymptotics techniques (see [16]) to the so-called effective Schrödinger equation that is equivalent to the Dirac system (see [17], [18]). This combinatorial procedure generates the WKB formulas that give the elements of the transfer matrix. This transfer matrix defines all the transmission and reflection coefficients in the scattering problems discussed here.

When the energies are positive around potential height $0.5U_0$, electronic incident, reflected and transmitted states occur outside the barrier. Underneath the barrier a hole state exists (n-p-n junction). The symmetrical nature of the barrier means that we see incident, reflected and transmitted hole states outside the barrier when the energies are negative and close to one-half of the potential height $U_0 < 0$. Thus, underneath the barrier there are electronic states (a p-n-p junction).

Incorporated into the semiclassical method is the assumption that all four turning points are spatially separated. Consequently, the transverse component of the momentum p_y is finite and there is a finite width to the total internal reflection zone. This results in a 1-D Fabry-Perot resonator, which is of great physical importance and may aid understanding in creating plasmonic devices that operate in the range of terahertz to infrared frequencies [19]. Quantum confinement effects are crucial at the nano-scale and plasmon waves can potentially be squeezed into much smaller volumes than noble metals. The basic description of propagating plasma modes is essentially the same in the 2-D electron gas as in graphene, with the notable exception of the linear electronic dispersion and zero band-gap in graphene [20]. Thus, we predicate that the methods applied here are also applicable to systems of 2-D electron gases, such as semiconductor superlattices. Due to the broad absorption range of graphene, nanoribbons as described here, or graphene islands of various geometries may also be incorporated in opto-electronic structures.

In our analysis, if $p_y \to 0$ then we have a quasi-normal incidence whereby and first two, x_1 and x_2, and the second two, x_3 and x_4, turning points coalesce. In the case of normal incidence, there is always total transmission through the barrier. The vital discovery in this form of analysis is that of the existence of modes that are localized in the bulk of the barrier. These modes decay exponentially as the proximity to the barrier decreases. These modes are two discrete, complex and real sets of energy eigen-levels that are determined by the Bohr-Sommerfeld quantization condition, above and below the cut-off energy, respectively. It is shown that the total transmission through the barrier takes place when the energy of an incident electron, which is above the cut-off energy, coincides with the real part of the complex energy eigen-level of one among the first set of modes localized within the barrier. These facts have been confirmed by numerical simulations for the reflection and transmission coefficients using finite elements methods (Comsol package).

In Part II we examine the high energy localized eigenstates in graphene monolayers and double layers. One of the most fundamental prerequisites for understanding electronic transport in quantum waveguide resonators is to be able to explain the nature of the conductance oscillations (see [25], [26], [27]). The inelastic scattering length of charge carriers is much larger than the size of modern electronic devices and consequently electronic motion is ballistic and resistance occurs due to scattering off geometric obstacles or features (e.g. the shape of a resonator micro or nano-cavity or the potential formed by a defect). It is an interesting area of development whereby defects are engineered deliberately into devices to generate a sought effect. In graphene, defects such as missing carbon atoms or the addition of adatoms can lead to interesting and novel effects, e.g magnetism or proximity effects. In the ballistic regime, conductance is analyzed by the total transmission coefficient and the Landauer formula for the zero temperature conductance of a structure (see monographs [25],

[26], [27], and papers [28], [29], [30], for example). The excitation of localized eigenmodes inside a quantum electronic waveguide has a massive effect on the conductivity because these modes could create an internal resonator inside the waveguide. This is a very good reason to research the role of localised eigenmodes for quantum resonator systems and 2-D electronic transport in quantum waveguides. Excitation of some modes could result in the the emergence of stop bands for electronic wave propagation in the dispersion characteristics of the system, whereby propagation through the waveguide is blocked entirely. Other modes will result in total transmission.

In this review, the semiclassical analysis of resonator eigenstates that are localized near periodic orbits is developed for a resonator of Fabry-Perot type. These are examined inside graphene monolayer nanoribbons in static magnetic fields and electrostatic potentials. The first results for bilayer graphene are also presented in parallel to this.

Graphene has generated a fervor throughout the scientific world and especially in the condensed matter physics community, with its unusual electronic properties in tunneling, charge carrier confinement and the appearance of the integer quantum Hall effect (see [31], [33], [34]), [35], [17])). Its low energy excitations are massless chiral Dirac fermion quasi-particles. The Dirac spectrum, that is valid only at low energies when the chemical potential crosses exactly at the Dirac point (see [31]), describes the physics of quantum electrodynamics for massless fermions, except that in graphene the Dirac electrons move with a Fermi velocity of $v_F = 10^6 m/s$. This is 300 times smaller than the speed of light. Graphene is a material that is easy to work with, it has a high degree of flexibility and agreeable characteristics for lithography. The unusual electronic properties of graphene and its gapless spectrum provide us with the ideal system for investigation of many new and peculiar charge carrier dynamical effects. It is also conceivable, if its promise is fulfilled, that a new form of carbon economy could emerge based upon exploitation of graphenes novel characteristics. The enhancements in devices are not just being found at the nano and micron-sized levels, though these hold the most potential (e.g. the graphene transistor, metamaterials etc), but in composites [36], electrical storage [37], solar harvesting [38] and many more applications. Following this train of thought, graphene is also a viable alternative to the materials normally used in plasmonics and nanophotonics. It absorbs light over the whole electromagnetic spectrum, including UV, visible and far-infrared wavelengths and as we have mentioned, it is capable of confining light and charge carriers into incredibly small volumes. Thus, there are a range of applications where band gap engineering is not required and it is satisfactory to directly use nanoribbons of graphene as optical-electronic devices.

In the analysis of graphene one also expects unusual Dirac charge carrier properties in the eigenstates of a Fabry-Perot resonator in a magnetic field. For example, two parts of the semiclassical Maslov spectral series with positive and negative energies, for electrons and holes, correspondingly, with two different Hamiltonian dynamics and families of classical trajectories are apparent. Semiclassical analysis can provide insight into the aforementioned physical systems and good quantitative predictions on quantum observables using classical insights. Application of semiclassical analysis in studying the quantum mechanical behavior of electrons has been demonstrated in descriptions of different nano-structures, electronic transport mechanisms in mesoscopic systems and, as another example, the quantum chaotic dynamics of electronic resonators [25], [26], [27], [39], [40], [41], [42] and many others.

However, it is important to state that the first semiclassical study on two-dimensional graphene systems only recently appeared in [43], [44], [45]. In [43] a semiclassical approximation for the Green's function in graphene monolayer and bilayers was discussed. In [44] and [45] bound states in inhomogeneous magnetic fields in graphene and graphene-based Andreev billiards were studied by semiclassical analysis, accordingly. This was carried out with one-dimensional WKB quantization due to total separation of variables.

In the second half of this review, the semiclassical Maslov spectral series of the proliferation of high-energy eigenstates (see [48], [49] [50]) of the electrons and holes for a resonator formed inside graphene mono and bilayer nanoribbons with zigzag boundary conditions, is specified. These states are localized around a stable periodic orbit (PO) under the influence of a homogeneous magnetic field and electrostatic potential. The boundaries of the nanoribbon act with perfect reflection to confine the periodic orbit to isolation. This system is a quantum electron-hole Fabry-Perot resonator of a type analogous to the "bouncing ball" high-frequency optical resonators found in studies of electromagnetics and acoustics. The asymptotic analysis of the high-energy localized eigenstates presented here is similar to ones used for optical resonators (see[50], [51], [54], and [55]). In this review, the semiclassical methods presented focus upon the stability of POs and electron and hole eigenstates that depend on the applied magnetic field.

We construct a solitary localized asymptotic solution to the Dirac system in the neighborhood of a classical trajectory called an electronic Gaussian beam (Gaussian wave package). In PO theory there are similarities between the asymptotic techniques used here and those used in the semiclassical analysis (see, for example, [27] (chapters 7, 8) or [39] and cited references). Further, the stability of a continuous family of closed trajectories in asymptotic proximity to a PO, confined between two reflecting interfaces, is studied. The classical theory of linear Hamiltonian systems with periodic coefficients gives the basis to study the stability using monodromy matrix analysis. The asymptotic eigenfunctions for electrons and holes are constructed only for the stable PO as a superposition of two Gaussian beams propagating in opposite directions between two reflecting points of the periodic orbit. A generalized Bohr-Sommerfeld quantization condition gives the asymptotic energy spectral series (see [46] and [47], [48], [49], [50], [51] and [55]). This work highlights that the single quantization condition derived herein for the quantum electron-hole graphene resonator fully agrees with the asymptotic quantization formula of a quite general type spectral problem in [51]. It is worth drawing attention to the fact that in a semiclassical approximation for the Green's function in a graphene monolayer and bilayer, the relationship between the semiclassical phase and the adiabatic Berry phase was discussed in the paper [43]. Our asymptotic solutions, for rays and Gaussian beams, possess the adiabatic phase introduced by Berry [64]. The importance of Berry-like and non-Berry-like phases in the WKB asymptotic theory of coupled differential equations and their roles in semiclassical quantization were discussed in [57], [58], [59].

Our results are a special class of POs that occur for graphene zigzag nanoribbons in a homogeneous magnetic field and piece-wise electrostatic potential that is embedded inside the nanoribbon. They are found by giving, to the leading order, a description of the general form of asymptotic solution of Gaussian beams in a graphene monolayer or bilayer. The key point in the asymptotic analysis is the quantization of the continuous one-parameter (energy) family of POs. For one subclass of lens-shaped POs, these localized eigenstates were evaluated against eigenvalues and eigenfunctions that have been computed by the finite element method using COMSOL. For a selectively chosen range of energy eigenvalues and eigenfunctions, agreement between the numerical results and those computed semiclassically is very good. In the graphene Fabry-Perot resonator, the electrostatic potential does not play a role of confinement, it behaves more like an inhomogeneity, but in some cases an electrostatic potential helps to make a family of POs stable.

In this chapter, we describe the tunneling through smooth potential barriers and the asymptotic solutions for a Dirac system in a classically allowed domain. This is done using WKB methods. We then go on to investigate the classically disallowed domain and tunneling through the smooth barrier. The asymptotic WKB solutions are presented for scattering and for quasi-bound states localized within the smooth barrier. The second part of the chapter, goes into detail about high energy localized eigenstates in monolayer and bilayer graphene. The graphene resonator is described when it is subjected to a magnetic field and ray asymptotic solutions outlined. Finally, the construction of periodic orbits, stability analyses and quantization conditions are thoroughly examined. A numerical analysis is given that compares the analytical techniques and results with those found using finite element methods.

PART I: Tunneling through graphene barriers

2. The rectangular barrier

In a conventional metal or semiconductor there are no propagating states connecting regions either side of the barrier (regions I and III). To get through the barrier an electron has to tunnel through the classically forbidden region and the tunneling amplitude depreciates exponentially as a function of the barrier width. Thus, transport between I and III is strongly suppressed. However, in each of the three regions of a barrier in a graphene system, the valence and conduction band touches, meaning that there are propagating states connecting I and III at all energies. There is no such suppression of the transport at energies incident and below the barrier. At normal incidence transmission is always perfect.

Potential barriers for single quasiparticle tunneling in graphene can be introduced by designing a suitable underlying gate voltage or even as a result of local uniaxial strain [68]. In the following we denote the angle of incidence with respect to the barrier to be θ_1. We are interested in the dependence of the tunneling transmission on this incidence angle. To illustrate quantum mechanical tunneling one must extract the transmission coefficient from the solution to the graphene barrier problem. The transmission coefficient is the ratio of the flux of the particles that penetrate the potential barrier to the flux of particles incident on the barrier. We demonstrate a rectangular barrier as described in detail in the Reviews by Castro Neto et al [60] and Pereira Jr et al [69]. The problem can be described by the following 2D Dirac system (see, for example, [60])

$$v_F(\bar{\sigma}, -i\hbar\nabla)\psi(\mathbf{x}) + U(\mathbf{x})\psi(\mathbf{x}) = E\psi(\mathbf{x}), \quad \psi(\mathbf{x}) = \begin{pmatrix} u \\ v \end{pmatrix}, \tag{1}$$

where $\mathbf{x} = (x, y)$ and u, v are the components of the spinor wave function describing electron localization on the sites of sublattice A or B of the honeycomb graphene structure, v_F is the

Fermi velocity, the symbol $(,)$ means scalar product, $\hbar$ is the Planck constant and $\bar{\sigma} = (\sigma_1, \sigma_2)$ with Pauli matrices

$$\sigma_1 = \begin{pmatrix} 0 & 1 \\ 1 & 0 \end{pmatrix}, \quad \sigma_2 = \begin{pmatrix} 0 & -i \\ i & 0 \end{pmatrix}.$$

If we assume that the potential representing the barrier does not depend on y, i.e. $U = U(x)$, then we can look for a solution in the form

$$\psi(\mathbf{x}) = e^{i\frac{p_y}{\hbar}y} \begin{pmatrix} \tilde{u}(x) \\ \tilde{v}(x) \end{pmatrix},$$

where p_y means value of the transverse momentum component describing the angle of incidence. Then, we obtain the Dirac system of two ODEs

$$\begin{pmatrix} U(x) - E & v_F[-i\hbar\partial_x - ip_y] \\ v_F[-i\hbar\partial_x + ip_y] & U(x) - E \end{pmatrix} \begin{pmatrix} \tilde{u} \\ \tilde{v} \end{pmatrix} = \begin{pmatrix} 0 \\ 0 \end{pmatrix}. \tag{2}$$

The particle incident with energy $E < U_0$ from the left of the barrier has wavevectors k_1, q, and k_2 to the left, in the barrier and to the right of the barier, respectively. These regions are denoted I, II and III, respectively. In the symmetric barrier $k_1 = k_2 = k$. Region II lies between $-L$ and L, where $\pm L$ defines the width of the barrier. The wave functions are defined for each of the three regions below:

$$\psi_I = \frac{a_1}{\sqrt{2}} \begin{pmatrix} 1 \\ e^{i\theta_1} \end{pmatrix} e^{i(k_x x + k_y y)} + \frac{a_2}{\sqrt{2}} \begin{pmatrix} 1 \\ -e^{-i\theta_1} \end{pmatrix} e^{i(-k_x x + k_y y)} \tag{3}$$

$$\psi_{II} = \frac{b_1}{\sqrt{2}} \begin{pmatrix} 1 \\ e^{i\theta_2} \end{pmatrix} e^{i(q_x x + k_y y)} + \frac{b_2}{\sqrt{2}} \begin{pmatrix} 1 \\ -e^{-i\theta_2} \end{pmatrix} e^{i(-q_x x + k_y y)} \tag{4}$$

$$\psi_{III} = \frac{c_1}{\sqrt{2}} \begin{pmatrix} 1 \\ e^{i\theta_1} \end{pmatrix} e^{i(k_x x + k_y y)} + \frac{c_2}{\sqrt{2}} \begin{pmatrix} 1 \\ -e^{-i\theta_1} \end{pmatrix} e^{i(-k_x x + k_y y)} \tag{5}$$

where we have introduced the wave function, as is done in [31]. The coefficients c_1, c_2 and a_1, a_2 are related by means of the transfer matrix, $c = Ta$. The transfer matrix has unique properties, which are demonstrated in Appendix B. In regions I and III the angle of incidence in momentum space is given by, $\theta_1 = \arctan(k_y/k_x)$ and in region II, $\theta_2 = \arctan(k_y/q_x)$. In regions I-III the valence and conduction bands touch. This allows propagating states to connect the regions at all energies and there is no suppression of transport at the energies below the height of the barrier. There is also perfect transmission at normal incidence. The graphene rectangular barrier can be thought of as a medium with a different refractive index to its surroundings. In an optical analogy, the refractive index of the barrier is $1 - U_0/E$ [8]. At the interface of the barrier the incidence angle splits into transmitted and reflected waves with the transmitted wave propagating with angle θ_2

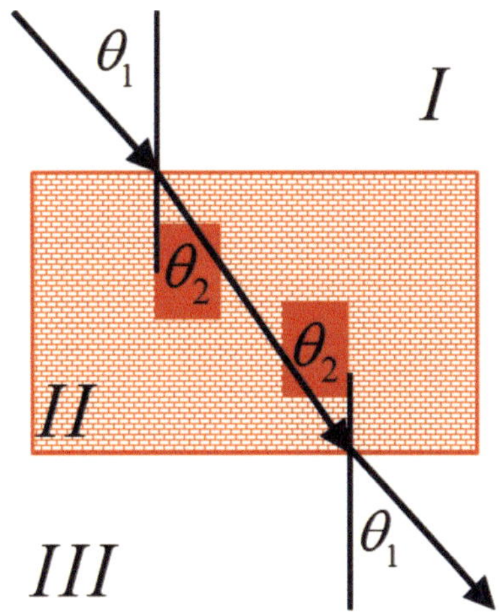

Figure 1. The angles related to the propagation of an electron through the rectangular barrier in the xy plane for a barrier of width W.

through the barrier. The wave inside the barrier is multiply reflected between $-L$ and L. The parallel wave vector inside the barrier is given by,

$$q_x = \frac{1}{\hbar}\sqrt{\frac{(E-U_0)^2}{v_F^2} - p_y^2}$$

and the wave vectors outside the barrier are defined as,

$$k_x = \frac{p_x}{\hbar} = \frac{1}{\hbar}\sqrt{\frac{E^2}{v_F^2} - p_y^2}$$

The wave functions in regions I and II are matched at $x = -L$. Likewise, the wavefunctions between regions II and III are matched at $x = L$. It is not necessary to match the derivatives, as is done in an analysis using the Schrödinger equation. One requires the wave functions to be continuous at the boundary of each region to generate relationships between the coefficients, $a_{1,2}$, $b_{1,2}$ and $c_{1,2}$. We seek solutions such that $|a_1|^2 - |a_2|^2 = |c_1|^2 - |c_2|^2$. The elements of the transfer matrix for the rectangular barrier are found to be,

$$T_{11} = \frac{e^{-ik_xL}}{4cos\theta_1 cos\theta_2}\left(e^{2iq_xL-ik_xL}|\alpha|^2 - e^{-2iq_xL-ik_xL}|\beta|^2\right), \tag{6}$$

$$T_{12} = \frac{e^{-ik_xL}}{4cos\theta_1 cos\theta_2}\left(e^{2iq_xL+ik_xL}\beta\bar{\alpha} - e^{-2iq_xL+ik_xL}\bar{\alpha}\beta\right), \tag{7}$$

$$T_{21} = \frac{e^{ik_xL}}{4cos\theta_1 cos\theta_2}\left(e^{-2iq_xL-ik_xL}\bar{\beta}\alpha - e^{2iq_xL-ik_xL}\alpha\bar{\beta}\right), \tag{8}$$

$$T_{22} = \frac{e^{ik_xL}}{4cos\theta_1 cos\theta_2}\left(e^{-2iq_xL+ik_xL}|\alpha|^2 - e^{2iq_xL+ik_xL}|\beta|^2\right). \tag{9}$$

where we make the substitutions $\alpha = e^{i\theta_1} + e^{-i\theta_2}$ and $\beta = e^{-i\theta_2} - e^{-i\theta_1}$ and their complex conjugate forms are denoted by $\bar{\alpha} = e^{i\theta_2} + e^{-i\theta_1}$ and $\bar{\beta} = e^{i\theta_2} - e^{-i\theta_1}$. If we assume that the incident wave approaches from the left, then $a_1 = 1$, $a_2 = r_1$ and $c_1 = t_1$, where r_1 is the reflection coefficient and t_1 is the transmission coefficient. If the incident wave approaches from the right then $c_1 = r_2$, $c_2 = 1$ and $a_2 = t_2$. We find that $t_1 = t_2 = t$ and the transmission coefficient is $t = 1/T_{22}$. The reflection coefficients are determined as $r_1 = -T_{21}/T_{22}$ and $r_2 = T_{12}/T_{22}$. The transmission probability is as usual given by $|t_1|^2$ with the definition $|t_1|^2 + |r_1|^2 = 1$. At normal incidence the carriers in graphene are transmitted completely through the barrier (Klein tunneling). However, the carriers can be reflected by a potential step when the angle of incidence increases and a non-zero momentum component parallel to the barrier ensues. Thus, the transmission of charge carriers through the potential barrier is anisotropic. When a beam of electrons is fired at an angle into the barrier, it splits up into transmitted and reflected beams, with multiple reflections occurring at the edges of the barrier. As is usual in quantum mechanics, the transmission is found by stipulating that there must be continuity between the wavefunctions. In the above this demand for continuity at the extremities of the barrier allowed us to find the coefficients of the wavefunctions. Thus, using these results and following the work of Castro Neto et al [60], the total transmission as a function of the incident angle is given by $T(\theta_1) = tt^*$:

$$T = \frac{16cos^2\theta_1 cos^2\theta_2}{|\alpha|^4 + |\beta|^4 - 2|\alpha|^2|\beta|^2 cos(4q_xL)}$$

When the tunneling resonance condition $2Lq_x = n\pi$ is met, where n is an integer, $T = 1$. This statement means that a half-integer amount of wavelengths will fit into the length of the potential barrier. The absolute transmission is the manifestation of Klein tunneling, which is unique for relativistic electrons, and it should occur when an incoming electron starts penetrating through a potential barrier of height, U_0 (which is far in excess of the electrons rest energy). The transport mechanism in a graphene tunneling structure is unique. This perfect transmission at incidence normal to the barrier is due to the pseudo-spin conservation that gives no backscattering. In order to attain an interference effect between the two interfaces an oblique incidence angle is required and it is under this prerequisite that multiple interference effects emerge. Thus, the potential barrier is analogous to two interfaces at $-L$ and L and also a Fabry-Perot interferometer [5]. The analogy of the graphene rectangular barrier to the Fabry-Perot resonator when $\theta_1 \neq 0$ extends to the potential barrier operating like an optical cavity. In region II the incoming wave can interfere with itself and with constructive interference, resonances will occur where $T(\theta_1 \neq 0) = 1$ [5]. The potential barriers for single quasiparticle tunneling in graphene are usually created by suitably changing the underlying gate voltage. In the next section we investigate the smooth barrier and expect that there will be similar scattering behavior as through the rectangular barrier. We seek to explore the similarities and the differences between the two.

3. The smooth barrier

Consider a scattering problem for the Dirac operator describing an electron-hole in the presence of a scalar potential representing a smooth localized barrier with the height U_0 (see Fig.2). It is convenient to use the dimensionless system

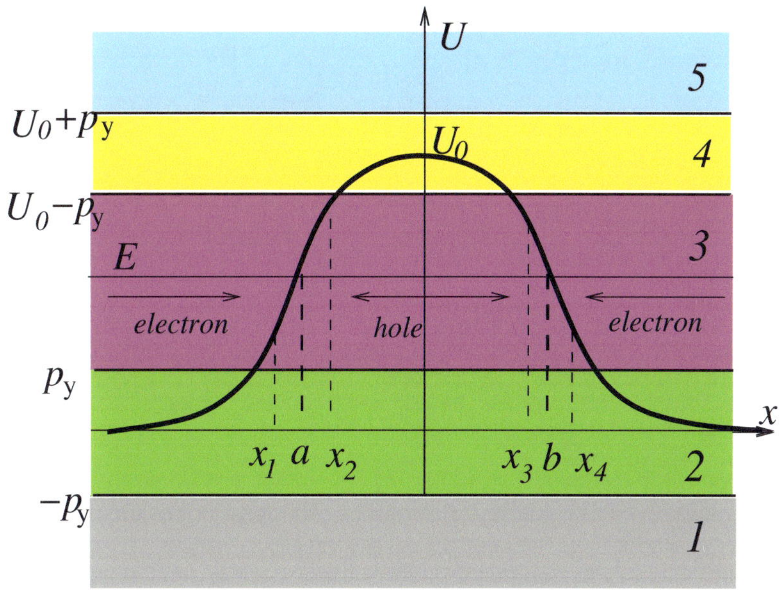

Figure 2. The generalization of a smooth potential barrier with Gaussian shape (we assume that $p_y > 0$). The Dirac electron and hole states arising in resonance tunneling are shown. The quasibound states are to be found above the green strip, $|E| < p_y$, where bound states are located. Quasibound (metastable) states are confined by two tunneling strips between x_1, x_2 and x_3, x_4, whereas the bound states are located between x_2 and x_3.

$$\begin{pmatrix} U(x) - E & -ih\partial_x - ip_y \\ -ih\partial_x + ip_y & U(x) - E \end{pmatrix} \begin{pmatrix} u \\ v \end{pmatrix} = \begin{pmatrix} 0 \\ 0 \end{pmatrix}, \tag{10}$$

in which we omitted the sign of tilde for brevity. In physical dimensions the energy is $U_0 E$, the potential is $U_0 U(x)$, the y-component of the momentum is $p_y U_0 / v_F$, and the dimensionless Planck constant (small WKB parameter) is given by $h = \hbar v_F / U_0 D$, where U_0 is the height of the potential barrier ($|U(x)| < 1$ for $x \in \mathbf{R}$) and D is a characteristic scale of the potential barrier with respect to the x-coordinate. Typical values of U_0 and D are within the ranges 10-100meV and 100-500nm. For example, for U_0=100meV, $D = 264nm$, we have $h = 0.025$ and also we assume that $p_y > 0$.

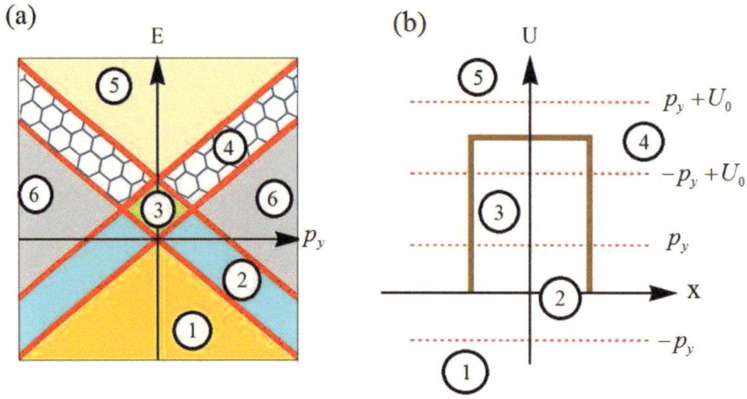

Figure 3. (a) The six different scattering regimes for smooth barrier tunneling. The six zones are separated by the four red diagonal lines, $E = \pm p_y$ and $E = \pm p_y + U_0$. We will now discuss the right hand side of this diagram. In zone 1 (orange shading), $E < -p_y$, there is total transmission and exponentially small reflection. The asymptotic solutions are of an oscillatory nature everywhere in this zone. In zone 2 (blue), $-p_y < E < p_y$ with the cut-off energy at $E = \pm p_y$. In zone 2 there is no propagation outside the barrier. However, there are oscillatory solutions within the barrier. Zone 3 (green) is the zone of Klein tunneling. Here, $p_y < E < U_0 - p_y$ and there are oscillatory solutions everywhere. Zone 4 (hexagons), $U_0 - p_y < E < U_0 + p_y$, is devoid of propagation through the barrier. There is total reflection and exponentially small transmission in this zone. The fifth zone (sand color) is limited to $E > U_0 + p_y$ and is characterized by total transmission, exponentially small reflection and the asymptotic solutions are oscillatory everywhere. The sixth zone is one of no propagation and exponentially decaying or growing asymptotic solutions, $U_0 < E < p_y$.

In Fig. 3, six zones (horizontal strips in Fig. 3b) are shown that illustrate different scattering regimes for the smooth barrier scattering problem. These six zones are exactly the same as for the rectangular barrier with the height U_0. In zone 1 $E < -p_y$, we have total transmission and exponentially small reflection, asymptotic solutions are of oscillatory type everywhere. In zone 2, $-p_y < E < p_y$ ($E = \pm p_y$ is the cut-off energy), there is no propagation outside the barrier, however there are oscillatory solutions within the barrier. In the zone 3, $p_y < E < U_0 - p_y$, there are oscillatory solutions everywhere (zone of the Klein tunneling). In zone 4, $U_0 - p_y < E < U_0 + p_y$, there is no propagation through the barrier, we have total reflection and exponentially small transmission. In zone 5 $E > U_0 + p_y$, we have total transmission and exponentially small reflection, asymptotic solutions are of oscillatory type everywhere. Finally, in the zone 6, $U_0 - p_y < E < p_y$, there is no propagation, everywhere asymptotic solutions are of exponential type, decaying or growing.

Firstly, we study the scattering problem for zone 3 (see Fig.2). In this case, there are 5 domains with different WKB asymptotic solutions: $\Omega_1 = \{x : -\infty < x < x_1\}$, $\Omega_2 = \{x : x_1 < x < x_2\}$, $\Omega_3 = \{x : x_2 < x < x_3\}$, $\Omega_4 = \{x : x_3 < x < x_4\}$ and $\Omega_5 = \{x : x_4 < x < +\infty\}$ and the turning points are x_i with $i = 1, 2, 3, 4$, are the roots of the equation $(E - U(x))^2 - p_y^2 = 0$. The regions Ω_1, Ω_3 and Ω_5, in which $(E - U(x))^2 - p_y^2 > 0$, will be referred to as classically allowed domains, whereas Ω_2 and Ω_4, in which $(E - U(x))^2 - p_y^2 < 0$, are classically disallowed domains. Note that as $p_y \to 0$ for fixed value of E, the turning points coalesce. We exclude this possibility in our analysis.

It is worth to remark that for fixed p_y, when E moves down from zone 3 to zone 2, the turning points x_1 and x_4 disappear ($x_1 \to -\infty$, $x_4 \to +\infty$) such that inside zone 2 we have only x_2 and x_3. When we move down from zone 2 to zone 1, the turning points x_2 and x_3 disappear. When E moves up from zone 3 to zone 4, the turning points x_2 and x_3 coalesce and disappear such that inside zone 4 we have only x_1 and x_4. When we move up from zone 4 to zone 5, the turning points x_1 and x_4 coalesce and disappear.

4. WKB asymptotic solution for Dirac system in classically allowed domain

The WKB oscillatory asymptotic solution to the Dirac system in the classically allowed domains is to be sought in the form (see [16]) with real $S(x)$

$$\psi = \begin{pmatrix} u \\ v \end{pmatrix} = e^{\frac{i}{h}S(x)} \sum_{j=0}^{+\infty} (-ih)^j \begin{pmatrix} u_j \\ v_j \end{pmatrix} = e^{\frac{i}{h}S(x)} \sum_{j=0}^{+\infty} (-ih)^j \psi_j(x). \tag{11}$$

Substituting this series into the Dirac system, and equating to zero corresponding coefficients of successive degrees of the small parameter h, we obtain a recurrent system of equations which determines the unknown $S(x)$ (classical action) and $\psi_j(x)$, namely,

$$(H - EI)\psi_0 = 0, \quad (H - EI)\psi_j = -R\psi_{j-1}, \ j > 0, \tag{12}$$

$$H = \begin{pmatrix} U(x) & p_x - ip_y \\ p_x + ip_y & U(x) \end{pmatrix}, \tag{13}$$

$$\widehat{R} = \begin{pmatrix} 0 & \partial_x \\ \partial_x & 0 \end{pmatrix}, \tag{14}$$

where I is the identity matrix and $S' = p_x$. The Hamiltonian H has two eigenvalues

$$h_{1,2} = U(x) \pm \sqrt{p_x^2 + p_y^2} \equiv U(x) \pm p$$

and

$$e_{1,2} = \frac{1}{\sqrt{2}} \begin{pmatrix} 1 \\ \pm \frac{p_x + ip_y}{p} \end{pmatrix}$$

with

$$p_x = \pm\sqrt{(E - U(x))^2 - p_y^2}.$$

From now on we will omit the dependence on x of U, S, and quantities derived from them. It turns out to be convenient to use different $e_{1,2}$ instead with

$$e_{1,2} = \frac{1}{\sqrt{2}} \begin{pmatrix} 1 \\ \pm e^{i\theta} \end{pmatrix}, \qquad e^{i\theta} = \frac{p_x + ip_y}{E - U}.$$

In this way we will be able to solve problems of electron and hole incidence on the barrier simultaneously. Note that, irrespective of whether $E > U$ or $E < U$,

$$He_1 = Ee_1, \qquad He_2 = (2U - E)e_2. \tag{15}$$

The classical action $S(x)$ is given by

$$S = \int p_x dx = \pm \int \sqrt{(E - U)^2 - p_y^2}\, dx, \tag{16}$$

the sign indicating the direction of the wave, with + corresponding to a wave traveling to the right.

For electrons and holes one can seek a solution to the Dirac system zero-order problem in the form

$$\psi_0 = \sigma^{(0)}(x)e_1 \tag{17}$$

with unknown amplitude $\sigma^{(0)}$. The solvability of the problem

$$(H - EI)\psi_1 = -\widehat{R}\psi_0$$

requires that the orthogonality condition

$$< e_1, \widehat{R}(\sigma^{(0)}e_1) >= 0$$

must hold, written as a scalar product implied with complex conjugation, and from this one obtains the transport equation for $\sigma^{(0)}$:

$$\frac{d\sigma^{(0)}}{dx}(e^{i\theta} + e^{-i\theta}) + \sigma^{(0)}\frac{de^{i\theta}}{dx} = 0. \tag{18}$$

It has a solution

$$\sigma^{(0)} = \left(\frac{c_0}{\sqrt{2\cos\theta}}\right)e^{-i\frac{\theta}{2}}$$

with $c_0 = const$, where a branch of the analytic function $\sqrt{z}$ is taken that satisfies the condition

$$Im(\sqrt{z}) \geq 0, \ z \in C.$$

For higher-order terms,

$$(H - EI)\psi_j = -\widehat{R}\psi_{j-1},$$

one can seek a solution to

$$(H - EI)\psi_j = -\widehat{R}\psi_{j-1}$$

$(j > 0)$ in the form

$$\psi_j = \sigma_1^{(j)} e_1 + \sigma_2^{(j)} e_2, \tag{19}$$

where from (15), $\sigma_2^{(j)}$ is given by

$$\sigma_2^{(j)} = \frac{< e_2, R(\psi_{j-1}) >}{2(E - U)}.$$

Then, from the orthogonality condition,

$$< e_1, R(\sigma_1^{(j)} \mathbf{e}_1 + \sigma_2^{(j)} e_2) >= 0,$$

one obtains

$$\sigma_1^{(j)} = \frac{e^{-i\theta/2}}{\sqrt{2\cos\theta}} \left(c_j - \int e^{i\theta/2}\sqrt{2\cos\theta} < e_1, R(\sigma_2^{(j)} e_2) > dx \right),$$

where $c_j = const$. Below we assume that $p_x > 0$, corresponding to a wave traveling in the positive x-direction. Thus, to the leading order we have

$$\psi = \begin{pmatrix} u \\ v \end{pmatrix} = \frac{e^{\pm \frac{i}{h} S_p(x,x_i)}}{\sqrt{J_p^{\pm}}} c_0 e_1^{\pm}(1 + O(h)), \tag{20}$$

$$S_p(x, x_i) = \int_{x_i}^{x} p_x dt, \quad J_p^{\pm} = 1 + e^{2i\theta^{\pm}}, \quad e_1^{\pm} = \frac{1}{\sqrt{2}} \begin{pmatrix} 1 \\ e^{i\theta^{\pm}} \end{pmatrix},$$

$$e^{i\theta^{\pm}} = \frac{\pm p_x + i p_y}{E - U}.$$

This asymptotic approximation is not valid near turning points where $S' = 0$ (see Fig. 1) at $x = x_j$, $j = 1, 2, 3, 4$ where $e^{i\theta} = \pm i$ and $\cos\theta = 0$, while at $x = a, b$ we have $E = U$. The WKB asymptotic solution, derived in this section, is valid for the domains Ω_i, $i = 1, 3, 5$.

5. WKB asymptotic solution for Dirac system in classically disallowed domain

The WKB asymptotic solution to the Dirac system in the classically disallowed domain is to be sought in the form

$$\psi = \begin{pmatrix} u \\ v \end{pmatrix} = e^{-\frac{1}{h}S(x)} \sum_{j=0}^{+\infty} (-ih)^j \begin{pmatrix} u_j \\ v_j \end{pmatrix} = e^{-\frac{1}{h}S(x)} \sum_{j=0}^{+\infty} (-ih)^j \psi_j(x), \tag{21}$$

with $S(x)$ real. As in section 4, we obtain a recurrent system of equations which determines the unknown $S(x)$ and $\psi_j(x)$, namely,

$$(H - EI)\psi_0 = 0, \quad (H - EI)\psi_j = -R\psi_{j-1}, \ j > 0, \tag{22}$$

$$H = \begin{pmatrix} U & i(q_x - p_y) \\ i(q_x + p_y) & U \end{pmatrix}, \tag{23}$$

where $S' = q_x$, and the matrix R is as in (14). The Hamiltonian H is not Hermitian. It has two eigenvalues and not orthogonal eigenvectors $Hl_{1,2} = h_{1,2}l_{1,2}$, where

$$h_{1,2} = U(x) \pm \sqrt{p_y^2 - q_x^2},$$

$$l_{1,2} = \begin{pmatrix} 1 \\ \pm i\sqrt{\frac{q_x + p_y}{p_y - q_x}} \end{pmatrix},$$

as we have

$$i\frac{q_x + p_y}{E - U} = \pm i\sqrt{\frac{q_x + p_y}{p_y - q_x}}$$

where

$$q_x = \pm\sqrt{p_y^2 - (E - U)^2}, \quad |q_x| < p_y.$$

Thus, the function $S(x)$ in a classically disallowed domain is given by

$$S = \int q_x dx = \pm \int \sqrt{p_y^2 - (E - U)^2} dx. \tag{24}$$

Again, for the sake of simplicity, we shall use different $l_{1,2}$

$$l_{1,2} = \frac{1}{\sqrt{1+\kappa^2}} \begin{pmatrix} 1 \\ \pm i\kappa \end{pmatrix} = \begin{pmatrix} \cos\phi \\ \pm i\sin\phi \end{pmatrix}, \tag{25}$$

where

$$\kappa = \frac{q_x + p_y}{E-U}, \quad \kappa = \tan\phi, \qquad -\frac{\pi}{2} < \phi < \frac{\pi}{2}.$$

For electrons and holes one can seek a solution to the Dirac system zero-order problem in the form

$$\psi_0 = \sigma^{(0)}(x) l_1 \tag{26}$$

with unknown amplitude $\sigma^{(0)}$. Solvability of the problem

$$(H - EI)\psi_1 = -\widehat{R}\psi_0$$

requires that the orthogonality condition must hold

$$< l_1^*, \widehat{R}(\sigma^{(0)} l_1) >= 0,$$

where

$$l_1^* = \frac{1}{\sqrt{1+\kappa^2}} \begin{pmatrix} \kappa \\ i \end{pmatrix} = \begin{pmatrix} \sin\phi \\ i\cos\phi \end{pmatrix}.$$

The vector l_1 is the eigenvector of H, whereas l_1^* is the eigenvector of H^*. From the orthogonality condition one obtains the transport equation for $\sigma^{(0)}$

$$\frac{d\sigma^{(0)}}{dx} - \sigma^{(0)} \tan 2\phi \frac{d\phi}{dx} = 0. \tag{27}$$

It has a solution

$$\sigma^{(0)} = \frac{c_0}{\sqrt{-\cos 2\phi}} = c_0 \sqrt{\frac{\kappa^2+1}{\kappa^2-1}}, \quad c_0 = const. \tag{28}$$

For higher-order terms, we have $(H - h_1 I)\psi_j = -\widehat{R}\psi_{j-1}$ and one should seek solution in the form

$$\psi_j = \sigma_1^{(j)} l_1 + \sigma_2^{(j)} l_2, \tag{29}$$

where $\sigma_2^{(j)}$ is given by

$$\sigma_2^{(j)} = \frac{< l_1^*, R(\psi_{j-1}) >}{2(E-U)}. \tag{30}$$

Then, from the orthogonality condition, $< l_1^*, \widehat{R}(\sigma_1^{(j)} l_1 + \sigma_2^{(j)} l_2) >= 0$ we obtain

$$\sigma_1^{(j)} = \frac{1}{\sqrt{-\cos 2\phi}} \left(c_j - \int \sqrt{-\cos 2\phi} < l_1^*, R(\sigma_2^{(j)} l_2) > dx \right), \quad c_j = const. \tag{31}$$

Below we assume that $q_x > 0$. Thus, to the leading order in classically disallowed domains we have

$$\psi = \frac{e^{\mp \frac{1}{h} S_q(x,x_i)}}{\sqrt{J_q^{\pm}}} l_1^{\pm} (1 + O(h)), \tag{32}$$

where

$$S_q(x, x_i) = \int_{x_i}^{x} q_x dt, \quad J_q^{\pm} = \pm((\kappa^{\pm})^2 - 1),$$

$$l_1^{\pm} = \begin{pmatrix} 1 \\ i\kappa^{\pm} \end{pmatrix}$$

and

$$\kappa^{\pm} = \frac{\pm q_x + p_y}{E - U}.$$

This asymptotic approximation is not valid near turning points $q_x = 0$. The WKB asymptotic solution, derived in this section, is valid for the domains Ω_i, $i = 2, 4$.

6. WKB asymptotic solution for scattering through the smooth barrier

Consider a problem of scattering through the smooth barrier (see Fig. 2) under the assumption that $|E| > |p_y|$ and all four turning points x_i, $i = 1,2,3,4$ are separated. In this case we have again 5 domains Ω_i, $i = 1, 2, ..., 5$ to describe 5 WKB forms of solution to the leading order. In considering a graphene system, if $E > 0$ we observe incident, reflected and transmitted electronic states at $x < a$ and $x > b$, whereas under the barrier $a < x < b$ we have a hole state (n-p-n junction, see Fig. 2).

To formulate the scattering problem for transfer matrix T, here we present the WKB solutions in the domains 1 and 5

$$\psi_1 = \frac{e^{\frac{i}{h} S_p(x,x_1)}}{\sqrt{J_p^{+}}} a_1 e_1^{+} + \frac{e^{-\frac{i}{h} S_p(x,x_1)}}{\sqrt{J_p^{-}}} a_2 e_1^{-}, \tag{33}$$

$$\psi_5 = \frac{e^{\frac{i}{h} S_p(x,x_4)}}{\sqrt{J_p^{+}}} d_1 e_1^{+} + \frac{e^{-\frac{i}{h} S_p(x,x_4)}}{\sqrt{J_p^{-}}} d_2 e_1^{-}. \tag{34}$$

The barrier is represented by the combination of the left and right slopes. The total transfer matrix T, that is $d = Ta$, is given by

$$T = T^R \begin{pmatrix} e^{\frac{i}{h}P} & 0 \\ 0 & e^{-\frac{i}{h}P} \end{pmatrix} T^L, \tag{35}$$

with T^R and T^L the transfer matrices of the right and left slopes (see formulas (137), (143) in Appendix C), respectively, and

$$P = \int_{x_2}^{x_3} \sqrt{(U(x) - E)^2 - p_y^2} dx.$$

The entries of the matrix T read (see formulas (121), (134), (144) in Appendix C)

$$T_{11} = e^{\frac{Q_1}{h} + \frac{Q_2}{h}} [s_1 s_2 e^{i(\theta_1 + \theta_2 + \frac{P}{h})} + e^{-i\frac{P}{h}}], \tag{36}$$

$$T_{22} = e^{\frac{Q_1}{h} + \frac{Q_2}{h}} [s_1 s_2 e^{-i(\theta_1 + \theta_2 + \frac{P}{h})} + e^{i\frac{P}{h}}], \tag{37}$$

$$T_{12} = -\mathrm{sgn}(p_y) e^{\frac{Q_1}{h} + \frac{Q_2}{h}} [s_2 e^{i(\theta_2 + \frac{P}{h})} + s_1 e^{-i(\theta_1 + \frac{P}{h})}], \tag{38}$$

$$T_{21} = -\mathrm{sgn}(p_y) e^{\frac{Q_1}{h} + \frac{Q_2}{h}} [s_2 e^{-i(\theta_2 + \frac{P}{h})} + s_1 e^{i(\theta_1 + \frac{P}{h})}], \tag{39}$$

where $s_i = \sqrt{1 - e^{-2Q_i/h}}, \quad i = 1, 2$. They satisfy the classical properties of the transfer matrix

$$T_{22} = T_{11}^*, \quad T_{21} = T_{12}^*, \quad \det T = 1,$$

and if $a_1 = 1, a_2 = r_1, d_1 = t_1, d_2 = 0$, then

$$t_1 = \frac{1}{T_{22}},$$

$$r_1 = -\frac{T_{21}}{T_{22}},$$

$$|t_1|^2 + |r_1|^2 = 1.$$

If $a_1 = 0, a_2 = t_2, d_1 = r_2, d_2 = 1$, then

$$t_2 = t_1 = t,$$

$$r_2(p_y) = \frac{T_{12}}{T_{22}},$$

$$|t_2|^2 + |r_2|^2 = 1$$

(see appendix B). Correspondingly, the unitary scattering matrix connecting

$$\begin{pmatrix} a_2 \\ d_1 \end{pmatrix} = \hat{S} \begin{pmatrix} a_1 \\ d_2 \end{pmatrix}$$

may be written as follows

$$\hat{S} = \begin{pmatrix} r_1 & t \\ t & r_2 \end{pmatrix}.$$

The transmission coefficient $t = 1/T_{22}$, looks exactly like the formula (131) in [18]. Total transmission takes place only for a symmetric barrier when $Q_2 = Q_1 = Q$ ($\theta_2 = \theta_1 = \theta$). Then

$$t = e^{i\theta}\left(\cos\left(\frac{P}{h} + \theta\right)(2e^{\frac{2Q}{h}} - 1) + i\sin\left(\frac{P}{h} + \theta\right)\right)^{-1}, \tag{40}$$

$$r_1(p_y) = \frac{2\text{sgn}(p_y)\cos\left(\frac{P}{h} + \theta\right)e^{\frac{2Q}{h} + i\theta}\sqrt{1 - e^{-2Q/h}}}{\cos\left(\frac{P}{h} + \theta\right)(2e^{\frac{2Q}{h}} - 1) + i\sin\left(\frac{P}{h} + \theta\right)} = -r_2(p_y). \tag{41}$$

However, it is worth noting that $r_1(p_y) = r_2(-p_y)$. It is clear that if

$$P(E) + h\theta = h\pi(n + \frac{1}{2}), \quad n = 0, 1, 2, \dots, \tag{42}$$

then we have total transmission $|t_1| = 1$.

7. WKB asymptotic solution for complex resonant (quasibound) states localized within the smooth barrier

Consider a problem of resonant states localized within the smooth barrier (see Fig. 2). In the first case when the energy of the electron-hole is greater than the cut-off energy ($E > E_c = |p_y|$), we have five domains Ω_i, $i = 1, 2, \dots, 5$ and five WKB forms of solution to the leading order. To determine the correct radiation conditions that are necessary for the localization, we present WKB solutions in the domains 1 and 5

$$\psi_1 = \frac{e^{-\frac{i}{h}S_p(x,x_1)}}{\sqrt{J_p^-}} a_2 e_1^-, \quad \psi_5 = \frac{e^{\frac{i}{h}S_p(x,x_4)}}{\sqrt{J_p^+}} d_1 e_1^+. \tag{43}$$

If $a_1 = 0, \ \ d_2 = 0$, then

$$T_{22}(E) = 0, \tag{44}$$

and as a result we obtain Bohr-Sommerfeld quantization condition for complex energy eigen-levels (quasi-discrete)

$$P(E) = h(\pi(n + \frac{1}{2}) - \theta - \frac{i}{2} \log (1 - e^{\frac{-2Q}{h}})), \qquad n = 0, 1, 2, ..., N_1 \tag{45}$$

for $|p_y| < E < U_0$. Solutions to this equation are complex resonances $E_n = Re(E_n) - i\Gamma_n$, where Γ_n^{-1} is the lifetime of the localized resonance. What is important is that the real part of these complex positive resonances is decreasing with n, thus showing off the anti-particle hole-like character of the localized modes. For these resonances we have $\Gamma_n > 0$. From (45), we obtain the important estimate

$$\Gamma_n = \frac{hw}{2\Delta t}, \qquad w = -\log (1 - e^{\frac{-2Q}{h}}), \qquad \Delta t = -P'(E_n). \tag{46}$$

that is the equivalent of the formula (14) in [35]. Namely, w is the transmission probability through the tunneling strip, Δt is the time interval between the turning points x_2 and x_3, and P' is the first derivative of P with respect to energy. If $p_y \to 0$, then $Q \to 0$, and $\Gamma_n \to +\infty$, that is opposite to [35] (to be exact, the estimate for Γ_n in [35] works only for a linear potential when p_y is not small).

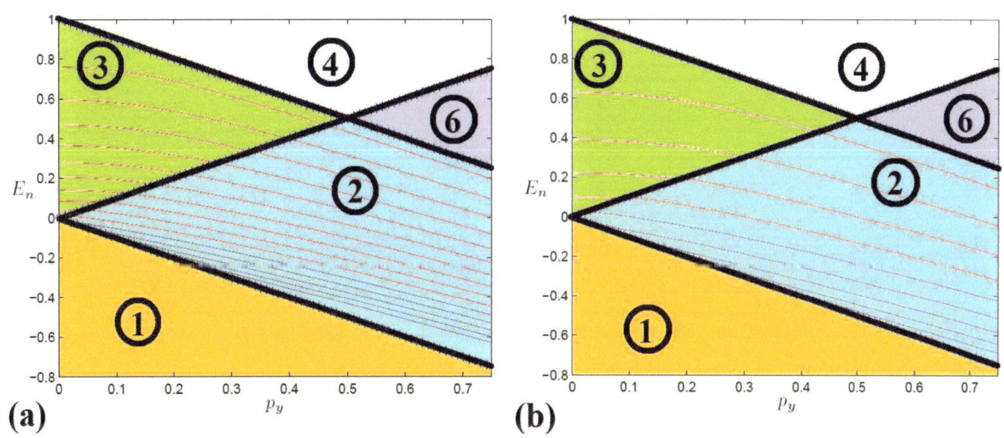

Figure 4. (a) The dispersion of energy levels $E_n(p_y)$ for complex and real bound states for $n = 0, 1, ..., 15$ are shown for $h = 0.1$ and $U = 1/cosh x$. (b) As in (a), except that $n = 0, 1, ..., 9$ and $U = 1/cosh 2x$. For complex resonant bound states the real part was taken. The energies $E = \pm p_y$ and $E = U_0 - p_y$ are shown with thick black lines. Semiclassical solutions are shown by the lines in zones 2 and 3. The upper and lower bounds for the dispersion branches are shown by the boundaries between zones 1, 4 and 6 with zones 2 and 3. The black line $p_y = E$, running between zones 2 and 3 is the upper bound for the bound states $\Gamma_n = 0$.

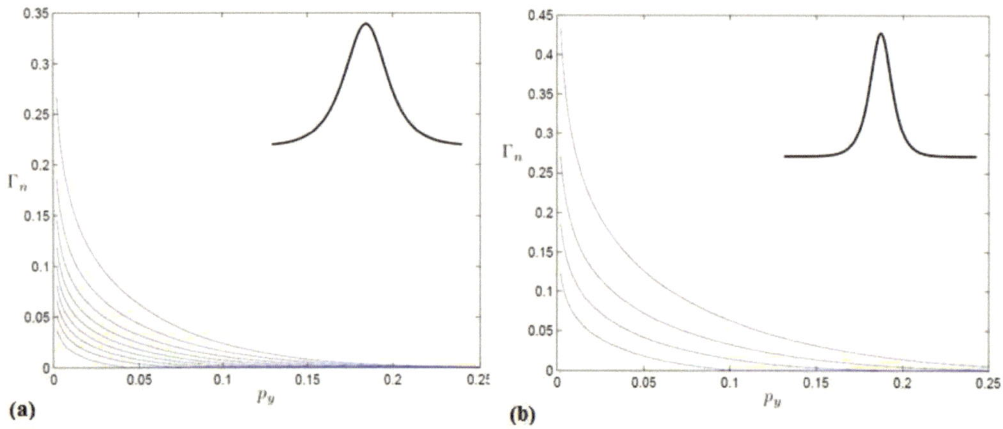

Figure 5. (a) The imaginary part Γ_n of the first nine quasi-bound eigenvalues. The semiclassical solutions are shown by the blue lines and the shape of the potential is shown in the inset ($U = 1/coshx$). (b) There are four quasibound states for Γ_n associated with a $U = 1/cosh2x$ potential. The narrower potential allows less complex bound states.

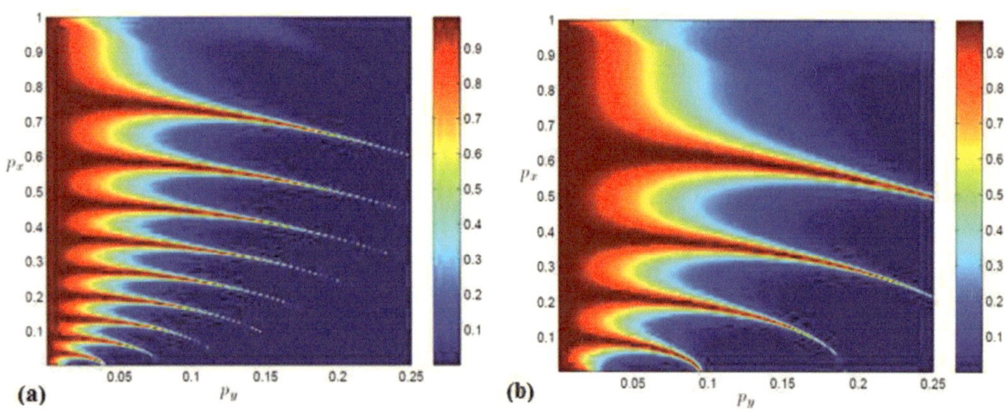

Figure 6. The transmission probability $|t|^2$ is shown in these colorbar diagrams with respect to dimensionless p_y and $p_x = \sqrt{E^2 - p_y^2}$ for the barriers (a) $U = 1/\cosh x$ and (b) $U = 1/\cosh 2x$.

For the second set of real resonances, when the energy of the electron-hole is smaller than the cut-off energy ($E < |p_y|$), we have 2 turning points x_2 and x_3. Between them there are oscillatory WKB solutions

$$\psi_1 = \frac{e^{\frac{i}{\hbar}S_p(x,x_2)}}{\sqrt{J_p^+}}\bar{d}_1 e_1^+ + \frac{e^{-\frac{i}{\hbar}S_p(x,x_2)}}{\sqrt{J_p^-}}\bar{d}_2 e_1^-, \tag{47}$$

or

$$\psi_1 = \frac{e^{\frac{i}{\hbar}S_p(x,x_3)}}{\sqrt{J_p^+}}\bar{a}_1 e_1^+ + \frac{e^{-\frac{i}{\hbar}S_p(x,x_3)}}{\sqrt{J_p^-}}\bar{a}_2 e_1^-, \tag{48}$$

and outside decaying solutions,

$$\psi_1 = \frac{e^{\frac{1}{h}S_q(x,x_2)}}{\sqrt{J_q^-}}\bar{c}_2 l_1^-, \quad x < x_2, \qquad \psi_3 = \frac{e^{-\frac{1}{h}S_q(x,x_3)}}{\sqrt{J_q^+}}\bar{c}_1 l_1^+, \quad x > x_3. \tag{49}$$

By gluing these WKB solutions together through the two boundary layers near x_2 and x_3, using the results in sections 5.1, 5.2 and the Appendix C, we eliminate $\bar{a}_{1,2}$ and $\bar{d}_{1,2}$ and obtain the homogeneous system

$$i\bar{c}_1 + \bar{c}_2 e^{\frac{i}{h}P} = 0,$$

$$i\bar{c}_1 - \bar{c}_2 e^{\frac{-i}{h}P} = 0.$$

Thus, we derive the Bohr-Sommerfeld quantization condition for real energy eigen-levels (bound states) inside the cut-off energy strip for $0 < E < |p_y|$.

$$P(E) = h\pi(n + \frac{1}{2}), \qquad n = N_1 + 1, ...N_2. \tag{50}$$

8. Numerical results

Based upon the analytical descriptions in the preceding sections for the smooth barrier, we present the results for the energy eigenvalues and eigenfunctions. These are shown in Fig's. 4-6 and compare favorably with those obtained through finite difference methods, as detailed in [71]. The energy dispersion curves, $E_n(p_y)$, are shown for the complex resonant and real bound states for $h = 0.1$ and potentials of different widths. In Fig. 4(a), the energy levels are illustrated for the potential, $U = 1/coshx$, with $n = 0, 1, ..., 15$. For complex resonant states the real parts are shown. It must be emphasized that in zone "3", which is restricted by $E < U_0 - p_y$ and $E = p_y$ with $p_y > 0$, the complex quasibound states reside. The bound states are located in zone "2", which lies between $E = \pm p_y$ and $E = U_0 - p_y$. In zone "3" there are nine complex resonances. In Fig. 4(b), the results for a narrower potential of $U = 1/cosh2x$ can be seen (all other parameters being the same as in Fig. 4 (a)). In this case, there are four complex resonances in zone "3" and $n = 0, 1, ..., 9$. The lifetimes of the local resonances, Γ_n, are shown in Fig. 5 (a) and (b) for the same two potentials as described in Fig. 4. The complex quasi-bound states that are witnessed in zone "3" in Fig. 4 are shown in Fig. 5 for Γ_n. The thinner potential allows less complex bound states. The bound states have infinite lifetimes. Both types of states are confined within the barrier in the x-direction, while the motion in the y-direction is controlled by the dispersion relations. In Fig. 6 we present the transmission probabilities $|t|^2$ for the two potentials. There are nine tunneling resonances, i.e. complex quasi-bound states within potential barrier defined as $U = 1/coshx$ that correlate with those shown in Fig. 4 in zone "3". Likewise, for the thinner barrier there are four complex quasi-bound states. These resonance states are a clear indication of the Fabry-Perot multiple interference effects inside the barrier.

PART II: High energy localized eigenstates in graphene monolayers and double layers

9. Graphene resonator in a magnetic field

We consider a spectral problem for the Dirac operator describing the electron-hole quantum dynamics in a graphene monolayer without a gap, in the presence of a homogeneous magnetic field **A** and arbitrary scalar potential $U(\mathbf{x})$ (see [31])

$$v_F < \bar{\sigma}, -ih\nabla + \frac{e}{c}\mathbf{A} > \psi(\mathbf{x}) + U(\mathbf{x})\psi(\mathbf{x}) = E\psi(\mathbf{x}), \quad \psi(\mathbf{x}) = \begin{pmatrix} u \\ v \end{pmatrix}, \tag{51}$$

where $\mathbf{x} = (x_1, x_2)$, and u, v are the components of the spinor wave function that describes electron localization on the sites of sublattice A or B of a honeycomb graphene structure. Here, e is the electron charge, c is the speed of light, **A** is magnetic potential in axial $\mathbf{A} = B/2(-x_2, x_1, 0)$ or Landau gauge $\mathbf{A} = B(-x_2, 0, 0)$ (magnetic field is directed along the x_3 axis), and v_F is the Fermi velocity. The symbol $<,>$ means a scalar product, and $\hbar$ is the Planck constant (which is a small parameter ($\hbar \to 0$) in semiclassical analysis). The vector $\bar{\sigma} = (\sigma_1, \sigma_2)$ with Pauli matrices corresponds to the K Dirac point of the first Brillouin zone (see [31]). The case of the second K' Dirac point with $\bar{\sigma}^* = (\sigma_1, -\sigma_2)$ is treated similarly and is not considered here.

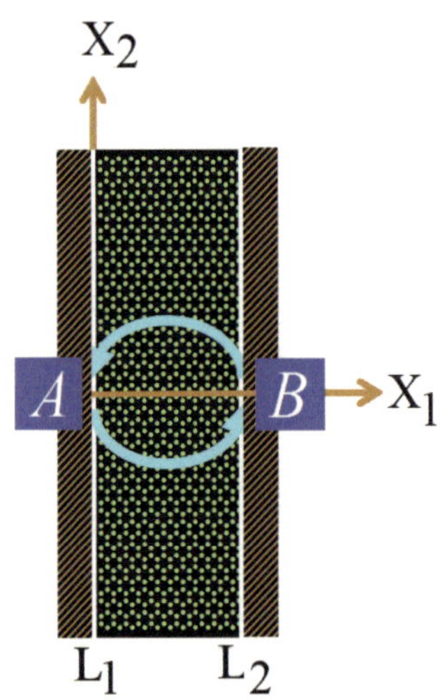

Figure 7. A periodic orbit inside the graphene nanoribbon resonator with magnetic field and electrostatic potential (electronic trajectory). Magnetic field is directed along the x_3 axis, the electrostatic field is piece-wise linear $U(x_2) = \beta|x_2|$.

We study the high energy spectral problem, using the semiclassical approximation, for a vertical graphene nanoribbon confined between two flat reflecting interfaces $L_{1,2}$ (see Fig.7). It is assumed that the spinor wave function satisfies zigzag boundary conditions on the interfaces $L_{1,2}$: $u|_{L_1} = 0, \quad v|_{L_2} = 0$. It will be discussed later that the electrostatic field $U(x_2) = \beta|x_2|$ makes the orbit shown in Fig. 7 periodic and stable. In the gener al case, as it was noted earlier for the Schrödinger operator (see [55]), if high-energy localised eigenstates are sought, which decay exponentially away from the resonator axis AB, the separation of variables will not help construct an exact solution due to the difficulty of satisfying the boundary conditions.

10. Ray asymptotic solution

The WKB ray asymptotic solution to the Dirac equation is sought through consideration of the eigenvalue problem associated with $H\phi = E\phi$. The magnetic vector potential $\mathbf{A} = B/2(-x_2, x_1, 0)$ is given in terms of the axial gauge for a magnetic field. The Hamiltonian of the Dirac system (see equations (2) and (10)) takes the form for monolayer graphene,

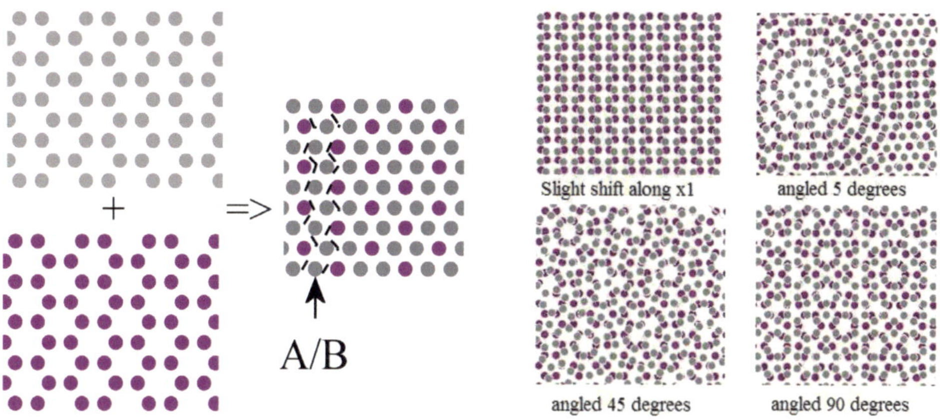

Figure 8. Bilayer graphene consists of two coupled graphene monolayers on top of one another. On the left-hand side, the planes of graphene have perfect Bernal stacking. The A_1 atoms of the sublattice of the top sheet overlap the B_2 atoms of the sublattice of the bottom sheet as is indicated. A triangular structure is seen when looking directly down upon the bilayers, as is schematically shown, and the A_2 and B_1 atoms are over the centers of the hexagonal structure of their opposite sheet. The group of four pictures on the right illustrate what happens if a slight shift of one of the graphene planes occurs. Going clockwise around this group, the first is when one graphene layer moves slightly from the ideal Bernal orientation along the x_1 direction. The second is with a 5° tilt and the third and fourth are with a tilt of the plane of 45° and 90°, respectively.

$$H = \begin{pmatrix} U(\mathbf{x}) & v_F[\hbar(-i\partial_{x_1} - \partial_{x_2}) - i\frac{\alpha x_1}{2} - \frac{\alpha x_2}{2}] \\ v_F[\hbar(-i\partial_{x_1} + \partial_{x_2}) + i\frac{\alpha x_1}{2} - \frac{\alpha x_2}{2}] & U(\mathbf{x}) \end{pmatrix},$$

In contrast, for bilayer graphene the Hamiltonian takes the form,

$$H = v_F \begin{pmatrix} g & p_x - ip_y & 0 & \zeta \\ p_x + ip_y & g & 0 & 0 \\ 0 & 0 & g & p_x - ip_y \\ \zeta & 0 & p_x + ip_y & g \end{pmatrix},$$

where $g \approx 0.4eV/v_F$ is the interlayer coupling. We consider the case when bilayer graphene has Bernal stacking as shown in Fig. 6. Bernal-stacked bilayer graphene occurs with half of the carbon atoms in the second layer sitting on top of the empty centers of hexagons in the first layer. An external electric field can tune its bandgap by up to $250meV$ [32]. This form of structure of bilayer graphene can be experimentally created using chemical vapor deposition [32]. Considering the low energy states of electrons, we can reduce the 4×4 matrix describing the bilayer graphene to a form similar to that of monolayer graphene [56]. The only difference from the monolayer form is the squaring of the off-diagonal entries and the inclusion of a band mass for bilayer electrons.

$$H = \begin{pmatrix} U(\mathbf{x}) & \frac{1}{2m}[\hbar(-i\partial_{x_1} - \partial_{x_2}) - i\frac{\alpha x_1}{2} - \frac{\alpha x_2}{2}]^2 \\ \frac{1}{2m}[\hbar(-i\partial_{x_1} + \partial_{x_2}) + i\frac{\alpha x_1}{2} - \frac{\alpha x_2}{2}]^2 & U(\mathbf{x}) \end{pmatrix}.$$

It is now convenient to introduce some dimensionless variables. The coordinate system $x \Rightarrow xD$, where D is a characteristic scale associated with a change in the potential (and correspondingly $U(x) \Rightarrow U(xD)$). Then, we write $\widetilde{U} = U(xD)/E_0$; where we define the characteristic energy scale as $E \Rightarrow E_0 E$. For single layers of graphene the small parameter, $h << 1$, is $h = v_F \hbar / U_0 D$. In double layer graphene it is slightly different; $h = \alpha D/\sqrt{2mU_0}$, with the magnetic length, as a function of the applied magnetic field, given to be $\widetilde{\alpha} = \alpha D/\sqrt{2mE_0}$ with $\alpha = eB/c$. We now write the dimensionless forms of the one and two layer graphene systems as (with the tildes omitted for brevity),

$$H = \begin{pmatrix} U(\mathbf{x}) & h(-i\partial_{x_1} - \partial_{x_2}) - i\frac{\alpha x_1}{2} - \frac{\alpha x_2}{2} \\ h(-i\partial_{x_1} + \partial_{x_2}) + i\frac{\alpha x_1}{2} - \frac{\alpha x_2}{2} & U(\mathbf{x}) \end{pmatrix} \quad (52)$$

and

$$H = \begin{pmatrix} U(\mathbf{x}) & (h(-i\partial_{x_1} - \partial_{x_2}) - i\frac{\alpha x_1}{2} - \frac{\alpha x_2}{2})^2 \\ (h(-i\partial_{x_1} + \partial_{x_2}) + i\frac{\alpha x_1}{2} - \frac{\alpha x_2}{2})^2 & U(\mathbf{x}) \end{pmatrix} \quad (53)$$

The solution for monolayer graphene is sought in the same form as equation (11). Substituting this series into the Dirac system, and equating to zero the corresponding coefficients of successive degrees of the small parameter h, we obtain a recurrent system of equations which determines the unknown $S(\mathbf{x})$ and $\psi_j(\mathbf{x})$.

The Hamiltonian H has two eigenvalues. In the domain $\Omega_e = \{\mathbf{x} : E > U(\mathbf{x})\}$, the Hamiltonian eigenvalue $h_1 = U(\mathbf{x}) + p$ on the level set $h_1 = E$ describes the dynamics of electrons. The corresponding classical trajectories can be obtained from the Hamiltonian system $\dot{x} = H^e_p, \quad \dot{p} = -H^e_x, \quad \mathbf{x} = (x_1, x_2), \ \mathbf{p} = (p_1, p_2)$, with an equivalent Hamiltonian (see [48])

$$H^e = \frac{1}{2}\Big((p_1 - \frac{\alpha x_2}{2})^2 + (p_2 + \frac{\alpha x_1}{2})^2 - (E - U(\mathbf{x}))^2 \Big)$$

on the level set $H^e = 0$ with $p_1 = S_{x_1}$ and $p_2 = S_{x_2}$. Opposite to this case, in the domain $\Omega_h = \{\mathbf{x} : E < U(\mathbf{x})\}$, the Hamiltonian eigenvalue $h_2 = U(\mathbf{x}) - p$ on the level set $h_2 = E$ describes the dynamics of holes. The corresponding classical trajectories can be obtained from the Hamiltonian system with Hamiltonian

$$H^h = \frac{1}{2}\Big(- (p_1 - \frac{\alpha x_2}{2})^2 - (p_2 + \frac{\alpha x_1}{2})^2 + (E - U(\mathbf{x}))^2 \Big)$$

on the level set $H^h = 0$. The Hamiltonian dynamics with $h_{1,2}$ or with $H^{e,h}$ are equivalent (see [48]).Classical action $S(x)$ satisfies the Hamilton-Jacobi equation in the monolayer case to be

$$[(S_{x_1} - \frac{\alpha x_2}{2}) + (S_{x_2} + \frac{\alpha x_1}{2})^2] - (E - U(\mathbf{x}))^2 = 0. \tag{54}$$

Likewise, in the case of bilayers,

$$H^e = \left[(p_1 - \frac{\alpha x_2}{2})^2 + (p_2 + \frac{\alpha x_1}{2})^2\right]^2 - (E - U(\mathbf{x}))$$

and

$$H^h = \left[(p_1 - \frac{\alpha x_2}{2})^2 + (p_2 + \frac{\alpha x_1}{2})^2\right]^2 + (E - U(\mathbf{x}))$$

The Hamiliton-Jacobi equation is satisfied in the two-layers of graphene case by,

$$[(S_{x_1} - \frac{\alpha x_2}{2})^2 + (S_{x_2} + \frac{\alpha x_1}{2})^2]^2 - (E - U(\mathbf{x}))^2 = 0. \tag{55}$$

The solutions of the Hamiltonian-Jacobi equations, for monolayer and bilayers, electrons and holes, are given by the following curvilinear integrals over classical trajectories connecting points $M^{(0)}$ and $M = (x_1, x_2)$ ($M^{(0)}$ is fixed and M is variable)

$$S(M) = \int_{M^{(0)}}^{M} [E - U(\mathbf{x}(s))]ds - \frac{e}{c}\int_{M^{(0)}}^{M} \mathbf{A}d\mathbf{x} = \tag{56}$$

$$\int_{M^{(0)}}^{M} \left([E - U(\mathbf{x}(t))]\sqrt{\dot{x}_1^2 + \dot{x}_2^2} - \frac{\alpha}{2}(-x_2\dot{x}_1 + x_1\dot{x}_2)\right)dt,$$

$$S(M) = \int_{M^{(0)}}^{M} \sqrt{E - U(\mathbf{x}(s))}ds - \frac{e}{c}\int_{M^{(0)}}^{M} \mathbf{A}d\mathbf{x} = \tag{57}$$

$$\int_{M^{(0)}}^{M} \left(\sqrt{E - U(\mathbf{x}(t))}\sqrt{\dot{x}_1^2 + \dot{x}_2^2} - \frac{\alpha}{2}(-x_2\dot{x}_1 + x_1\dot{x}_2)\right)dt,$$

for mono and bi-layer, respectfully, where s is the arc length. This representation is correct in the neighbourhood of a regular family of classical trajectories emanating from $M^{(0)}$. For electrons and holes one can seek solution to the Dirac system zero-order problem in the form

$$\psi_0 = \sigma^{(0)}(\mathbf{x})\mathbf{e}_1 \tag{58}$$

with unknown amplitude $\sigma^{(0)}(\mathbf{x})$. Solvability of the problem

$$(H - EI)\psi_1 = -\widehat{R}\psi_0, \qquad E = h_{1,2}$$

requires that the orthogonality condition with complex conjugation

$$< \mathbf{e}_1, \widehat{R}(\sigma^{(0)}(\mathbf{x})\mathbf{e}_1) >= 0$$

must hold, where

$$Monolayer: \ \ \widehat{R} = \begin{pmatrix} 0 & \partial_x - i\partial_y \\ \partial_x + i\partial_y & 0 \end{pmatrix},$$

$$Bilayer: \widehat{R} = \begin{pmatrix} 0 & Y_1 \\ Y_1^* & 0 \end{pmatrix},$$

where,

$$Y_1 = 2\left(S_{x_1} - iS_{x_2} - \frac{i\alpha x_1}{2} - \frac{i\alpha x_2}{2}\right)\left(\partial_{x_1} - i\partial_{x_2}\right) + S_{x_1x_1} - 2iS_{x_1x_2} - S_{x_2x_2}$$

Using the basic elements of the techniques described in [48], from the orthogonality condition one obtains the transport equation for $\sigma^{(0)}(\mathbf{x})$. The geometrical spreading for an electron or hole with respect to the Hamiltonian system with $h_{1,2} = U \pm v_F p$ has a solution

$$Monolayer: \ \ \sigma^{(0)} = \frac{c_0}{\sqrt{J}} e^{-i\frac{\theta}{2}}, \tag{59}$$

$$Bilayer: \ \ \sigma^{(0)} = \frac{c_0}{\sqrt{J}} e^{-i\theta}, \quad c_0 = const, \tag{60}$$

where

$$J(t, \gamma) = \left| \frac{\partial(x_1, x_2)}{\partial(t, \gamma)} \right| \tag{61}$$

where we have introduced θ, which is the angle made by the velocity of the particle trajectory with the x_1 axis:

$$\left(p_1 - \frac{\alpha x_2}{2}\right) + i\left(p_2 - \frac{\alpha x_1}{2}\right) = pe^{i\theta} \tag{62}$$

Here $-\theta/2$ is the adiabatic phase for monolayer graphene, as introduced by Berry [64]. Chirality results in a Berry phase of θ in bilayer graphene and the confinement of electronic states. Conservation of chirality in monolayer graphene means that the particles cannot backscatter and this leads to normal incidence transmission equal to unity. This is not the case in bilayer graphene and backscattering can occur.

http://dx.doi.org/10.5772/52267

11. Construction of eigenfunctions, periodic orbit stability analysis and quantization conditions

Let $\mathbf{x}_0 = (x_1(s), x_2(s))$ be a particle (electron or hole) classical trajectory, where s is the arc length measured along the trajectory. Consider the neighborhood of the trajectory in terms of local coordinates s, n, where n is the distance along the vector, normal to the trajectory, such that

$$\mathbf{x} = \mathbf{x}^{(0)}(s) + \mathbf{e_n}(s)n, \tag{63}$$

where $\mathbf{e_n}(s)$ is the unit vector normal to the trajectory. Introducing $\nu = n/\sqrt{\hbar} = O(1)$, we seek an asymptotic solution to the Dirac system related to (2) where $S_0(s)$ and $S_1(s)$ are chosen similar to [55], [66] as they give a linear approximation for solution to the Hamilton-Jacobi equation (55) (see [55], [66])). The parameter for monolayers

$$a(s) = E - U_0(s)$$

and for Bernal bilayers,

$$a(s) = \sqrt{E - U_0(s)}$$

is obtained from the approximation

$$U(\mathbf{x}) = U_0(s) + U_1(s)n + U_2(s)n^2 + \ldots .$$

In the following $\gamma_i(s),\ \ i = 1,2$ are the Cartesian components of $\mathbf{e_n}(s)$. Following [54], [55], [65], and [66], we apply the asymptotic boundary-layer method to the Dirac system (2). We allow that the width of the boundary layer is determined by $|n| = O(\sqrt{\hbar})$ as $\hbar \to 0$. We assume that we deal with a continuous family of POs symmetric with respect to both axes (see Fig. 4). Thus, the trajectory of the PO consists of two symmetric parts between two reflection points A and B. We seek the asymptotic solution of the eigenfunction for electrons and holes localized in the neighborhood of a PO as a combination of two Gaussian beams

$$\psi(\mathbf{x}, E) = \psi_1(\mathbf{x}, E) + \widehat{R}\psi_2(\mathbf{x}, E),$$

described by

$$\psi_{1,2}(\mathbf{x}, E) = \exp\left(\frac{i}{\hbar}(S_0^{(1,2)}(s) + S_1(s)n + \frac{1}{2}\frac{p_{1,2}(s)}{z_{1,2}(s)}n^2)\right)$$

$$e^{-i\theta/2}\frac{Q_m(z_{1,2}(s), \nu)}{\sqrt{z_{1,2}(s)}}\mathbf{e_1}\left(1 + O(\hbar^{1/2})\right),$$

where

$$S_0^{(1)}(s) = \int_0^s \left(a(s') + \frac{\alpha}{2}(x_1^{(0)}\gamma_1 + x_2^{(0)}\gamma_2)\right)ds', \quad 0 < s < s_0,$$

$$S_0^{(2)}(s) = S_0^{(1)}(s_0) + \int_{s_0}^{s} \left(a(s') + \frac{\alpha}{2}(x_1^{(0)}\gamma_1 + x_2^{(0)}\gamma_2) \right) ds', \quad s_0 < s < 2s_0.$$

$$S_1(s) = \frac{\alpha}{2}\left(x_2^{(0)}\gamma_1 - x_1^{(0)}\gamma_2 \right), \tag{64}$$

where we have defined the Berry phase to be, $e^{i\theta} = \gamma_2 - i\gamma_1$, and $\mathbf{e}_{1,2} = (1, e^{i\theta})/\sqrt{2}$. Here, each beam is related to the corresponding part of the periodic orbit. Namely, ψ_1 is determined by $z = z_1(s)$, $p = p_1(s)$ for $0 < s < s_0$, describing the electrons propagation along the lower part of the orbit from A to B, whereas ψ_2 is determined by $z = z_2(s)$, $p = p_2(s)$ for $s_0 < s < 2s_0$, for the electrons propagation along the upper part of the orbit from B to A. The complete derivation of the electronic Gaussian beams for monolayer graphene can be found in the work of Zalipaev [66]. Following the methodology developed in the previous work [66], we state the problem in terms of the function $z(s)$ and write the Hamiltonian in its terms,

$$\dot{z} = \frac{p}{a(s)}, \quad \dot{p} = -a(s)d(s)z \tag{65}$$

with the Hamiltonian,

$$H(z,p) = a(s)\frac{\dot{z}^2}{2} - \frac{z^2}{2} \tag{66}$$

The above are the same for both mono and bilayer graphene, but with different $d(s)$ (and $a(s)$, as mentioned above),

$$Monolayer: d(s) = \frac{2}{a(s)}(U_2 - \frac{U_1}{\rho}) + \frac{\alpha}{\rho a(s)}$$

$$Bilayer: d(s) = \frac{U_2}{E - U_0} + \frac{U_1^2}{4(E-U_0)^2} - \frac{U_1}{\rho(E-U_0)^2} - \frac{\alpha}{\rho a(s)}$$

where $\rho(s)$ is the radius of curvature of a trajectory. Thus, $(z_1(s), p_1(s))$ and $(z_2(s), p_2(s))$ define $(z(s), p(s))$ for $0 < s < 2s_0$. The asymptotic localized solution of Gaussian beam $\psi(s,n)$ is constructed in an asymptotically small neighbourhood of the PO. This solution is to be periodic with respect to $s \in \mathbf{R}$ with the period $2s_0$, and satisfies the zigzag boundary conditions. The reflection coefficient R is derived in the short-wave approximation, and given by

$$Monolayer: R = i\exp\left[i(2\gamma + \frac{\Delta}{2})\right], \tag{67}$$

where γ is the angle of incidence, and $\delta_1 = \theta(s_0 + 0) - \theta(s_0 - 0)$. In the bilayer graphene system the reflection coefficient is,

$$Bilayer: R = i\exp\left[i(4\gamma + \Delta)\right], \tag{68}$$

The localized solution can be constructed if $z(s)$, $p(s)$ is a complex (in the complex phase space $C^2_{z,p}$) quasi-periodic Floquet solution of Hamiltonian system (65) with periodic coefficients (see [50], [54]). Namely, for the monodromy 2×2 matrix M, describing the mapping for the period $2s_0$,

$$\begin{pmatrix} z(s+2s_0) \\ p(s+2s_0) \end{pmatrix} = \begin{pmatrix} M_{11} & M_{12} \\ M_{21} & M_{22} \end{pmatrix} \begin{pmatrix} z(s) \\ p(s) \end{pmatrix}, \tag{69}$$

a Floquet solution for arbitrary s is defined as

$$M \begin{pmatrix} z(s) \\ p(s) \end{pmatrix} = \lambda \begin{pmatrix} z(s) \\ p(s) \end{pmatrix}. \tag{70}$$

The structure of the monodromy matrix M is given by the following product

$$M = M_2 R^B M_1 R^A, \qquad \mathbf{det} M = 1, \tag{71}$$

where $M_1 = M_1(s_0)$ and $M_2 = M_2(2s_0)$ are fundamental matrices of the system (65) describing the evolution $(z(s), p(s))$ for $0 < s < s_0$ and $s_0 < s < 2s_0$, correspondingly.

The reflection matrices at points A and B (see Fig. 6), R^A and R^B are given by

$$\begin{pmatrix} z_1(0) \\ p_1(0) \end{pmatrix} = R^A \begin{pmatrix} z_2(2s_0) \\ p_2(2s_0) \end{pmatrix}, \qquad R^A = \begin{pmatrix} -1 & 0 \\ R^A_{21} & -1 \end{pmatrix},$$

$$\begin{pmatrix} z_2(s_0) \\ p_2(s_0) \end{pmatrix} = R^B \begin{pmatrix} z_1(s_0) \\ p_1(s_0) \end{pmatrix}, \qquad R^B = \begin{pmatrix} -1 & 0 \\ R^B_{21} & -1 \end{pmatrix},$$

$$R^A_{21} = R^B_{21} = 2\alpha \tan(\gamma),$$

where γ is the angle of incidence of the trajectory at the points A, B. To attain R^A and R^B, the classical action S of the phase function at the reflecting boundary requires continuity to be set between the incident and the reflected beams (see [50], [54], [51]).

In a general case, the entries of $M_{1,2}$ are to be determined numerically as the Hamiltonian system has variable coefficients. It is worth to remark that all the multipliers in (71) are symplectic matrices. Thus, the monodromy matrix M is symplectic.

The classical theory of linear Hamiltonian systems with periodic coefficients states that, if $|TrM| < 2$, we have a stable PO (elliptic fixed point, for example, see [27]), and $||M^n|| < const$ for arbitrary $n \in \mathbf{N}$. Then, there exist two bounded complex Floquet's solutions for $-\infty < s < +\infty$, namely, $(z(s), p(s))$ and $(\bar{z}(s), \bar{p}(s))$ with Floquet's multipliers $\lambda_{1,2} = e^{\pm i\varphi}$ $(0 < \varphi < \pi)$, which are solutions of

$$\lambda^2 - TrM\lambda + 1 = 0.$$

These solutions $(z(s), p(s))$ and $(\bar{z}(s), \bar{p}(s))$ may be obtained as follows. If $|TrM| < 2$, the monodromy matrix has complex eigenvectors $w = (w_z, w_p)$ and $\bar{w} = (\bar{w}_z, \bar{w}_p)$

$$M\begin{pmatrix} w_z \\ w_p \end{pmatrix} = e^{i\varphi}\begin{pmatrix} w_z \\ w_p \end{pmatrix}, \quad M\begin{pmatrix} \bar{w}_z \\ \bar{w}_p \end{pmatrix} = e^{-i\varphi}\begin{pmatrix} \bar{w}_z \\ \bar{w}_p \end{pmatrix}.$$

Then, the first solution is determined by

$$M_1(s)R^A\begin{pmatrix} w_z \\ w_p \end{pmatrix} = \begin{pmatrix} z(s) \\ p(s) \end{pmatrix}, \qquad 0 \leq s \leq s_0,$$

$$M_2(s)R^B\begin{pmatrix} z(s_0-0) \\ p(s_0-0) \end{pmatrix} = \begin{pmatrix} z(s) \\ p(s) \end{pmatrix}, \qquad s_0 \leq s \leq 2s_0.$$

Satisfaction of the solution to the zigzag boundary condition at the interface L_1 ($u|_{L_1} = 0$), as well as the requirement that the solution should be periodic with respect to $s \in \mathbf{R}$ with the period $2s_0$ lead to a generalized Bohr-Sommerfeld quantization condition determining semiclassical asymptotics of the high energy spectrum. Namely, after the integration around the closed loop of PO, the total variation of the classical action S and the phase of the amplitude of ψ_0 must be equal to $2\pi m_1$. Thus, we obtain the quantization condition for electrons and holes in the form

$$\int_0^{2s_0} a(s)ds - \alpha A = h[\pm 2\pi m_1 + (m_2 + 1/2)\varphi + \Delta], \tag{72}$$

$$Monolayer: \Delta = \pi - \frac{\gamma}{2},$$

$$Bilayer: \Delta = -\gamma$$

where $m_{1,2} \in \mathbf{N}$ are the longitudinal and the transversal quantization indexes, and for electrons we have $+$, for holes $-$. The index m_2 and factor Δ appear due to the variation of the phase of $\sigma_{m_2}(s, \nu)$ (see the formulas in [66]). Here in the left-hand side in (72)

$$A = \frac{1}{2}\int_0^{2s_0}((x_1^{(0)}\gamma_1 + x_2^{(0)}\gamma_2))ds$$

is the area encircled by PO.

Assuming the presence of a continuous family of POs that depend on E, the quantization condition is satisfied only for a discrete set of energy levels $E = E_{m_1,m_2}$. It is clear that the quantization condition may be fulfilled only if the longitudinal index m_1 is positive and large as $h \to 0$. At the same time, the transversal index $m_2 = 0, 1, 2, ...$ should be of the order 1 as very large values of m_2 would lead to the asymptotic solution $\psi = \psi_1 + R\psi_2$ becoming not localized.

12. Numerical results

In this section we concentrate on the example for monolayer graphene with piece-wise linear potential $U(x_2) = \beta|x_2|$. The numerical techniques used in this section are described in [55]. We deal with the Dirac system (2) by incorporating the Landau gauge $\mathbf{A} = B(-x_2, 0, 0)$. Thus, using dimensionless U, E, α and dimensionless coordinates, the Dirac system is written in the following form

$$\begin{pmatrix} U(x_2) - E & h(-i\partial_x - \partial_y) - \alpha x_2 \\ h(-i\partial_x + \partial_y) - \alpha x_2 & U(x_2) - E \end{pmatrix} \begin{pmatrix} u \\ v \end{pmatrix} = \begin{pmatrix} 0 \\ 0 \end{pmatrix}. \tag{73}$$

The energy in eV is given by $U_0 E$, where $U_0 = v_F \hbar / hD = 6.59 meV/h$ is the characteristic scale of the potential U. Here we assume that $D = 10^{-7} m$. A new small dimensionless parameter h $(0 < h << 1)$ is supposed to be predetermined. The magnetic induction amplitude is given by $B = \alpha c U_0 / v_F e D = \alpha / h\ 6.5910^{-2} T$. Consider, as an example, a family of continuous POs which are symmetric with respect to both axes, with two reflection points A, B (see Fig. 1). The formulas describing electronic POs as solutions of the corresponding integrable system with the Hamiltonian in the Landau gauge

$$H = \frac{1}{2}((p_1 - \alpha x_2)^2 + p_2^2 - (E - U(x_2))^2), \tag{74}$$

on the level set $H = 0$, are easily obtained and given by

$$x_1 = f_1(t, \pi_1, \pi_2, \beta) = (\pi_1 - \alpha\frac{\alpha\pi_1 + E\beta}{\Omega^2})t + \frac{\pi_2\alpha}{\Omega^2}(\cos\Omega t - 1) + \alpha\frac{\alpha\pi_1 + E\beta}{\Omega^3}\sin\Omega t,$$

$$x_2 = f_2(t, \pi_1, \pi_2, \beta) = \frac{\alpha\pi_1 + E\beta}{\Omega^2}(1 - \cos\Omega t) + \frac{\pi_2}{\Omega}\sin\Omega t, \tag{75}$$

for the lower part $0 < t < t_0$, and

$$x_1 = D + f_1(t, -\pi_1, -\pi_2, -\beta),$$

$$x_2 = f_2(t, -\pi_1, -\pi_2, -\beta), \tag{76}$$

for the upper part $0 < t < t_0$. Here π_1 and π_2 are the initial values of the components of momentum p_1 and p_2 at the point A, and $\Omega = \sqrt{\alpha^2 - \beta^2}$. It is important that $\alpha > \beta$. In this case a drift motion of electrons and holes takes place in the positive direction of the x_1 axis, from the point A to the point B (see Fig. 6). This fact helps to construct POs. We assume that everywhere in a domain, in which we construct asymptotic solutions for electronic eigenfunctions, that the inequality $E > U(x_2)$ holds. In equations (75-76) π_1 is a fixed parameter, whereas π_2 and t_0 as functions of π_1 are determined uniquely by the equations $f_1(t_0,\pi_1,\pi_2,\beta) = D, \quad f_2(t_0,\pi_1,\pi_2,\beta) = 0$. The formulas, describing PO holes with Hamiltonian

$$H = \frac{1}{2}\big(- (p_1 - \alpha x_2)^2 - p_2^2 + (E - U(x_2))^2 \big), \tag{77}$$

on the level set $H = 0$, are given by

$$x_1 = g_1(t,\pi_1,\pi_2,\beta) = -(\pi_1 - \alpha\frac{\alpha\pi_1 - E\beta}{\Omega^2})t + \frac{\pi_2\alpha}{\Omega^2}(\cos\Omega t - 1) - \alpha\frac{\alpha\pi_1 - E\beta}{\Omega^3}\sin\Omega t,$$

$$x_2 = g_2(t,\pi_1,\pi_2,\beta) = \frac{\alpha\pi_1 - E\beta}{\Omega^2}(1 - \cos\Omega t) - \frac{\pi_2}{\Omega}\sin\Omega t, \tag{78}$$

for the upper part $0 < t < t_0$, and

$$x_1 = D + g_1(t,-\pi_1,-\pi_2,-\beta),$$

$$x_2 = g_2(t,-\pi_1,-\pi_2,-\beta), \tag{79}$$

for the lower part $0 < t < t_0$. It is worth to remark that holes move along a clockwise direction of PO whereas electrons run counter-clockwise around the PO contour. Thus, we have for electrons and holes the continuous family of POs with respect to parameter π_1 which look like lens-shaped contours. As soon as the parameter π_1 has been determined from the generalized Bohr-Sommerfeld quantization condition (72), the semiclassical energy levels for electrons and holes are computed by

$$E = \pm\sqrt{\pi_1^2 + \pi_2^2}. \tag{80}$$

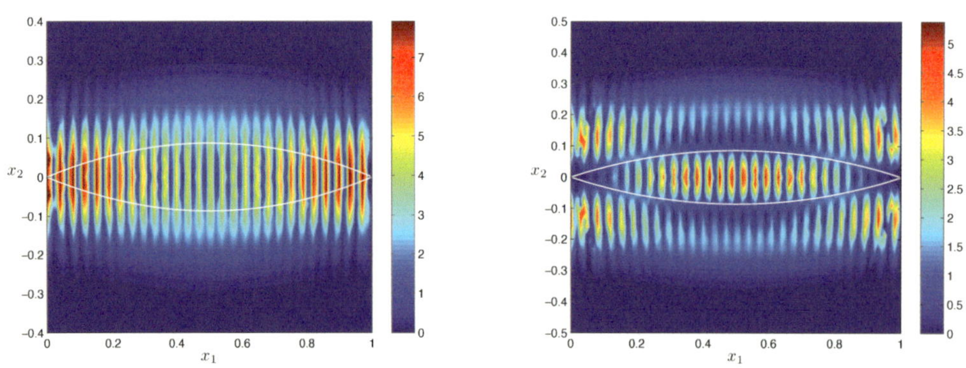

(a) (b)

Figure 9. Electronic eigenfunction density component $|u|^2$ computed by semiclassical analysis for the state $m_1 = 27, m_2 = 0$ with $E = 2.2538$ - (a) and $m_1 = 27.1, m_2 = 1$ with $E = 2.2812$ - (b), for $\alpha = 1$, $\beta = 0.5$, $h = 0.025$.

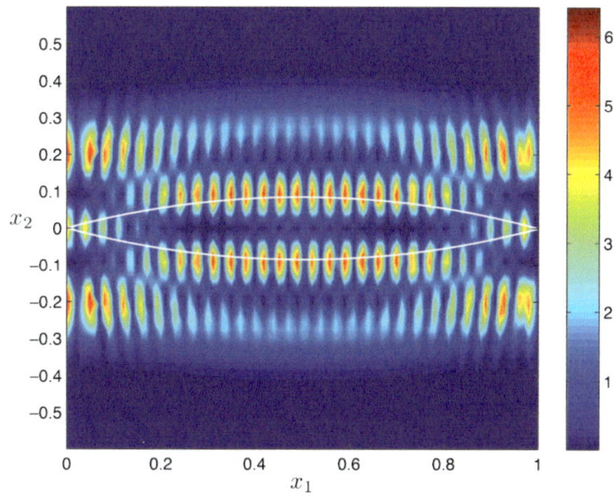

Figure 10. Electronic eigenfunction density component $|u|^2$ computed by semiclassical analysis for the state $m_1 = 27.2, m_2 = 2$ with $E = 2.3078$ for $\alpha = 1$, $\beta = 0.5$, $h = 0.025$.

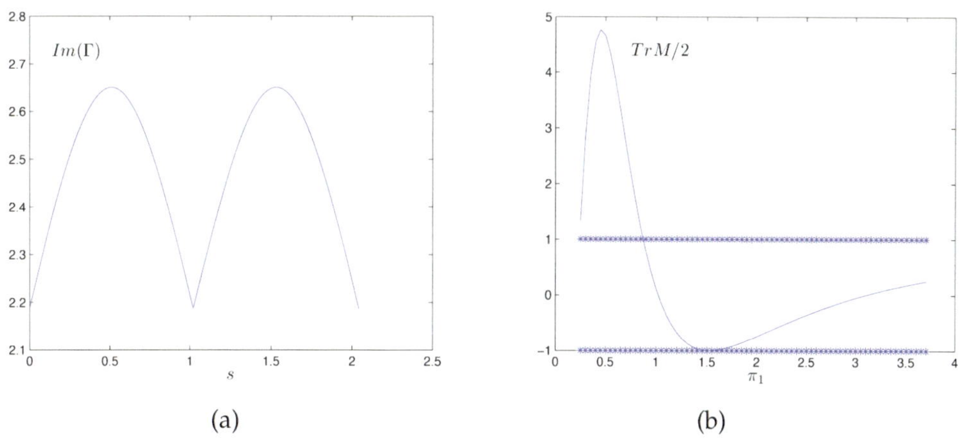

Figure 11. Dependence of $TrM/2$ on π_1 - (a) and dependence of $Im(\Gamma)$ on s - (b), for the state $m_1 = 27, m_2 = 0$ with $E = 2.2538$ for $\alpha = 1$, $\beta = 0.5$, $h = 0.025$.

For the lens-shaped class of POs the high-energy semiclassical localized eigenstates were tested successfully against the energy eigenvalues and the eigenfunctions computed by the finite element method using COMSOL (see [70]). The boundary conditions

$$u|_{x_1=D} = 0, \qquad v|_{x_1=0} = 0.$$

were used in the following numerical experiments.

In Fig's. 9 and 10, the electronic eigenfunction density component $|u|^2$ is shown that was computed semiclassically for the the states $m_1 = 27, m_2 = 0, 1, 2$ with $E = 2.2538,\ 2.2812,\ 2.3078$ for $h = 0.025$, $\alpha = 1.0$, $\beta = 0.5$. It is worth to remark that in this case the localization of eigenfunction density components takes place in close neighborhood of PO. In all shown figures computed by semiclassical analysis one can easily see a white contour of PO. In Fig. 9 (a) dependence of $TrM/2$ on π_1 - (a) and dependence of $Im(\Gamma)$ on s - (b), for the state $m_1 = 27, m_2 = 0$ with $E = 2.2538$ $\alpha = 1$, $\beta = 0.5$, $h = 0.025$ are shown.

13. Conclusion

In this review we have outlined our work on the semiclassical analysis of graphene structures and introduced some new results for monolayer and bilayer graphene. We have outlined a range of new asymptotic methods and a semiclassical analysis of Dirac electron-hole tunneling through a Gaussian shaped barrier that represents an electrostatic potential. We started by analyzing the rectangular barrier and have found some important differences between it and the smooth barrier. Namely, the smooth barrier exhibits a quasi-discrete spectrum and complex bound states that do not exist for the rectangular barrier (in zone "3" in Fig. 3). In both types of barrier Klein tunneling occurs. The WKB approximation deals with the asymptotic analysis of matched asymptotic techniques and boundary layers for the turning points in the barrier. The main results of this work are eloquent WKB formulas for the entries in a smooth barrier transfer matrix. This matrix explains the mechanism of total transmission through the barrier for some resonance values of energy of skew incident electrons or holes. Moreover, it has been shown that the existence of modes localized within the barrier, and exponentially decaying away from it, for two discrete complex and real sets of energy eigenlevels can be determined by the Bohr-Sommerfeld quantization condition. It was shown that the total transmission through the barrier takes place when the energy of an incident electron or hole coincides with a real part of the complex energy eigenlevel of one among the set of localized modes. These facts were confirmed by numerical simulations done by finite elements methods and have been found to also be in excellent agreement with the results found using finite difference methods as in [71].

We have also applied the Gaussian beam methods, originated by Popov [51] and expanded by Zalipaev [66] to quantum problems, to describe monolayer and bilayer graphene. We have constructed eigenfunctions and defined the stable periodic orbit conditions and the quantization conditions. The reflection and transmission coefficients of monolayer and bilayer graphene have been derived within the context of semiclassical physics in full. It is clear that these methods can offer good insights into the behavior of the graphene Fabry-Perot resonator.

Such systems will find applications in plasmonics, and nanoribbon heterostructures made from graphene are promising to emerge. The kind of bilayer structure analyzed here can be created by chemical vapor deposition [32] and this opens up the road to a flurry of geometrically optimized graphene resonator systems, whether acting in isolation or as part of a composite, or array.

14. Appendix A. Transfer and scattering matrix properties for a smooth step

Let us formulate this scattering problem in terms of transfer matrix T for the left slope of the entire barrier (see [67])

$$\psi_1 = \frac{e^{\frac{i}{h}S_p(x,x_1)}}{\sqrt{J_p^+}} a_1 e_1^+ + \frac{e^{-\frac{i}{h}S_p(x,x_1)}}{\sqrt{J_p^-}} a_2 e_1^-, \qquad x \in \Omega_1, \tag{81}$$

$$\psi_3 = \frac{e^{\frac{i}{h}S_p(x,x_2)}}{\sqrt{J_p^+}} d_1 e_1^+ + \frac{e^{\frac{-i}{h}S_p(x,x_2)}}{\sqrt{J_p^-}} d_2 e_1^-, \qquad x \in \Omega_3, \tag{82}$$

and

$$d = Ta, \quad T = \begin{pmatrix} T_{11} & T_{12} \\ T_{21} & T_{22} \end{pmatrix}, \quad d = \begin{pmatrix} d_1 \\ d_2 \end{pmatrix}, \quad a = \begin{pmatrix} a_1 \\ a_2 \end{pmatrix}. \tag{83}$$

Taking into account the conservation of the x-component of the probability density current (see equation (8) in [17] or (18) in [18])

$$J_x = \bar{v}u + \bar{u}v, \tag{84}$$

we obtain that

$$|a_1|^2 - |a_2|^2 = |d_2|^2 - |d_1|^2. \tag{85}$$

Thus, for the slope transfer matrix T it holds that

$$\bar{T}_{21}T_{22} - \bar{T}_{11}T_{12} = 0, \quad |T_{21}|^2 - |T_{11}|^2 = 1, \quad |T_{22}|^2 - |T_{12}|^2 = -1, \tag{86}$$

or

$$T^+ \begin{pmatrix} -1 & 0 \\ 0 & 1 \end{pmatrix} T = \begin{pmatrix} 1 & 0 \\ 0 & -1 \end{pmatrix}. \tag{87}$$

As a result we have $|T_{11}| = |T_{22}|$, $|T_{12}| = |T_{21}|$, $|det(T)| = 1$. For the scattering matrix

$$S = \begin{pmatrix} S_{11} & S_{12} \\ S_{21} & S_{22} \end{pmatrix} \tag{88}$$

we have

$$S \begin{pmatrix} a_1 \\ d_1 \end{pmatrix} = \begin{pmatrix} a_2 \\ d_2 \end{pmatrix}, \tag{89}$$

and

$$|a_1|^2 + |d_1|^2 = |d_2|^2 + |d_2|^2. \tag{90}$$

From (90) we obtain that

$$S^+S = SS^+ = I, \tag{91}$$

thus, the scattering matrix is unitary. If the entries of S are known, then,

$$T = \begin{pmatrix} -S_{11}/S_{12} & 1/S_{12} \\ S_{21} - S_{11}S_{22}/S_{12} & S_{22}/S_{12} \end{pmatrix}, \quad det(T) = -\frac{S_{21}}{S_{12}}. \tag{92}$$

Time-reversal symmetry in scattering through the graphene barrier would mean that

$$(\sigma_3\psi_1)^* = \frac{e^{-\frac{i}{\hbar}S_p(x,x_1)}}{\sqrt{J_p^-}} a_1^* e_1^- + \frac{e^{\frac{i}{\hbar}S_p(x,x_1)}}{\sqrt{J_p^+}} a_2^* e_1^+, \quad x \in \Omega_1, \tag{93}$$

$$(\sigma_3\psi_3)^* = \frac{e^{-\frac{i}{\hbar}S_p(x,x_2)}}{\sqrt{J_p^-}} d_1^* e_1^- + \frac{e^{\frac{i}{\hbar}S_p(x,x_2)}}{\sqrt{J_p^+}} d_2^* e_1^+, \quad x \in \Omega_3, \tag{94}$$

are both asymptotic solutions to the Dirac system, and

$$\sigma_3 = \begin{pmatrix} 1 & 0 \\ 0 & -1 \end{pmatrix}.$$

Thus, we have

$$S \begin{pmatrix} a_2^* \\ d_2^* \end{pmatrix} = \begin{pmatrix} a_1^* \\ d_1^* \end{pmatrix}, \tag{95}$$

$$T \begin{pmatrix} a_2^* \\ a_1^* \end{pmatrix} = \begin{pmatrix} d_2^* \\ d_1^* \end{pmatrix}. \tag{96}$$

In what follows that

$$S = S^T, \quad \begin{pmatrix} 0 & 1 \\ 1 & 0 \end{pmatrix} T \begin{pmatrix} 0 & 1 \\ 1 & 0 \end{pmatrix} = T^*. \tag{97}$$

Thus, $S_{12} = S_{21}$,

$$detT = -1, \tag{98}$$

and

$$T = \begin{pmatrix} T_{11} & T_{12} \\ T_{12}^* & T_{11}^* \end{pmatrix}. \tag{99}$$

If $a_1 = 1, \ a_2 = r_1, \ d_1 = 0, \ d_2 = t_1$, then

$$t_1 = \frac{1}{T^L_{12}}, \quad r_1 = -\frac{T^L_{11}}{T^L_{12}}, \quad |r_1|^2 + |t_1|^2 = 1. \tag{100}$$

If $a_1 = 0, \;\; a_2 = t_2, \;\; d_1 = 1, \;\; d_2 = r_2$, then

$$t_2 = \frac{1}{T^L_{12}}, \quad r_2 = \frac{T^L_{22}}{T^L_{12}}, \quad |r_2|^2 + |t_2|^2 = 1. \tag{101}$$

15. Appendix B. Transfer and scattering matrix properties for a smooth barrier

Let us formulate this scattering problem in terms of transfer matrix T for the entire barrier. The definition of T is given by (81),(82), and looks the same $Ta = d$. However, for the barrier we have

$$|a_1|^2 - |a_2|^2 = |d_1|^2 - |d_2|^2, \tag{102}$$

and

$$T^+ \begin{pmatrix} 1 & 0 \\ 0 & -1 \end{pmatrix} T = \begin{pmatrix} 1 & 0 \\ 0 & -1 \end{pmatrix}. \tag{103}$$

For the scattering matrix S we have

$$S \begin{pmatrix} a_1 \\ d_2 \end{pmatrix} = \begin{pmatrix} a_2 \\ d_1 \end{pmatrix}, \tag{104}$$

and

$$|a_1|^2 + |d_2|^2 = |a_2|^2 + |d_1|^2. \tag{105}$$

From (105) we obtain that

$$S^+ S = SS^+ = I. \tag{106}$$

If the entries of S are known, then,

$$T = \begin{pmatrix} S_{21} - S_{11}S_{22}/S_{12} & S_{22}/S_{12} \\ -S_{11}/S_{12} & 1/S_{12} \end{pmatrix}, \quad det(T) = \frac{S_{21}}{S_{12}}. \tag{107}$$

Taking into account the time-reversal symmetry in scattering through the graphene barrier, we obtain $S = S^T$, and

$$T = \begin{pmatrix} T_{11} & T_{12} \\ T^*_{12} & T^*_{11} \end{pmatrix}, \quad detT = 1. \tag{108}$$

16. Appendix C. WKB asymptotic solution for tunneling through a smooth step.

16.1. Left slope tunneling

Let us assume that $E > E_c$, where $E_c = |p_y|$ is the cut-off energy. In the case $|E| < E_c$ there is no wave transmission through the barrier. On the other side, we assume that $E < U_0 - \delta E$, and δE is chosen such as to avoid coalescence of all four turning points. Consider a scattering problem through a smooth step that is the left slope of the barrier. Assume that the right slope in Fig. 1 does not exist, that is $U(x) = U_0$ if $x > x_{max}$. In this case we have three domains Ω_i, $i = 1,2,3$ with the only difference for $\Omega_3 = \{x : x_2 < x < +\infty\}$. Thus, to the leading order, in the domain Ω_1 we have a superposition of waves traveling to the left and to the right.

$$\psi_1 = \frac{e^{\frac{i}{\hbar}S_p(x,x_1)}}{\sqrt{J_p^+}} a_1 e_1^+ + \frac{e^{-\frac{i}{\hbar}S_p(x,x_1)}}{\sqrt{J_p^-}} a_2 e_1^- . \tag{109}$$

In the domain Ω_2 we have exponentially decaying and growing contributions. In the domain Ω_3 we have

$$\psi_3 = d_1 \frac{e^{\frac{i}{\hbar}S_p(x,x_2)}}{\sqrt{J_p^+}} e^+ + d_2 \frac{e^{\frac{-i}{\hbar}S_p(x,x_2)}}{\sqrt{J_p^-}} e^- . \tag{110}$$

where $d = T^L a$. It is worth remarking that the electron state in $x < a$ transfers into a hole state for $x > a$.

To determine the unknown entries of the transfer matrix T^L (see Appendix A), we have to match the principal terms of all asymptotic expansions by gluing them through asymptotically small boundary layers at $x = x_1$ and $x = x_2$. To perform matching asymptotics techniques in this case we introduce a new variable $U(x) - E = \xi$ and derive an effective Schrödinger equation. Then, we have

$$\begin{pmatrix} \xi & -i\hbar\alpha\partial_\xi - ip_y \\ -i\hbar\alpha\partial_\xi + ip_y & \xi \end{pmatrix} \begin{pmatrix} u \\ v \end{pmatrix} = \begin{pmatrix} 0 \\ 0 \end{pmatrix}, \tag{111}$$

where $\alpha = \alpha(\xi) = d\xi/dx > 0$. Changing u, v as follows

$$W = \frac{(u+v)}{2}, \quad V = \frac{(u-v)}{2},$$

we obtain the system of

$$-i\hbar\alpha W_\xi + ip_y V + \xi W = 0,$$

and

$$i\hbar\alpha V_\xi - ip_y W + \xi V = 0.$$

Next, differentiating the first equation with respect to ξ and eliminating V, we obtain a second order ODE for W

$$h^2 W'' + W(\frac{\xi^2 - p_y^2}{\alpha^2} + \frac{ih}{\alpha}) + h^2 \frac{\alpha'}{\alpha} W' = 0. \tag{112}$$

Then, after we have found W, we have

$$V = \frac{ih\alpha W_\xi - \xi W}{ip_y}. \tag{113}$$

Both boundary layers for two turning points $\xi = -|p_y|$ and $\xi = |p_y|$ are determined by following scale, well-known in WKB asymptotics for turning points in 1D Schrödinger equation as $h \to 0$ (see for example [16]),

$$\xi + |p_y| = O(h^{2/3}), \quad \xi - |p_y| = O(h^{2/3}).$$

On the other side, this scattering problem for the equation (112) written as effective Schrödinger equation

$$h^2 w'' + w(\frac{\xi^2 - p_y^2}{\alpha^2} + \frac{ih}{\alpha}) = 0, \quad w = W\sqrt{\alpha}, \tag{114}$$

may be represented to leading order as follows

$$w = \frac{1}{2(\xi^2 - p_y^2)^{1/4}} \sqrt{\frac{|p_y|\alpha}{2}} (a_1 \mathrm{sgn}(p_y) \frac{e^{\frac{i}{h}\Phi^-(\xi)}}{\sqrt{D_\xi^-}} - ia_2 e^{\frac{-i}{h}\Phi^-(\xi)} \sqrt{D_\xi^-}), \tag{115}$$

for $\xi < -|p_y|$, where

$$\Phi^\pm(\xi) = \int\limits_{\pm|p_y|}^{\xi} \sqrt{q_0(t)}dt, \quad q_0(t) = \frac{t^2 - p_y^2}{\alpha^2(t)}, \quad D_\xi^\pm = \frac{\xi + \sqrt{\xi^2 - p_y^2}}{\pm|p_y|},$$

and

$$w = \frac{1}{2(\xi^2 - p_y^2)^{1/4}} \sqrt{\frac{|p_y|\alpha}{2}} (- id_1 \frac{e^{\frac{i}{h}\Phi^+(\xi)}}{\sqrt{D_\xi^+}} - d_2 \mathrm{sgn}(p_y) e^{\frac{-i}{h}\Phi^+(\xi)} \sqrt{D_\xi^+}) \tag{116}$$

for $\xi > |p_y|$. According to the method of comparison equations described in [61] and [62], we seek asymptotic solutions, uniform with respect to $|p_y|$, as follows

$$w = \frac{\sqrt{|p_y|}}{2} h^{\nu/2} \left(\frac{\frac{z^2}{4} - a^2)}{q(\xi)} \right)^{1/4} \left(b_1 D_\nu (h^{-1/2} z) + b_2 D_{-\nu-1} (ih^{-1/2} z) \right), \tag{117}$$

and the function $z(\xi)$ is determined by

$$z'^2 (a^2 - z^2/4) = q(\xi). \tag{118}$$

where

$$q = \frac{\xi^2 - p_y^2}{\alpha^2} + \frac{ih}{\alpha}, \qquad a^2 = h(\nu + \frac{1}{2});$$

The asymptotics include the parabolic cylinder function $D_\nu(z)$ that is a solution to

$$h^2 y_{zz} + (h(\nu + 1/2) - z^2/4)y = 0.$$

From (124) we obtain

$$i \int_{\xi_1}^{\xi_2} \sqrt{q(\xi)} d\xi = \int_{-2a}^{2a} \sqrt{a^2 - \frac{z^2}{4}} dz = \pi a = \pi h(\frac{1}{2} + \nu), \tag{119}$$

where $\xi_{1,2}$ are the complex roots of $q(\xi) = 0$. Using the estimate

$$\frac{1}{h} \int_{\xi_1}^{\xi_2} \sqrt{-q(t)} dt = \frac{1}{h} \int_{-p_y}^{p_y} \sqrt{-q_0(t)} dt - \frac{i\pi}{2} + O(h),$$

we obtain

$$i \int_{-p_y}^{p_y} \sqrt{-q_0(\xi)} d\xi + \frac{\pi h}{2} = \pi a = \pi h(\frac{1}{2} + \nu), \tag{120}$$

where,

$$q_0 = \frac{\xi^2 - p_y^2}{\alpha^2}$$

Thus,

$$\nu = \frac{iQ_1}{\pi h}, \qquad Q_1 = \int_{-|p_y|}^{|p_y|} \sqrt{-q_0(\xi)}d\xi. \tag{121}$$

For $\xi > |p_y|$ we have

$$i\int_{\xi_2}^{\xi} \sqrt{q(\xi)}d\xi = \int_{2a}^{z} \sqrt{z^2/4 - a^2},$$

whereas for $\xi < -|p_y|$ we have

$$i\int_{\xi}^{\xi_1} \sqrt{q(\xi)}d\xi = \int_{z}^{-2a} \sqrt{z^2/4 - a^2}.$$

In case of the linear potential with constant α we obtain the substitute

$$z = \sqrt{\frac{2}{\alpha}}e^{i\pi/4}\xi, \ \xi \in \mathbf{R} \tag{122}$$

from (124) (see [18]). For a general case, we assume that our z belongs to a sector of the complex plane based on this central line (122), where $\alpha = \alpha_0$ is evaluated at the point $x = a$, $(E = U(a))$. Thus, we assume that the asymptotic expansions for the parabolic cylinder functions in (123) are applied in a way similar to the case of the linear potential.
Then, the following important techniques for matching asymptotic estimates may be obtained for $\xi << -|p_y|$

$$\frac{z^2}{4h} - \frac{a^2}{h}log(-z) \approx \frac{i}{h}\int_{\xi}^{-|p_y|} \sqrt{q_0(\xi)}d\xi - \frac{1}{2}log\frac{2\xi}{-|p_y|} + \gamma \tag{123}$$

and for $\xi >> |p_y|$

$$\frac{z^2}{4h} - \frac{a^2}{h}log(z) \approx \frac{i}{h}\int_{|p_y|}^{\xi} \sqrt{q_0(\xi)}d\xi - \frac{1}{2}log\frac{2\xi}{|p_y|} + \gamma \tag{124}$$

where

$$\gamma = a^2/2h(1 - \log a^2) + 1/2(\nu + 1/2)\log\nu + 1/2/\nu - 1/4$$

$$= 1/2(\nu + 1/2)(1 - \log(h\nu)) - 1/4.$$

Using the asymptotic expansions of the parabolic cylinder functions for large argument (see the appendices in [9]), we obtain for $\xi << -|p_y|$

$$w \sim \frac{1}{2(\xi^2 - p_y^2)^{1/4}} \sqrt{\frac{|p_y|}{2}} \Big(b_1 [e^{-z^2/4h} z^\nu h^{-\nu/2} - e^{z^2/4h - i\pi\nu} z^{-\nu-1} h^{\nu/2+1/2} \frac{\sqrt{2\pi}}{\Gamma(-\nu)}] + \tag{125}$$

$$b_2 e^{z^2/4h - i\frac{\pi}{2}(\nu+1)} z^{-\nu-1} h^{\nu/2+1/2} \Big),$$

where $\Gamma(z)$ is the Gamma function. For $\xi >> |p_y|$ we have

$$w \sim \frac{1}{2(\xi^2 - p_y^2)^{1/4}} \sqrt{\frac{|p_y|}{2}} \Big(b_1 e^{-z^2/4h} z^\nu h^{-\nu/2} + \tag{126}$$

$$b_2 [e^{z^2/4h - i\frac{\pi}{2}(\nu+1)} z^{-\nu-1} h^{\nu/2+1/2} + e^{-z^2/4h - i\pi\nu/2} z^\nu h^{-\nu/2} \frac{\sqrt{2\pi}}{\Gamma(-\nu)}] \Big).$$

Matching these two asymptotic expansions with the asymptotics, correspondingly, leads to the following system

$$\begin{cases} a_1 sgn(p_y) = b_1 (-1)^\nu e^{-\gamma}, \\ -ia_2 = (-b_1 e^{-i\pi\nu} \frac{\sqrt{2\pi}}{\Gamma(-\nu)} + b_2 e^{-i\frac{\pi}{2}(\nu+1)}) h^{\nu+1/2} (-1)^{-\nu-1} e^{\gamma}, \\ -id_1 = b_2 e^{\gamma - i\frac{\pi}{2}(\nu+1)} h^{\nu+1/2}, \\ -d_2 sgn(p_y) = (b_1 + b_2 e^{-i\frac{\pi}{2}\nu} \frac{\sqrt{2\pi}}{\Gamma(\nu+1)}) e^{-\gamma}. \end{cases} \tag{127}$$

Let us introduce new notations

$$t = -(-1)^{-\nu} = -e^{i\pi\nu} = -e^{-\frac{Q_1}{h}}, \tag{128}$$

$$r_1 = isgn(p_y) e^{i\pi\nu + 2\gamma} \frac{\sqrt{2\pi}}{\Gamma(-\nu)} h^{\nu+1/2}, \quad r_2 = -sgn(p_y) e^{-2\gamma} \frac{\sqrt{2\pi}}{\Gamma(1+\nu)} h^{-\nu-1/2}. \tag{129}$$

Then, the system (127) reads

$$\begin{cases} a_1 \mathrm{sgn}(p_y) = -b_1 \frac{e^{-\gamma}}{t}, \\ -ia_2 = b_1 \frac{ir_1 sgn(p_y)}{t} e^{-\gamma} - ib_2 e^{\gamma - i\frac{\pi}{2}\nu} h^{\nu+1/2} t, \\ -id_1 = -ib_2 e^{\gamma - i\frac{\pi}{2}\nu} h^{\nu+1/2}, \\ -d_2 \mathrm{sgn}(p_y) = b_1 e^{-\gamma} + b_2 e^{\gamma + i\frac{\pi}{2}\nu} h^{\nu+1/2} \frac{r_2 sgn(p_y)}{t}. \end{cases} \tag{130}$$

Eliminating b_1 and b_2 from the system (130), we obtain the relations determining the transfer matrix T^L

$$\begin{cases} d_1 = -a_1 \frac{r_1}{t} + a_2 \frac{1}{t}, \\ d_2 = a_1 (t - \frac{r_1 r_2}{t}) + a_2 \frac{r_2}{t}, \end{cases} \tag{131}$$

that is

$$T^L_{11} = -\frac{r_1}{t}, \quad T^L_{12} = \frac{1}{t}, \quad T^L_{21} = t - \frac{r_1 r_2}{t}, \quad T^L_{22} = \frac{r_2}{t}. \tag{132}$$

The expressions for r_1 and r_2 can be simplified as follows

$$r_1 = -i \operatorname{sgn}(p_y) \nu \exp\big(i\pi\nu + (\nu + \frac{1}{2})(1 - \log(h\nu)) - \frac{1}{2}\big) \frac{\sqrt{2\pi}}{\Gamma(1-\nu)} h^{\nu+1/2}$$

$$= -i \operatorname{sgn}(p_y) \nu \exp\big(- \frac{Q_1}{2h} + i\frac{Q_1}{\pi h}(1 - \log(\frac{Q_1}{\pi h})) - \frac{1}{2}\log\nu \big) \frac{\sqrt{2\pi}}{\Gamma(1-\nu)}.$$

Using the properties of the Gamma function (see [63])

$$|\Gamma(1 \mp \nu)| = \sqrt{\frac{\pi\nu}{\sin(\pi\nu)}} = \sqrt{\frac{2Q_1}{h(e^{Q_1/h} - e^{-Q_1/h})}},$$

we derive

$$r_1 = \operatorname{sgn}(p_y) e^{i\theta_1} \sqrt{1 - e^{-2Q_1/h}}, \tag{133}$$

where

$$\theta_1 = \theta(Q_1) = \frac{Q_1}{\pi h}(1 - \log(\frac{Q_1}{\pi h})) - \frac{\pi}{4} - \arg\Gamma(1 - i\frac{Q_1}{\pi h}). \tag{134}$$

Similarly, taking into account that $\arg\Gamma(1+\nu) = -\arg\Gamma(1-\nu)$, we obtain

$$r_2 = -\operatorname{sgn}(p_y) e^{-i\theta_1} \sqrt{1 - e^{-2Q_1/h}}. \tag{135}$$

Hence, for the left slope transfer marix we obtain

$$T^L(Q_1) = \begin{pmatrix} -\frac{r_1}{t} & \frac{1}{t} \\ \frac{1}{t} & \frac{r_2}{t} \end{pmatrix} = \tag{136}$$

$$\begin{pmatrix} \operatorname{sgn}(p_y) e^{i\theta_1 + Q_1/h} \sqrt{1 - e^{-2Q_1/h}} & -e^{\frac{Q_1}{h}} \\ -e^{\frac{Q_1}{h}} & \operatorname{sgn}(p_y) e^{-i\theta_1 + Q_1/h} \sqrt{1 - e^{-2Q_1/h}} \end{pmatrix} \tag{137}$$

This is the main result of this section and this formula was originally stated in [18]. It is clear that the transfer matrix for the left slope satisfies all the properties in the Appendix A, namely $T^L_{22} = (T^L_{11})^*$, $T^L_{12} = (T^L_{21})^*$, $\det T_L = -1$. Now it clear that the quantities $r_{1,2}$ mean the corresponding reflection cefficients, t is the transmission coefficient. It is worth to remark that due to the asymptotics as $x \to +\infty$,

$$Im \log(\Gamma(-ix)) = \pi/4 + x(1 - \log x) + O(1/x),$$

if $Q_1/h \to +\infty$ (the turning points $\xi = \pm|p_y|$ do not coalesce), we observe that

$$\arg\Gamma(1 - iQ_1/h\pi) = \arg(-iQ_1/h\pi) + \arg\Gamma(-iQ_1/h\pi)$$

$$= -\pi/4 + Q_1/h\pi(1 - \log Q_1/h\pi),$$

and, consequently from (134), we obtain that $\theta_1 \to 0$.

16.2. Right slope tunneling

Now let us formulate the scattering problem with transfer matrix T^R for the right slope. Taking into account that $\alpha = |\frac{d\xi}{dx}|$, the problem for transfer matrix written in terms of solution to the effective Schrödinger equation

$$h^2 W'' + w\left(\frac{\xi^2 - p_y^2}{\alpha^2} - \frac{ih}{\alpha}\right) = 0, \tag{138}$$

may be represented as follows

$$w = \frac{1}{2(\xi^2 - p_y^2)^{1/4}}\sqrt{\frac{|p_y|}{2}}\left(-ia_1\frac{e^{-\frac{i}{h}\Phi^+(\xi)}}{\sqrt{D^+_\xi}} - a_2\text{sgn}(p_y)e^{\frac{i}{h}\Phi^+(\xi)}\sqrt{D^+_\xi}\right) \tag{139}$$

for $\xi > |p_y|$,

$$w = \frac{1}{2(\xi^2 - p_y^2)^{1/4}}\sqrt{\frac{|p_y|}{2}}\left(d_1\text{sgn}(p_y)\frac{e^{-\frac{i}{h}\Phi^-(\xi)}}{\sqrt{D^-_\xi}} - id_2e^{\frac{i}{h}\Phi^-(\xi)}\sqrt{D^-_\xi}\right) \tag{140}$$

for $\xi < -|p_y|$. If w is a solution to (138), then w^* is the solution to (114). Thus, the coefficients from (139), (140) are connected by

$$\begin{pmatrix} -a_1^* \\ a_2^* \end{pmatrix} = T^L(Q_2)\begin{pmatrix} d_1^* \\ -d_2^* \end{pmatrix}. \tag{141}$$

Hence, we have

$$\begin{pmatrix} d_1 \\ d_2 \end{pmatrix} = \begin{pmatrix} 1 & 0 \\ 0 & -1 \end{pmatrix} \left((T^L(Q_2))^{-1} \right)^* \begin{pmatrix} -1 & 0 \\ 0 & 1 \end{pmatrix} \begin{pmatrix} a_1 \\ a_2 \end{pmatrix}. \tag{142}$$

Since $d = T^R a$, we obtain

$$T^R = \begin{pmatrix} 1 & 0 \\ 0 & -1 \end{pmatrix} \begin{pmatrix} -\frac{r_2^*}{t} & \frac{1}{t} \\ \frac{1}{t} & \frac{r_1^*}{t} \end{pmatrix} \begin{pmatrix} -1 & 0 \\ 0 & 1 \end{pmatrix} = \begin{pmatrix} -\frac{r_1}{t} & \frac{1}{t} \\ \frac{1}{t} & \frac{r_2}{t} \end{pmatrix} = T^L(Q_2). \tag{143}$$

where

$$Q_2 = \int_{-p_y}^{p_y} \frac{\sqrt{p_y^2 - \xi^2}}{\alpha(\xi)} d\xi = \int_{x_3}^{x_4} q_x(x) dx, \qquad \theta_2 = \theta(Q_2). \tag{144}$$

It is worth to remark that Q_1 and Q_2 differ as the function $\alpha(\xi)$ behave differently for the same segment $\xi \to (-|p_y|, |p_y|)$ for left and right slopes of non-symmetric barrier.

Acknowledgments

The authors would like to thank Prof J.Ferapontov, Dr A.Vagov and Dr D. Maksimov for constructive discussions and valuable remarks. DMF would like to acknowledge support from the EPSRC (KTA) for funding under the Fellowship: "Developing prototypes and a commercial strategy for nanoblade technology".

Author details

V.V. Zalipaev, D.M. Forrester,
C.M. Linton and F.V. Kusmartsev

School of Science, Loughborough University, Loughborough, UK

17. References

[1] Heersche, H.B., Jarillo-Herrero, P., Oostinga, J.B.,Vandersypen, L.M.K., Morpurgo,A.F., Solid State Communications. 143, 72 (2007).

[2] Cheianov, V.V, Fal'ko, V., Altshuler, B. L., Science. 315, 1252 (2007).

[3] Huard, B., Sulpizio, J.A., Stander, Todd, K., Yang, B. and Goldhaber-Gordon, D., PRL 98, 236803 (2007).

[4] Tzalenchuk, A., Lara-Avila, S., Cedergren, K., Syvajarvi, K., Yakimova, R., Kazakova, O., Janssen, T.J.B.M, Moth-Poulsen, K. Bjornholm, T., Kopylov, S. Fal'ko, V. Kubatkin, S., Solid State Communications 115 1094 (2011).

[5] Allain, P.E. and Fuchs, J.N., Eur.Phys.J. B 83, 301 (2011).

[6] Young A. F. and Kim, P. , Nat. Phys. 5, 222-6 (2009)

[7] Shytov, A. V., Rudner, M.S, and Levitov, L.S., Phys. Rev. Lett. 101, 156804 (2008).

[8] Ramezani Masir, M., Vasilopoulos, P.,Peeters, F. M., Phys. Rev. B. 82, 115417 (2010).

[9] Yanqing Wu, Vasili Perebeinos, Yu-ming Lin, Tony Low, Fengnian Xia, and Phaedon Avouris, Nano Lett., 12 (3), pp 1417-1423, (2012). DOI: 10.1021/nl204088b

[10] P. Barbara et al, (2012) unpublished.

[11] J. Velasco Jr et al, New J. Phys. 11, 095008 (2009).

[12] Liang W J, Bockrath M, Bozovic D, Hafner J H, Tinkham M and Park H, Nature, 411, 665-9 (2001).

[13] Miao F, Wijeratne S, Zhang Y, Coskun U, Bao W and Lau C N, Science 317 1530-3 (2007).

[14] O'Hare, A.O. Kusmartsev, F.V, Kugel, K.I. Nano Letters 12, 1045 (2012).

[15] Williams, J.R., Low, T., Lundstrom, M.S and Marcus, C.M., Nature Nanotechnology 6, 222 (2011).

[16] Bender, C.M., Orszag, S.A. 1978, *Advanced Mathematical Methods For Scientists and Engineers*, McGraw-Hill, Inc.

[17] Sonin, E.B. Phys.Rev. B. 79, 195438 (2009).

[18] Tudorovskiy, T., Reijnders, K.J.A., Katsnelson, M.I. Phys. Scr. T146, 014010, 17pp, (2012).

[19] Rodriquez-Sensale, B., Yan, R., Kelly, M.M, Fang, T., Tahy, K., Hwang, W.S., Jena, D., Liu, L. and Xing, H.G Nature Communications. 3 780 doi:10.1038/ncomms1787 (2012).

[20] Crassee, I., Orlita, M. Potemski, M., Walter, A.L., Ostler, M., Seyller, Th. Gaponenko, I., Chen, J. and Kuzmenko, A.B. Nano Letters 12 2470 (2012).

[21] L. V. Berlyand and S. Yu. Dobrokhotov. Operator separation of variables in problems of short-wave asymptotics for differential equations with rapidly oscillating coefficients. Doklady Akad. Nauk SSSR, V32, p.714, (1987).

[22] V. V. Belov, S. Yu. Dobrokhotov, and T. Ya. Tudorovskiy. Operator separation of variables for adiabatic problems in quantum and wave mechanics. Journal of Engineering Mathematics, 55(1-4):183-237, (2006).

[23] V. P. Maslov. Perturbation Theory and Asymptotic Methods. Dunod, Paris, (1972).

[24] V. P. Maslov and M. V. Fedoryuk. Semi-Classical Approximation in Quantum Mechanics. Reidel, Dordrecht, (1981).

[25] Datta, S., *Electronic transport in mesoscopic systems*, Cambridge University Press, Cambridge, (1995) .

[26] Mello, P. A., Kumar, N., 2004 *Quantum Transport in Mesoscopic systems*, Oxford University Press, New York.

[27] Stockmann, H. J., *Quantum Chaos. An Introduction*, Cambridge University Press, Cambridge, (2000).

[28] Schwieters, C.D., Alford, J.A. and Delos, J.B. Phys.Rev.B. 54, N15, 10652 (1996).

[29] Beenaker, C.W.J. Rev. Mod. Phys. 69, 731 (1997).

[30] Blomquist, T. and Zozoulenko, I. V. Phys.Rev.B. 61, N3, 1724 (2000).

[31] Castro Neto, A.H., Guinea, F., Peres, N.M.R., Geim, A.K., and Novoselov, K.S. Rev. Mod. Phys. 81, Jan-March, 109-162, (2009).

[32] Yan, K., Peng, H., Zhou, Y. Li, H., Liu, Z. Nano Letters 11, 1106 (2011)

[33] Cheianov, V.V., Fal'ko, V.I. Phys.Rev. B. 74, 041403(R)(2006).

[34] Fistul, M.V., Efetov, K.B. PRL 98, 256803 (2007).

[35] Silvestrov, P.G., Efetov, K.B. PRL 98, 016802 (2007).

[36] Rafiq, R., Cai, D., Jin, J. and Song, M. Carbon, 48, 4309 (2010)

[37] Zhu, Y., Murali, S., Stoller, M.D., Ganesh, K.J., Cai, W., Ferreira, P.J., Pirkle, A., Wallace, R.M., Cychosz, K.A., Thommes, M., Su, D., Stach, E.A., Ruoff, R.S. Science 332(6037) 1537 (2011)

[38] Echtermeyer, T.J., Britnell, L., Jasnos, P.K., Lombardo, A., Gorbachev,R.V., Grigorenko,A.N., Geim, A.K., Ferrari, A.C., Novoselov, K.S., Nature Communications 2, 458 doi:10.1038/ncomms1464 (2011).

[39] Brack, M. and Bhaduri R.K., 1997, *Semiclassical Physics. Frontiers in Physics*, Vol. 96, Addison-Wesley, Reading, MA.

[40] Belov, V.V, Dobrokhotov, S.Yu., and Tudorovskii, T. Ya., Theoretical and Mathematical Physics 141(2), 1562-1592 (2004).

[41] Bruning, J., Dobrokhotov, S., Sekerzh-Zenkovich, S., and Tudorovskii, T. , Russian Journal of Mathematical Physics 13(4), 380-396 (2006).

[42] Bruning, J., Dobrokhotov, S., Nekrasov, R., and Tudorovskii, T., Russian Journal of Mathematical Physics 15, 1, 1-16 (2008).

[43] Carmier, P., Ullmo, D. Phys.Rev. B. 77, N24, 245413 (2008).

[44] Kormanyos, A., Rakyta, P., Oroszlany, L, and Cserti, J. Phys.Rev. B. 78, 045430 (2008).

[45] Cserti, J., Hagymasi, I., Kormanyos, A., Phys.Rev. B. 80, 073404 (2009).

[46] Keller, J.B. Corrected Borr-Sommerfeld quantum conditions for non-separable systems. Ann. Phys., 4, N12, 180-188 (1958).

[47] Keller, J.B. Rubinov, S. Asymptotic solution of eigenvalue problems. Ann. Phys., 9, N1, 24-75 (1960).

[48] Maslov, V. P. and M.V.Fedoriuk, 1981, *Semiclassical approximation in quantum mechanics*, Reidel, Dordrecht.

[49] Maslov, V. P., 1977, *Complex WKB method in nonlinear equations*, Nauka, Moscow (in Russian).

[50] Babich, V. M., Buldyrev, V.S. 1991, *Asymptotic methods in shortwave diffraction problems*, Springer-Verlag, New York.

[51] Popov, M.M., The asymptotic behaviour of certain subsequences of eigenvalues of boundary problems for the Helmholtz equation in higher dimensions, English Translation of Soviet Physics, Doklady, vol.14, pp.108-110,(1969).

[52] Dubnov, V.L., Maslov, V.P., Nazaikinskii, V.E. The complex lagrangian germ and the canonical operator. Russsian Journal of Mathematical Physics. 3, 2, pp.141-190, (1995).

[53] Dobrokhotov S Yu, Martinez-Olive V, Shafarevich A I, Closed trajectories and two-dimensional tori in the quantum problem for a three-dimensional resonant anharmonic oscillator. Russsian Journal of Mathematical Physics. 3, 1, pp.133-138, (1995).

[54] Babich, V. M., Kirpichnikova, N. Ya., 1979, *The boundary-layer method in diffraction problems*, Springer-Verlag, Berlin.

[55] Zalipaev,V.V., Kusmartsev,F. V. and Popov,M. M., J. Phys. A: Math. Theor. 41, 065101 (2008).

[56] McCann, E., Fal'ko,V.I., Phys.Rev. Letters. 96, 086805 (2006).

[57] Yabana, K., Horiuchi, H., Prog. Theor. Phys. 75, 592 (1986).

[58] Duncan, K.P., Gyorffy, B.L., Ann. Phys.(N.Y.) 298, 273 (2002).

[59] Keppeler, S., Ann. Phys.(N.Y.) 304, 40 (2003).

[60] Castro Neto, A.H., Guinea, F., Peres, N.M.R., Geim, A.K., and Novoselov, K.S. Rev. Mod. Phys. 81, Jan-March, (2009).

[61] F. W. J. Olver. *Asymptotics and Special Functions*, Academic Press, 1974.

[62] M. V. Fedoryuk. Asymptotic Analysis: Linear Ordinary Differential Equations. Springer, 1993.

[63] M. Abramowitz and I. Stegun, editors. *Handbook on Mathematical Functions with Formulas, Graphs, and Mathematical Tables*. Dover, New York, 1965. Chapter 19 on parabolic cylinder functions.

[64] Berry, M.V. Proc. R. Soc. Lond. A. 392, 45 (1984).

[65] Popov M M 2002 *Ray Theory and Gaussian Beam Method for Geophysicists*, Edufba, Salvador-Bahia, Brazil.

[66] V. V. Zalipaev, J.Phys. A: Math. Theor. 42, 205302, 14pp, (2009).

[67] P. Markos and C Soukoulis. Wave Propagation. From electrons to photonic crystals and left-handed materials. Priceton University Press, 2008.

[68] Pelligrino, F. M. D., Angilella, G. G. N., Pucci, R., Phys. Rev. B. 84, 195404 (2011).

[69] Pereira Jr,J.M., Peeters,F.M., Chaves,A., Farias, G.A., Semicond. Sci. Technol 25, 033002, (2010).

[70] Zalipaev, V.V., High-energy localized eigenstates in Fabry-Perot graphene resonator in a magnetic field, J.Phys.A.: Math. Theor 45, 215306, 20pp (2012).

[71] Zalipaev, V.V., Maksimov, D.N., Linton. C.M., Kusmartsev, F.V., submitted to Phys. Let. A (2012)

Chapter 3

Electronic Properties of Deformed Graphene Nanoribbons

Guo-Ping Tong

Additional information is available at the end of the chapter

http://dx.doi.org/10.5772/51348

1. Introduction

As early as 1947, the tight-binding electronic energy spectrum of a graphene sheet had been investigated by Wallace (Wallace, 1947). The work of Wallace showed that the electronic properties of a graphene sheet were metallic. A better tight-binding description of graphene was given by Saito et al. (Saito et al., 1998). To understand the different levels of approximation, Reich et al. started from the most general form of the secular equation, the tight binding Hamiltonian, and the overlap matrix to calculate the band structure (Reich et al., 2002). In 2009, a work including the non-nearest-neighbor hopping integrals was given by Jin et al. (Jin et al., 2009).

It is common knowledge that a perfect grphene sheet is a zero-gap semiconductor (semimetal) that exhibits extraordinarily high electron mobility and shows considerable promise for applications in electronic and optical devices, high sensitivity gas detection, ultracapacitors and biodevices. How to open the gap of graphene has become a focus of the study. Early in 1996, Fujita et al. started to study the electronic structure of graphene ribbons (Fujita et al., 1996; Nakada et al., 1996) by the numerical method. The armchair shaped edge ribbons can be either semiconducting (n=3m and n=3m+1, where m is an integer) or metallic (n=3m+2) depending on their widths, i. e., on their topological properties. First-principles calculations showed that the origin of the gaps for the armchair edge nanoribbons arises from both quantum confinement and the deformation caused by edge dangling bonds (Son et al., 2006; Rozhkov et al., 2009). This result implies that the energy gap can be changed by deformation. In 1997, Heyd et al. studied the effects of compressive and tensile, unaxial stress on the density of states and the band gap of carbon nanotubes (Heyd et al., 1997). Applying mechanical force (e.g., nanoindentation) on the graphene can lead to a strain of about 10%(Lee et al., 2008). Xiong et al. found that engineering the strain on the graphene planes forming a channel can drastically change the interfacial friction of water transport through it (Xiong et al., 2011). Density functional perturbation theory is a well-tested *ab initio* method for accurate phonon calculations. Liu et al. (Liu et al., 2007) studied the phonon spectra of graphene as a function of uniaxial tension by using this theory. Edge stresses and edge energies of the armchair and zigzag edges in graphene also were studied by means of the theory (Jun, 2008). Jun found that both edges are under compression along the edge and the magnitude of compressive edge stress of armchair edge is larger than that of zigzag edge. By simulations of planar graphene undergoing in-plane deformations, Chung (Chung, 2006) found that crystal structures are different from the usual hexagonal configuration. The thermodynamic or kinetic character of the rearrangement was found to depend on the macroscopic straining direction. Neek-Amal et al. (Neek-Amal et al., 2010) simulated the bending of rectangular graphene nanoribbons subjected to axial stress both for free boundary and supported boundary conditions. Can et al. (Can et al., 2010) applied density-functional theory to calculate the equilibrium shape of graphene sheets as a function of temperature and hydrogen partial pressure. Their results showed that the edge stress for all edge orientations is compressive. Shenoy et al. (Shenoy et al., 2008) pointed out that edge stresses introduce intrinsic ripples in freestanding graphene sheets even in the absence of any thermal effects. Compressive edge stresses along zigzag and armchair edges of the sheet cause out-of-plane warping to attain several degenerate mode shapes and edge stresses can lead to twisting and scrolling of nanoribbons as seen in experiments. Marianetti et al. (Marianetti et al., 2010) reveals the mechanisms of mechanical failure of pure graphene under a generic state of tension at zero temperature. Their results indicated that finite wave vector soft modes can be the key factor in limiting the strength of monolayer materials. In the chemical activity of graphene, de Andres et al. (de Andres et al., 2008) studied how tensile stress affects $\sigma$and π bonds and pointed out that stress affects more stronglyπbonds that can become chemically active and bind to adsorbed species more strongly. Kang et al. (Kang et al., 2010) performed a simulation study on strained armchair graphene nanoribbons. By comparison, those with strained wide archair nanoribbons can achieve better device performance. By combining continuum elasticity theory and tight-binding atomistic simulations, Cadelano et al. (Cadelano et al., 2009) worked out the constitutive nonlinear stress-strain relation for graphene stretching elasticity and calculated all the corresponding nonlinear elastic moduli. Gui et al. (Gui et al., 2008) found that graphene with a symmetrical strain distribution is always a zero band-gap semiconductor and its pseudogap decreases linearly with the strain strength in the elastic regime. For asymmetrical strain distributions the band gaps were opened at the Fermi level. This is because small number of k points is chosen (Farjam et al., 2009). We also investigated the energy spectrum and gap of wider graphene ribbons under a tensile force (Wei et al., 2009) and found that the tensile force can have the gap of the ribbon opened.

In this Chapter, we focus on the effects of deformed graphene sheets and nanoribbons under unaxial stress on the electronic energy spectra and gaps based on the elasticity theory. Meanwhile, the energy spectrum of the curved graphene nanoribbons with the tubular warping is studied by the tight-binding approach. The energy spectrum of deformed graphene sheets subjected to unaxial stress is given in Section 2. In Section 3, we discuss the electronic properties of graphene nanoribbons under unaxial stress. The tubular warping deformation of graphene nanoribbons is presented in last Section.

http://dx.doi.org/10.5772/51348

2. Graphene under uniaxial stress

2.1. Elasticity theory

Since graphene is a monolayer structure of carbon atoms, when a force is exerted on it parallel to its plane, the positions of the atoms will change with respect to some origin in space. Let the x-axis be in the direction of the armchair edge of graphene and the y-axis in that of the zigzag edge, as shown in Fig. 1. Let $\boldsymbol{R}$ and $\boldsymbol{R}'$ denote the positions of a carbon atom before and after deformation, respectively. According to the theory of elasticity, the relation between the positions can be written in the form

$$\begin{pmatrix} R_x' \\ R_y' \end{pmatrix} = \begin{pmatrix} 1+\delta_1 & 0 \\ 0 & 1+\delta_2 \end{pmatrix} \begin{pmatrix} R_x \\ R_y \end{pmatrix} \tag{1}$$

where$\delta_1 = +\delta(\text{or} - \delta)$ is the tensile (or compression) stress along the x-direction and$\delta_2$is the stress in the y-direction and small compared toδ_1, approximately equal to$\delta_1/6$. When the deformation of graphene occurs, the bond length between the carbon atoms changes and which leads to the change of electronic hopping energies.

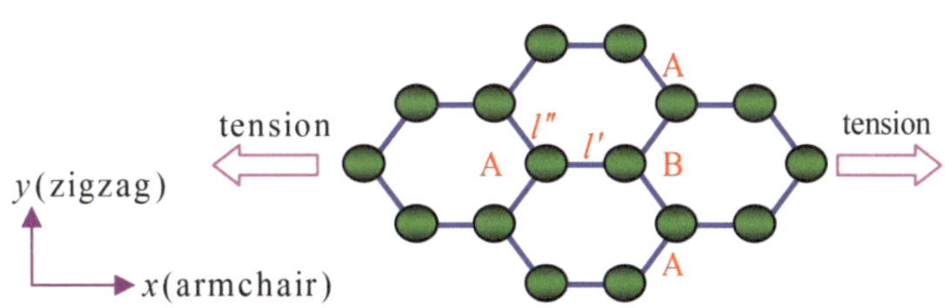

Figure 1. Graphene sheet subjected to the tensile stress in the x-direction. Symbols A and B denote sublattices with two kinds of different carbon atoms, respectively. $l'$and l''denote the bond lengths between tow adjacent carbon atoms after deformation, respectively.

According to Harrison's formula (Harrison, 1980), the hopping energy after deformation is expressed as follows

$$t' = \left(\frac{l_0}{l'}\right)^2 t_0 \tag{2}$$

where$l_0$and$t_0$denote the bond length and the hopping energy before deformation, respectively. $l'$and$t'$are the bond length and the hopping energy after deformation, respectively. From Fig. 1 and Eq. (1), the bond lengths between atoms A and B can be obtained

$$l' = R'_x = l_0(1+\delta),$$
$$l'' = \sqrt{R'^2_x + R'^2_y} = l_0\sqrt{1+\frac{1}{4}\delta+\frac{13}{48}\delta^2} \tag{3}$$

The nearest neighbour hopping integrals associated with the bond lengths are

$$t' = \frac{t_0}{(1+\delta)^2}$$
$$t'' = \frac{t_0}{\left(1+\frac{1}{4}\delta+\frac{13}{48}\delta^2\right)} \tag{4}$$

If graphene is subject to a tensile force in the y-direction, the hopping energies are given by

$$t' = \frac{t_0}{\left(1-\frac{\delta}{6}\right)^2}$$
$$t'' = \frac{t_0}{\left(1+\frac{17}{12}\delta+\frac{109}{144}\delta^2\right)} \tag{5}$$

2.2. The tight-binding energy spectrum

Let us now consider the band structure from the viewpoint of the tight-binding approximation. The structure of graphene is composed of two types of sublattices A and B as shown in Fig. 1. If $\varphi(r)$is the normalized orbital $2p_z$ wave function for an isolated carbon atom, then the wave function of graphene has the form

$$|\psi\rangle = C_A|\psi_A\rangle + C_B|\psi_B\rangle \tag{6}$$

where

$$|\psi_A\rangle = \frac{1}{\sqrt{N}}\sum_A e^{ik\cdot R_A}|\varphi(r-R_A)\rangle$$

and

$$|\psi_B\rangle = \frac{1}{\sqrt{N}}\sum_B e^{i\boldsymbol{k}\cdot\boldsymbol{R}_B}|\varphi(\boldsymbol{r}-\boldsymbol{R}_B)\rangle \tag{7}$$

The first sum is taken over A and all the lattice points generated from it by primitive lattice translation; the second sum is similarly over the points generated from B. Here$C_A$andC_B are coefficients to be determined, $R_A$and$R_B$are the positions of atoms A and B, respectively, and $N$is the number of the unit cell in graphene. Substituting Eq. (6) in

$$H|\psi\rangle = E|\psi\rangle \tag{8}$$

we obtain the secular equation

$$\begin{vmatrix} H_{AA} - E & H_{AB} \\ H_{AB}^{*} & H_{AA} - E \end{vmatrix} = 0 \tag{9}$$

For the tensile stress in the x-direction, the solution to Eq.(9) is

$$E(k_x, k_y) = \pm \left[t'^2 + 4t't'' \cos\left(\frac{\sqrt{3}}{2} k_y R'_y\right) \cos\left(\frac{3}{2} k_x R'_x\right) + 4t''^2 \cos^2\left(\frac{\sqrt{3}}{2} k_y R'_y\right) \right]^{\frac{1}{2}} \tag{10}$$

where$t'$and$t''$are the nearest-neighbour hopping integrals after deformation, given by Eq.(4).

Fig. 2 shows the electronic energy spectra of deformed graphene sheets for some high symmetric pointsΓ, M, and K under uniaxial stress. Because of uniaxial stress, the hexagonal lattice is distorted and the shape of the first Brillouin zone changes accordingly as the stress upon the lattice. Six "saddle" points on the boundary in the first Brillouin zone can be divided into two groups: M and M′. At the same time, Dirac point K will drift towards the saddle point M and is accompanied by a small angle. For the convenience of comparison, we give the spectrum of undeformed graphene in Fig. 2(a). From Fig. 2 (c) and (d), we see that tension along the armchair shape edge can reduce the band width at point $\Gamma$and increase the bandwidth at point M, and the result of compression is just opposite to that of tension. Fig. 2(e) tells us that tension along the zigzag shape edge can not only narrow the bandwidth at Γpoint but decrease the bandwidth at M point as well. On the contrary, compression can simultaneously increase the bandwidth at$\Gamma$and M. Moreover, it may be seen from Fig. 2 that whether the tensile stress or compressive stress, the result of the high symmetric point M′ is always opposite to that of the point M and the energy gap cannot be opened at Dirac point (K). On the other hand, we see yet that the energy band curves between M and M′ for the graphene without stress are a straight line, but for the graphene with stress the curves are not. It appears to graphene that the uniaxial stress does not open the energy gap at Dirac point. When graphene is compressed along the armchair shape edge or extended along the zigzag shape edge a small energy gap is opened at K point, which is approximately equal to 0.1eV as the stress parameter takes to be 12%. From this reason, the graphene under uniaxial stress still is a semiconductor with the zero-energy gaps.

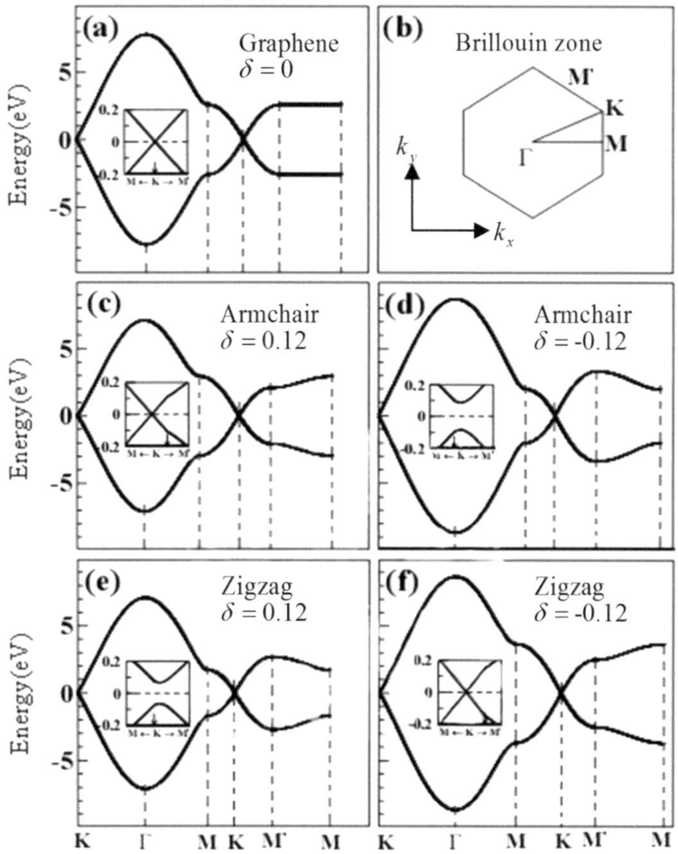

Figure 2. Electronic energy spectra of graphene under uniaxial tress for some high symmetric points.

3. Graphene nanoribbon under uniaxial stress

As mentioned in Section 2, for a graphene sheet subject to uniaxial stress there are no energy gaps at Dirac point. How to open the energy gaps of graphene? Studies showed that we can realize this goal by deducing the size of graphene, i.e., changing its toplogical propertiy(Son et al., 2006). On the other hand, the band gaps of graphene nanoribbons can be mamipulated by changing the bond lengths between carbon atoms, i.e., changing the hopping integrals, by exerting a strain force (Sun et al., 2008). The nearest-neighbor energy spectrum of an armchair nanoribbon was given by the tight-binding approach and using the hard-wall aboundary condition (Zheng et al., 2007). In the non-nearest-neighbor band structure of the nanoribbon was given by Jin et al (Jin et al., 2009). In this section we use the tight-binding approach to study the energy spectrum and gap of the nanoribbon under uniaxial stress along the length direction, i.e., x-direction, of the nanoribbon, as shown in Fig. 3.

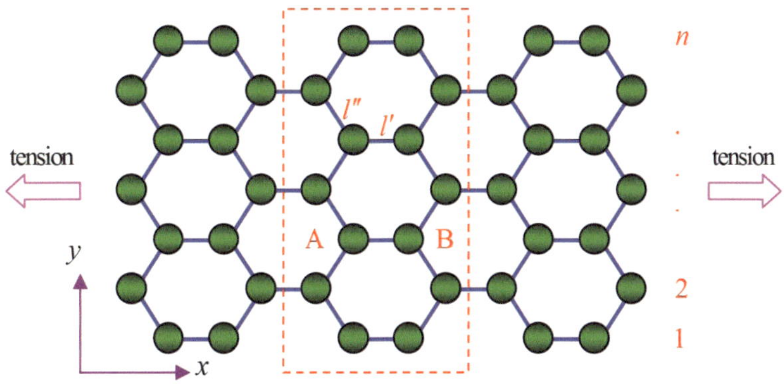

Figure 3. Structure of an armchair graphene nanoribbon with sublattices *A* and *B*. The tension is exerted on the nanoribbon along the x-axis. Symbol *n* denotes the width of the nanoribbon. There are *n* sublatices *A* or *B* in a unit cell.

Since the unit cell of the nanoribbon has the translational symmetry in the *x*-direction, we can choose the plane-wave basis in the *x*-direction and take the stationary wave in the *y*-direction. For the armchair nanoribbon there are two kinds of sublattices *A* and *B* in a unit cell. Therefor, the wave functions of *A* and *B* sublattices in hard-wall conditions can be written as

$$\begin{aligned} |\psi_A(k_x,q)\rangle &= \frac{1}{N_A}\sum_{j=1}^{n}\sum_{x_{A_j}} e^{ik_x x_{A_j}} \sin\left(\frac{\pi q}{(n+1)} j\right) |\varphi(\boldsymbol{r}-\boldsymbol{R}_{A_j})\rangle, \quad (q=1,2,\cdots,n) \\ |\psi_B(k_x,q)\rangle &= \frac{1}{N_B}\sum_{j=1}^{n}\sum_{x_{B_j}} e^{ik_x x_{B_j}} \sin\left(\frac{\pi q}{(n+1)} j\right) |\varphi(\boldsymbol{r}-\boldsymbol{R}_{B_j})\rangle, \quad (q=1,2,\cdots,n) \end{aligned} \tag{11}$$

where $N_A$and$N_B$are the normalized coefficients,$q=1,2,...,n$ is the quantum number associated with the wave vector k_y, which denotes the discrete wave vector in the y-direction. When a graphene nanoribbon is subject to uniaxial stress, Eq.(11) still is available. For a nanoribbon, as long as the wave vector$k_y$in Eq.(10) is replaced by the discrete wave vector$k_y(q)$, we can obtain the energy dispersion relation of the form

$$E(k_x,q) = \pm\left[t'^2 + 4t''^2\cos^2\left(\frac{q\pi}{n+1}\right) + 4t't''\cos\left(\frac{q\pi}{n+1}\right)\cos\left(\frac{3}{2}k_x R'_x\right)\right]^{\frac{1}{2}} \tag{12}$$

Since the electronic energy spectrum of the perfect armchair nanoribbon depends strongly on the width of the nanoribbon, the different width has the different spectrum. For instance, the nanoribbon with widths *n*=3*m*+2 (*m* is an integer) is metallic and others are insulating. When we exert a tensile (or compressive) force on the nanoribbon along the x-axis, the metal nanoribbon is converted into an insulator or semiconductor.

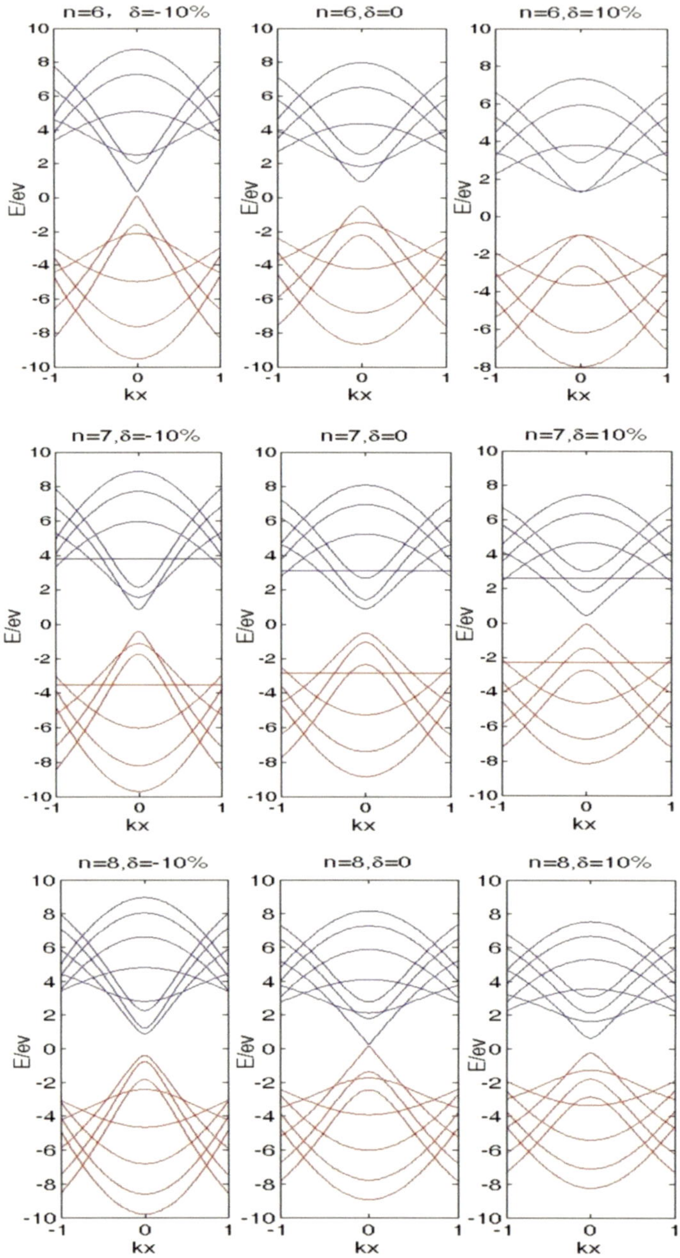

Figure 4. Band structures of armchair graphene nanoribbons under unaxial stress with widths *n*=6, 7, 8. The stress parameterδis taken to be - 0.1, 0, and 0.1 respectively.

Fig. 4 shows the energy spectra of three kinds of the nanoribbons under uniaxial stress, and in which the next-nearest neighbor hopping integrals are taken into account. In order to facilitate comparison, the energy spectrum of the undeformed nanoribbon is given in Fig. 4. When width n=6, the tensile stress can make the energy gap increase and the bandwidth decrease slightly. On the contrary, the compressive stress can decrease the gap and make the bandwidth widen. It is obvious that the energy band corresponding to quantum number q=n–1=5 plays an important role in the change of the band gap. When n=7, the tensile stress can make the gap narrow and the compressive stress has larger influence on the energy bandwidth, but not obvious on the gap. It can be seen that the energy band with quantum number q=n-1=6 contributes to the gap under compressive stress and which is clearly different from the tensile situation, where q=n-2=5. As for n=8, whether it is tensile or compression can open the gap and the energy band contributing to the gap belongs to quantum number q=n-2=6. It fallows from this that either tension or compression can change the gap and the bandwidth. Therefore, the electronic properties of armchair nanoribbons can be controlled by uniaxial stress.

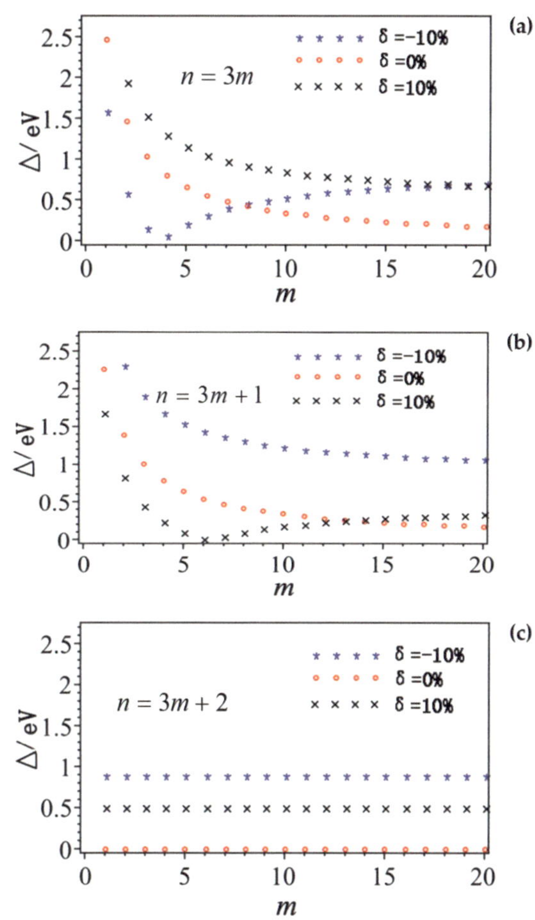

Figure 5. Energy gaps of deformed armchair nanoribbons as a function of the width n (m). SymbolΔdenotes the energy gap. The stress parameter$\delta$is taken to be -0.1, 0, and 0.1.

When the stress is constant, three graphs of the energy gaps with the width of the nanoribbon changes are shown in Fig. 5, where (a) the width n is equal to $3m$, (b) n=3m+1, and (c) n=3m+2. The results shown in Fig. 5 are inclusive of the nearest-neighbor hopping integrals. We see from Fig. 5(a) that the compressive stress can make an inflection point of the band gap minimum for the 3m-type nanoribbon and the width corresponding with the inflection point is about 12, and the tensile stress can not make a minimum value of the gap. For the 3m+1-type nanoribbon, the result shown in Fig. 5(b) tells us that the tensile stress also can produce the minimum value of the gap and the corresponding width is 19. In the case of the 3m+2-type nanoribbon, tensile or compression does not change the energy gap (see Fig. 5(c)). Furthermore, we found by calculations that with the inclusion of the next-nearest neighbor and the third neighbor respectively, the minimum point of the gap moves toward the direction of the origin of coordinates (zero width), i.e., the width of the non-nearest-neighbor hopping is less than that of the nearest neighbor.

On the other hand, in order to make certain of the relationship between the gap and the stress, the curves of the gap versus the stress are given in Fig. 6. As shown in Fig. 6, the gap increases as the stress increases for the 3m- and 3m+1-type nanoribbons and changes in the V-shaped curve for the 3m+2-type nanoribbon.

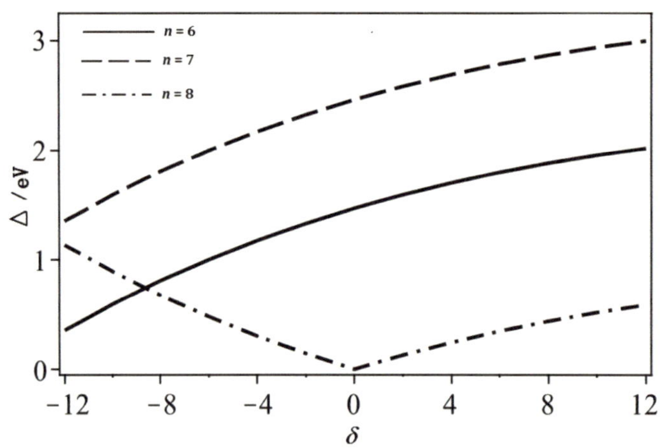

Figure 6. Energy gaps of deformed armchair nanoribbons as a function of stressδ. SymbolΔdenotes the energy gap. Solid, dashed, and dotted lines denote widths n=6, n=7, and n=8, respectively.

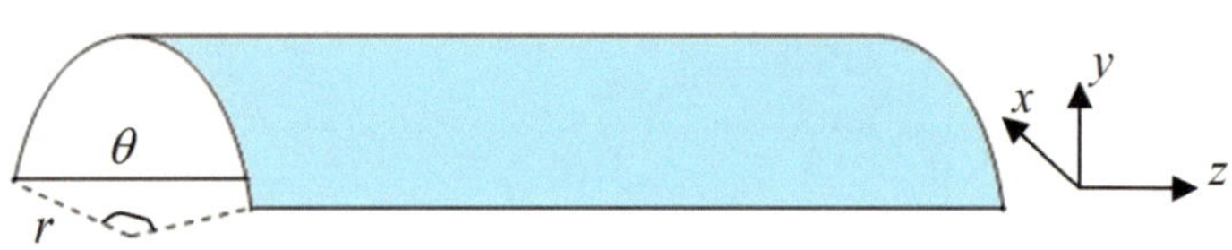

Figure 7. A curved armchair graphene nanoribbon with the tubular shape. $\theta$is the central angle and $r$is the curved radius of the nanoribbon.

http://dx.doi.org/10.5772/51348

4. The tubular warping graphene nanoribbon

4.1. Theoretical Model

In this section we choose an armchair ribbon as an example and which is bent into the tubular shape (cylindrical shape), as shown in Fig. 7. This tubular ribbon still has the periodicity in its length direction, but its dimensionality has changed. The consequence of such a dimension change is to lead to the change of the electronic energy dispersion relation. This is because the sp^2 hybridization of a flat ribbon turns into the sp^3 hybridization of a curved ribbon, i.e., the curvature of graphene nanoribbons will result in a significant rehybridization of the π orbitals (Kleiner et al., 2001). From this reason, the *s*-orbital component must be taken into account in calculating electronic energy bands.

Because of the curl of the ribbon, the wavefunction of π electrons should be composed of the *s*- and *p*-orbital components. The wavefunctions of sublattices *A* and *B* in cylindrical coordinates are written then

$$|\psi_A\rangle = \frac{1}{N_A}\sum_{j=1}^{n}\sum_{z_{A_j}} e^{ik_z z_{A_j}} \sin\left(j\frac{\sqrt{3}}{2}k_\phi a\right)\left[\sqrt{c_j}\,|s_{A_j}\rangle + \sqrt{1-c_j}\,|p_{zA_j}\rangle\right] \tag{13}$$

$$|\psi_B\rangle = \frac{1}{N_B}\sum_{j=1}^{n}\sum_{z_{B_j}} e^{ik_z z_{B_j}} \sin\left(j\frac{\sqrt{3}}{2}k_\phi a\right)\left[\sqrt{c_j}\,|s_{B_j}\rangle + \sqrt{1-c_j}\,|p_{zB_j}\rangle\right] \tag{14}$$

where c is the s-orbital component of electrons, given by (Huang et al., 2006; 2007)

$$c = \frac{2\sin^2\beta}{1-\sin^2\beta} \tag{15}$$

and

$$\beta = \frac{a}{4\sqrt{3}r} \tag{16}$$

Here β is a small inclined angle (Kleiner et al., 2001) between the p_z orbital and the normal direction of the cylindrical surface, r is the radius of the cylindrical surface, and a is the distance between two adjacent carbon atoms.

4.2. Results and Discussion

To clearly understand the effect of curvature, we choose the width *n*=6, 7, and 8 respectively as examples to show the characteristics of their electronic energy spectra. On the other hand, in order to compare with the ideal flat nanoribbon, the results of the ideal ribbon along with the tubular warping ribbon are also given in Fig. 8, where black lines denote the ideal ribbon and red lines are the tubular warping ribbon.

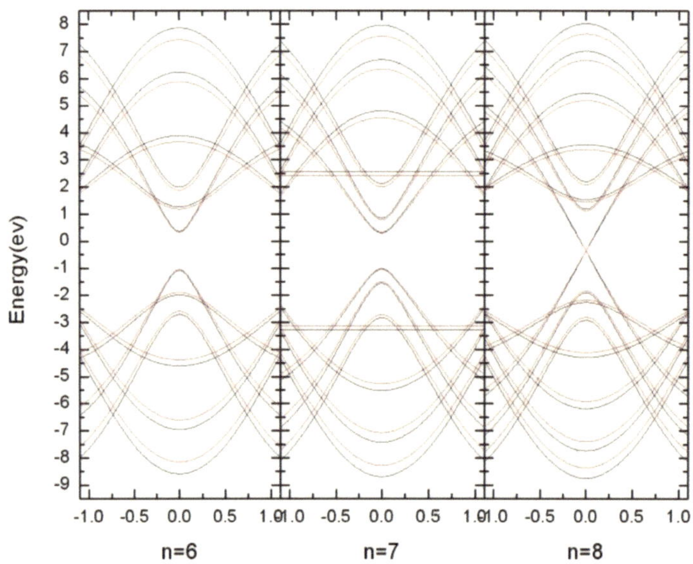

Figure 8. Band structures of the curved armchair nanoribbons with widths $n = 6$, $n = 7$, and $n = 8$, respectively. Black lines are the energy band of a perfect nanoribbon and red lines denote the band of a curved nanoribbon with the tubular shape.

By comparison, we found that the energy bandwidths become narrowed obviously for the widths n=6, 7, 8 and then this bending does not nearly influence on the energy gaps. This is because the localization of electrons is enhanced from two-dimensional plane to three-dimensional curved surface. When n=6, the increment of the gap with respect to the flat ribbon is equal to 0.074eV. When n=7, the change of the gap is 0.065eV. As for n=8, its metallic behavior does not change as the ribbon is rolled up. Fig. 9 illustrates the density of states of the warping ribbons with widths n=6, 7, 8. The meaning of the black and red lines in Fig. 9 is the same as in Fig. 8. From Fig. 9, we see that the tubular warping is responsible for the energy bandwidth narrowing. The density of states of both the top of the valence band and the bottom of the conduction band does not nearly change. It follows that this warping ribbon still keeps all the characteristics of the flat ribbon, especially for n=7. This means that the change of this dimension does not affect the electronic structure seriously. This is why we usually use a graphene sheet to study the electronic structure of a carbon nanotube. In addition, in order to show the effect of the curvature on the energy gap, a graph of the gap varying with the central angle is plotted in Fig. 10. It is apparent that for a fixed width the gap has a maximum value as the increasing of the central angle. When n=6, the central angle corresponding to the maximum value is between $5\pi/4$and$3\pi/2$. When n=7, this angle approximately equals$3\pi/2$. As the central angle is equal to zero, the warping ribbon becomes a flat ribbon and as the central angle goes to2π, the warp-

http://dx.doi.org/10.5772/51348

ing ribbon becomes a carbon nanotube. Fig. 10 also shows such a fact that when a graphene nanoribbon is bent into a nanotube, its energy gap is increased.

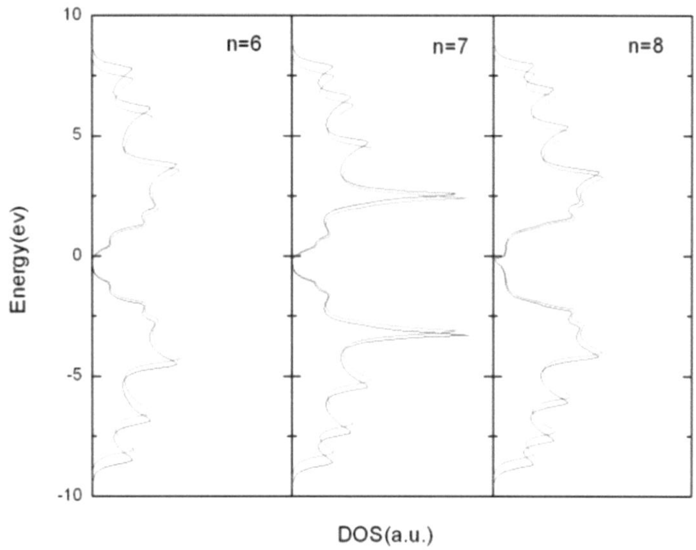

Figure 9. Density of states of tubular warping armchair nanoribbons. Black and red lines are the flat and warping nanoribbons, respectively.

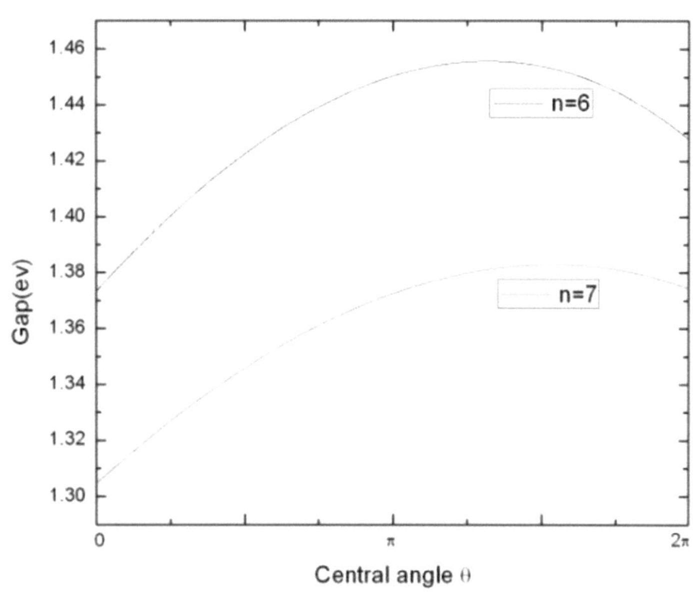

Figure 10. Energy gaps as a function of the central angle (or curvature).

5. Graphene nanoribbon modulated by sine regime

A free standing graphene nanoribbon could have out-of-plane warping because of the edge stress (Shenoy et al., 2008). This warping will bring about a very small change of the electronic energy spectrum. An ideal graphene nanoribbon only has periodicity in the direction of its length and there is no periodicity in the y-direction. To show the periodic effect in the y-direction, we modulate it with the aid of a sine periodic function

$$z = A\sin(fy) \tag{17}$$

where A is the modulation amplitude and f denotes the modulation frequency, i.e., modulated number per unit length. The modulated graph of an armchair graphene nanoribbon is shown in Fig. 11.

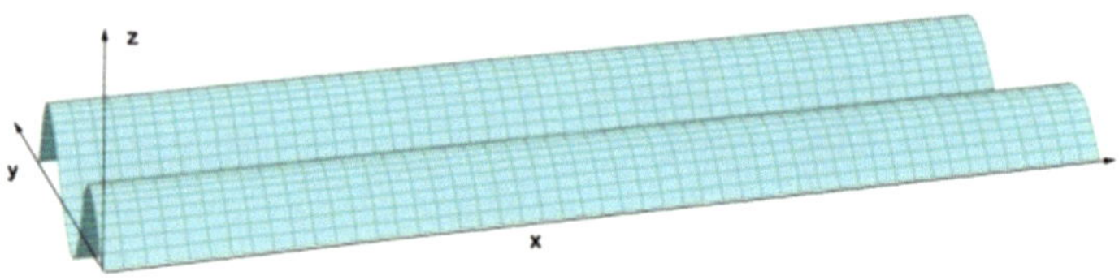

Figure 11. Graphene nanoribbon modulated by sine regime in the direction of the width.

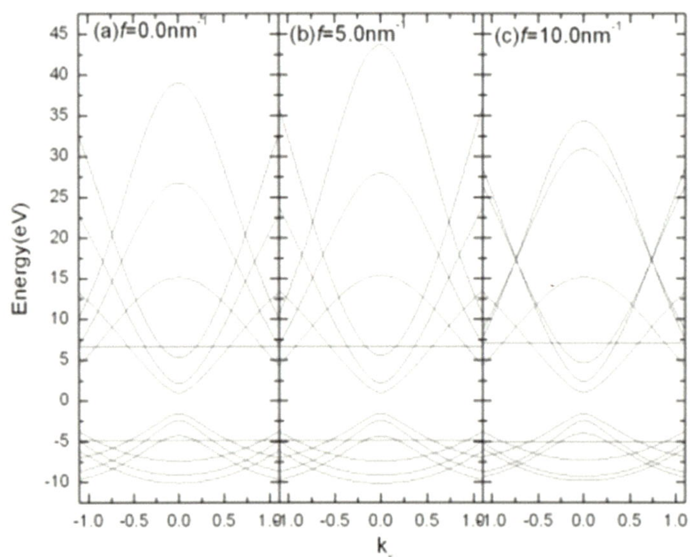

Figure 12. Energy spectra of graphene nanoribbons with width n=7 modulated by sine regime in the direction of the width. f is the modulation frequency.

By numerical calculations, we found that different modulation frequencies have different electronic band structures, i.e., the energy band structures depend strongly on the modulation frequency f and the modulation amplitude A. We take the width n=7 as an example to calculate the electronic energy spectrum. When the amplitude A is fixed, the energy band structures with different frequencies are shown in Fig. 12. It may be seen from Fig. 12 that this periodic modulation does not damage the Dirac cones, i.e., the topological property of armchair graphene nanoribbons is not destroyed. On the other hand, Fig. 12 tells us that this periodic modulation can change the energy band structure, i.e., both the bandwidth and band gap can be controlled by the modulation frequency. The density of states of electrons for an armchair graphene nanoribbon with n=7, modulated by using a sine function along the width direction, is plotted in Fig. 13.

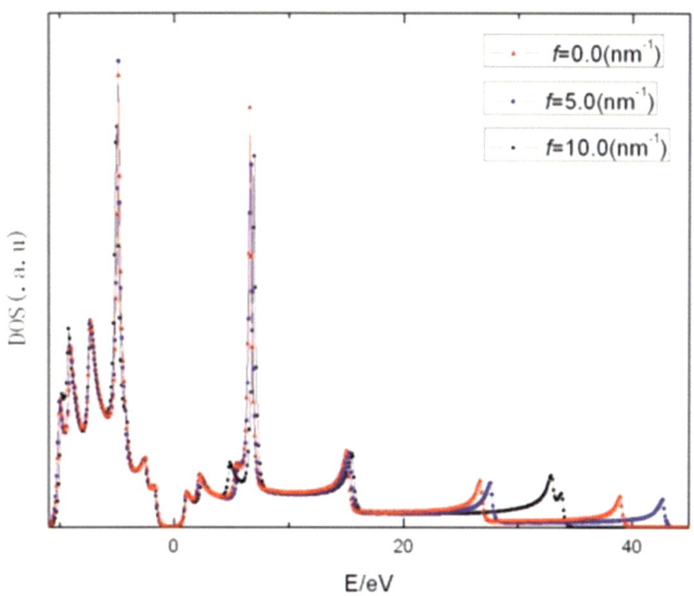

Figure 13. Density of states of graphene nanoribbons with width n=7 modulated by sine regime in the direction of the width.

When the modulation amplitude, taken to be 0.1nm, is fixed, different modulation frequencies have slightly different densities of states of electrons. The main difference between the frequencies $0.0nm^{-1}$, $5.0nm^{-1}$, and $10.0nm^{-1}$ is in the conduction band and the density of states of the valence band is the same nearly. It follows that the modulation along the width direction of the ribbon makes a notable impact for the density of states of the conduction band, especially for the high energy band corresponding to the standing wave of the smaller quantum number. In order to reveal the effect of the modulation amplitude on the electronic properties, the energy bands for the different amplitudes are calculated under certain frequency. Fig. 14 shows the band structures of the different amplitudes A=0.0nm, A=0.05nm, and A=0.1nm for an armchair nanoribbon with n=7, where the frequency f is taken to be

10nm^{-1}. When the modulation amplitude A=0.1nm, the band gaps corresponding to frequencies f=0.0nm^{-1}, 5nm^{-1}, and 10nm^{-1} are 2.580eV, 2.600eV, and 2.666eV, respectively. It seems that the band gaps linearly increase as the frequency increases. In fact, the inflection point of the smallest gap appears at f=6.02nm^{-1}, where the gap is equal to 2.571eV. There are other inflection points of the gap as the frequency increases, but the gaps of these points are big compared to that of the lowest inflection point (see Fig. 15(b)).

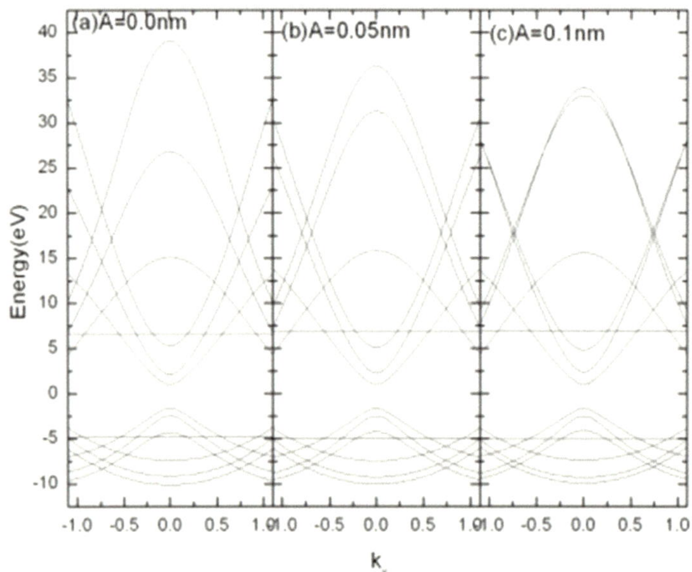

Figure 14. Energy spectra of graphene nanoribbons with width n=7 modulated by sine regime in the direction of the width. The modulation frequency f is taken to be 10nm^{-1}and A is the modulation amplitude.

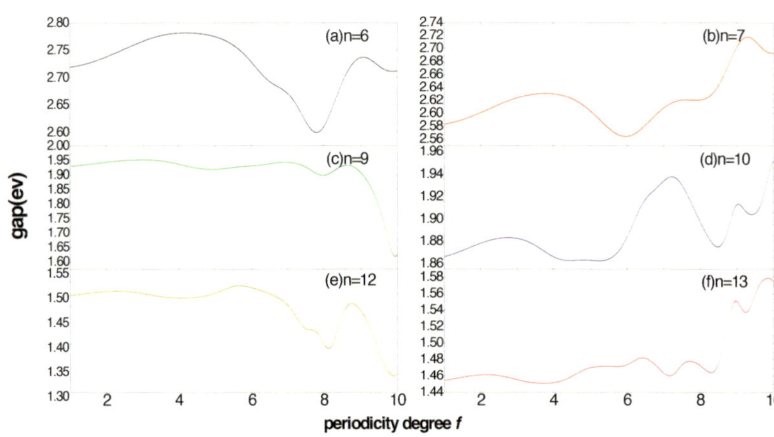

Figure 15. Energy gaps as a function of the modulation frequency f.

6. Conclusion

We investigated the electronic energy spectra of graphene and its nanoribbon subject to unaxial stress within the tight-binding approach. The unaxial stress can not open the energy gap of graphene at Dirac point K. But compression along the armchair shape edge or extension along the zigzag shape edge will make a small energy gap opened at K point. From this reason, the graphene subject to uniaxial stress still is a semiconductor with the zero-energy gaps. The position of Dirac point will vary as the stress. For the armchair graphene nanoribbon, the tensile or compressive stress not only can transfer the metallicity into the semiconductor, but also have the energy gap increased or decreased and the energy bandwidth widened or narrowed. Therefore, we can use the unaxial stress to control the electronic properties of armchair graphene nanoribbons. In addition, the tubular warping deformation of armchair nanoribbons does not nearly influence on the energy gap, but it is obvious to effect on the bandwidth. In addition, we also studied the periodic modulation of the shape of armchair nanoribbons by sine regime. This modulation can change its electronic properties. For the other modulation manner, we no longer discuss it here.

The advantage of the tight-binding method is that the physical picture is clearer and the calculating process is simpler compared to the first-principles calculations. This method is suitable only for narrow energy bands. Because graphene nanoribbons are the system of wider energy bands, this method has its limitation.

Author details

Guo-Ping Tong*

Address all correspondence to: tgp6463@zjnu.cn

Zhejiang Normal University, China

References

[1] Cadelano, E., Palla, P. L., Giordano, S., & Colombo, L. (2009). Nonlinear elasticity of monolayer graphene. *Phys. Rev. Lett.*, 102(23), 235502-4.

[2] Can, C. K., & Srolovitz, D. J. (2010). First-principles study of graphene edge properties and flake shapes. *Phys. Rev. B*, 81(12), 125445-8.

[3] Chung, P. W. (2006). Theoretical prediction of stress-induced phase transformations of the second kind in graphene. *Phys. Rev. B*, 73(7), 075433-5.

[4] de Andres, P. L., & Vergés, J. A. (2008). First-principles calculation of the effect of stress on the chemical activity of graphene. *Applied Physics Letters*, 93(17), 171915-3.

[5] Farjam, M., & Rafii-Tabar, H. (2009). Comment on "Band structure engineering of graphene by strain: First-principles calculations". *Phys. Rev. B*, 80(16), 167401-3.

[6] Feyd, R., Charlier, A., & Mc Rae, E. (1997). Uniaxial-stress effects on the electronic properties of carbon nanotubes. *Phys. Rev. B*, 55(11), 6280-6824.

[7] Fujita, M., Wakabayashi, K., Nakada, K., & Kusakabe, K. (1996). Peculiar localized state at zigzag graphite edge. *J. Phys. Soc. Jpn.*, 65(7), 1920-1923.

[8] Gui, G., Li, J., & Zhong, J. X. (2008). Band structure engineering of graphene by strain: First-principles calculations. *Phys. Rev. B*, 78(7), 075435-6.

[9] Harrison, W. A. (1980). *Electronic structure and properties of solids*, San Francisco, Freeman.

[10] Huang, Q. P., Yin, H., & Tong, G. P. (2006). Effect of π orbital orientations on the curvature and diameter of single-wall carbon nanotubes. *Journal of Atomic and Molecular Physics*, 23(4), 704-708.

[11] Huang, Q. P., Tong, G. P., & Yin, H. (2007). Calculation of the hybridization orbital of single-wall carbon nanotubes. *Journal of Atomic and Molecular Physics*, 24(1), 45-50.

[12] Jin, Z. F., Tong, G. P., & Jiang, Y. J. (2009). Effect of the non-nearest-neighbor hopping on the electronic structure of armchair graphene nanoribbons. *Acta Physica Sinica*, 58(12), 8537-8543, 1000-3290.

[13] Jun, S. (2008). Density-functional study of edge stress in graphene. *Phys. Rev. B*, 78(7), 073405-4.

[14] Kang, J. H., He, Y., Zhang, J. Y., Yu, X. X., Guan, X. M., & Yu, Z. P. (2010). Modeling and simulation of uniaxial strain effects in armchair graphene nanoribbon tunneling field effect transistors. *Applied Physics Letters*, 96(25), 252105-3.

[15] Kleiner, A., & Eggert, S. (2001). Curvature, hybridization, and STM images of carbon namotubes. *Phys. Rev. B*, 64(11), 113402-4.

[16] Lee, C., Wei, X., Kysar, J. W., & Hone, J. (2008). Measurement of the elastic properties and intrinsic strength of monolayer graphene. *Science*, 321(5887), 385-388.

[17] Liu, F., Ming, P. B., & Li, J. (2007). Ab initio calculation of ideal strength and phonon instability of graphene under tension. *Phys. Rev. B*, 76(6), 064120-7.

[18] Marianetti, C. A., & Yevick, H. G. (2010). Failure mechanisms of graphene under tension. *Phys. Rev. Lett.*, 105(24), 245502-4.

[19] Nakada, K., Fujita, M., Dresselhaus, G., & Dresselhaus, M. S. (1996). Edge state in graphene ribbons: Nanometer size effect and edge shape dependence. *Phys. Rev. B*, 54(24), 17954-17961, 1550-235X.

[20] Neek-Amal, M., & Peeter, F. M. (2010). Graphene nanoribbons subjected to axial stress. *Phys. Rev. B*, 82(8), 085432-6.

[21] Reich, S., Maultzsch, J., & Thomsen, C. (2002). Tight-binding description of graphene. *Phys. Rev. B*, 66(3), 035412-5.

[22] Rozhkov, A. V., Savel'ev, S., & Nori, F. (2009). Electronic properties of armchair graphene nanoribbons. *Phys. Rev. B*, 79(12), 125420-10.

[23] Saito, R., Dresselhaus, D., & Dresselhaus, M. S. (1998). *Physical Properties of Carbon Nanotubes*, Imperial College Press, 1-86094-093-5, London.

[24] Shenoy, V. B., Reddy, C. D., Ramasubramaniam, A., & Zhang, Y. W. (2008). Edge-stress- induced warping of graphene sheets and nanoribbons. *Phys. Rev.Lett.*, 101(24), 245501-4.

[25] Son, Y. W., Cohen, M. L., & Louie, S. G. (2006). Energy gaps in graphene nanoribbons. *Phys. Rev. Lett.*, 97(21), 216803-4.

[26] Sun, L., Li, Q., Ren, H., Su, H., Shi, Q., & Yang, J. (2008). Strain effect on electronic structures of graphene nanoribbons : A first-principles study. *The Journal of Chemical Physics*, 129(7), 074704.

[27] Wallace, P. R. (1947). The band theory of graphite. *Phys. Rev.*, 71(9), 622-634.

[28] Wei, Y., & Tong, G. P. (2009). Effect of the tensile force on the electronic energy gap of graphene sheets. *Acta Physica Sinica*, 58(3), 1931-1934.

[29] Xiong, W., Zhe, Liu. J., , M., Xu, Z. P., Sheridan, J., & Zheng, Q. S. (2011). Strain engineering water transport in graphene nanochannels. *Phys. Rev. E*, 84(5), 056329-7.

[30] Zheng, H. X., Wang, Z. F., Luo, T., Shi, Q. W., & Chen, J. (2007). Analytical study of electronic structure in armchair graphene nanoribbons. *Phys. Rev. B*, 75(16), 165411-6.

Chapter 4

The Cherenkov Effect in Graphene-Like Structures

Miroslav Pardy

Additional information is available at the end of the chapter

http://dx.doi.org/10.5772/50296

1. Introduction

More than 70 years ago, Peierls [29] and Landau [16] performed a proof that the 2-dimensional crystal is not thermodynamically stable and cannot exist. They argued that the thermodynamical fluctuations of such crystal leads to such displacements of atoms that these displacements are of the same size as the distances between atoms at the any finite temperature. The argument was extended by Mermin [21] and it seemed that many experimental observations supported the Landau-Peierls-Mermin theory. So, the "impossibility" of the existence of graphene was established.

In 2004, Andre Geim, Kostia Novoselov [13, 22, 23] and co-workers at the University of Manchester in the UK by delicately cleaving a sample of graphite with sticky tape produced a sheet of crystalline carbon just one atom thick, known as graphene. Geims group was able to isolate graphene, and was able to visualize the new crystal using a simple optical microscope. Nevertheless, Landau-Peierls-Mermin proof remained of the permanent historical and pedagogical meaning.

At present time, there are novel methods how to create graphene sheet. For instance, Dato et al. [5] used the plasma reactor, where the graphene sheets were synthesized by passing liquid ethanol droplets into an argon plasma.

Graphene is the benzene ring (C_6H_6) stripped out from their H-atoms. It is allotrope of carbon because carbon can be in the crystalline form of graphite, diamond, fullerene (C_{60}), carbon nanotube and glassy carbon (also called vitreous carbon).

Graphene unique properties arise from the collective behavior of so called pseudoelectrons with pseudospins, which are governed by the Dirac equation in the hexagonal lattice.

The Dirac fermions in graphene carry one unit of electric charge and so can be manipulated using electromagnetic fields. Strong interactions between the electrons and the honeycomb lattice of carbon atoms mean that the dispersion relation is linear and given by $E = vp$, where v is so called the Fermi-Dirac velocity, p is momentum of a pseudoelectron.

The linear dispersion relation follows from the relativistic energy relation for small mass together with approximation that the Fermi velocity is approximately only about 300 times less than the speed of light.

The pseudospin of the pseudoelectron follows from the graphene structure. The graphene is composed of the system of hexagonal cells and it means that graphene is composed from the systems of two equilateral triangles. If the wave function of the first triangle sublattice system is φ_1 and the wave function of the second triangle sublattice system is φ_2, then the total wave function of the electron moving in the hexagonal system is superposition $\psi = c_1\varphi_1 + c_2\varphi_2$, where c_1 and are c_2 appropriate functions of coordinate $\mathbf{x}$ and functions φ_1, φ_2 are functions of wave vector $\mathbf{k}$ and coordinate $\mathbf{x}$. The next crucial step is the new spinor function defined as [19].

$$\chi = \left\{ \begin{matrix} \varphi_1 \\ \varphi_2 \end{matrix} \right\} \tag{1}$$

and it is possible to prove that this spinor function is solution of the Pauli equation in the nonrelativistic situation and Dirac equation of the generalized case. The corresponding mass of such effective electron is proved to be zero.

The introduction of the Dirac relativistic Hamiltonian in graphene physics is the description of the graphene physics by means of electron-hole medium. It is the analogue of the description of the electron-positron vacuum by the Dirac theory of quantum electrodynamics. The pseudoelectron and pseudospin are not an electron and the spin of quantum electrodynamics (QED), because QED is the quantum theory of the interaction of real electrons and photons where mass of an electron is the mass defined by classical mechanics and not by collective behavior in hexagonal sheet called graphene.

The graphene can be considered as the special form the 2-dimensional graphene-like structures, where for instance silicene has the analogue structure as graphene [8]. The band structure of a free silicene layer resembles the band structure of graphene. The Fermi velocity v of electrons in silicene is lower than that in graphene.

If we switch on an electric field, the symmetry between the A and B sublattices of silicene's honeycomb structure breaks and a gap Δ is open in the band structure at the hexagonal Brillouin zone (BZ) points K and K'. In the framework of a simple nearest-neighbor tight-binding model, this manifests itself in the form of an energy correction to the on-site energies that is positive for sublattice A and negative for B. This difference in on-site energies $\Delta = E_A - E_B$ leads to a spectrum with a gap for electrons in the vicinity of the corners of the BZ with $E_\pm = \pm\sqrt{(\Delta/2)^2 + |v\mathbf{p}|^2}$, where $\mathbf{p}$ is the electron momentum relative to the BZ corner. Opening a gap in graphene by these means would be impossible because the A and B sublattices lie in the same plane [7].

So, silicene consists of a honeycomb lattice of silicon atoms with two sublattices made of A sites and B sites. The states near the Fermi energy are orbitals residing near the K and K' points at opposite corners of the hexagonal Brillouin zone. While silicon is dielectric medium, silicene is the conductive medium with Hall effect and it is possible to study the Mach cone generated by motion of a charged particle through the silicene sheet.

On the other side, there are amorphous solids - glasses, the atomic structure of which lack any long range translational periodicity. However, due to chemical bonding characteristics, glasses do possess a high degree of short-range order with respect to local atomic polyhedra.

http://dx.doi.org/10.5772/50296

In other words such structures can be considered as the graphene-like structures with the appropriate index of refraction, which is necessary for the the existence of Čerenkov effect.

The last but not least graphene-like structure can be represented by graphene-based polaritonic crystal sheet [2] which can be used to study the Čerenkov effect.

We derive in this chapter the power spectrum of photons generated by charged particle moving in parallel direction and perpendicular direction to the graphene-like structure with index of refraction n. The Graphene sheet is conductive contrary to some graphene-like structures, for instance graphene with implanted ions, which are dielectric media and it means that it enables the experimental realization of the Čerenkov radiation. We calculate it from the viewpoint of the Schwinger theory of sources [24–27, 30–32].

To be pedagogically clear we introduce the quantum theory of the index of refraction (where the dipole polarization of matter is the necessary condition for its existence), the classical and quantum theory of Čerenkov radiation and elements of the Schwinger source theory formalism for electrodynamic effect in dielectric medium. We involve also the Čerenkov effect with massive photons.

2. The quantum theory of index of refraction

The quantum theory of dispersion can be derived in the framework of the nonrelativistic Schrödinger equation [33] for an electron moving in dielectric medium and in the field with the periodic force

$$F_x = -eE_0 \cos \omega t, \quad F_y = F_z = 0. \tag{2}$$

Then, the corresponding potential energy is

$$V' = -exE_0 \cos \omega t \tag{3}$$

and this potential energy is the perturbation energy in the Schrödinger equation

$$\left(i\hbar \frac{\partial}{\partial t} - H_0 - V' \right) \psi_k(t) = 0, \tag{4}$$

where for $V' = 0$ $\psi_k(t) \to \psi_k^0(t)$ and

$$\psi_k^0(t) = \psi_k^0 e^{-\frac{i}{\hbar}E_k t} = \psi_k^0 e^{-i\omega_k t}, \tag{5}$$

where ψ_k^0 is the solution of the Schrödinger equation without perturbation, or,

$$\left(i\hbar \frac{\partial}{\partial t} - H_0 \right) \psi_k^0(t) = 0. \tag{6}$$

We are looking for the solution of the Schrödinger equation involving the perturbation potential in the form

$$\psi_k(t) = \psi_k^0(t) + \psi_k^1(t), \tag{7}$$

where $\psi_k^1(t)$, is the perturbation wave function correction to the non-perturbation wave function.

After insertion of formula (7) to eq. (4), we get

$$\left(i\hbar\frac{\partial}{\partial t} - H_0\right)\psi_k^1(t) = \frac{1}{2}exE_0\psi_k^0\left(e^{-it(\omega_k-\omega)} + e^{-it(\omega_k+\omega)}\right). \tag{8}$$

Let us look for the solution of eq. (8) in the form:

$$\psi_k^1(t) = ue^{-it(\omega_k-\omega)} + ve^{-it(\omega_k+\omega)}. \tag{9}$$

After insertion of (9) into (8), we get two equations for u and v:

$$(\hbar(\omega_k - \omega) - H_0)\,u = \frac{1}{2}exE_0\psi_k^0, \tag{10}$$

$$(\hbar(\omega_k + \omega) - H_0)\,v = \frac{1}{2}exE_0\psi_k^0. \tag{11}$$

Then, using the formal expansion

$$u = \sum_{k''} C_{k''}\psi_{k''}^0, \tag{12}$$

we get from eq.

$$(E_{k''} - H_0)\,\psi_{k''}^0 = 0 \tag{13}$$

the following equation

$$\hbar\sum_{k''} C_{k''}\left(\omega_{kk''} - \omega\right)\psi_{k''}^0 = \frac{exE_0}{2}\psi_k^0 \tag{14}$$

with

http://dx.doi.org/10.5772/50296

$$\omega_{kk''} = \frac{E_k - E_{k''}}{\hbar}. \tag{15}$$

Using the orthogonal relation

$$\int \psi_{k'}^{0*} \psi_{k''}^{0} d^3x = \delta_{k'k''}, \tag{16}$$

we get the following relation for C_k and u as follows:

$$C_k = -\frac{eE_0}{2\hbar} \cdot \frac{x_{k'k}}{\omega_{k'k} + \omega}, \tag{17}$$

$$u = \sum_{k'} \left(-\frac{eE_0}{2\hbar} \right) \cdot \frac{x_{k'k}}{\omega_{k'k} + \omega} \psi_{k'}^{0} \tag{18}$$

and $v = u(-\omega)$, or

$$v = \sum_{k'} \left(-\frac{eE_0}{2\hbar} \right) \cdot \frac{x_{k'k}}{\omega_{k'k} - \omega} \psi_{k'}^{0} \tag{19}$$

and

$$x_{k'k} = \int \psi_{k'}^{0*} x \psi_{k}^{0} d^3x. \tag{20}$$

The general wave function can be obtained from eqs. (7), (9), (18) and (19) in the form:

$$\psi_k(t) = e^{-i\omega_k t} \left\{ \psi_k^0 - \frac{eE_0}{\hbar} \sum_{k'} \frac{x_{k'k}}{\omega_{k'k}^2 - \omega^2} \psi_{k'}^0 \left[\omega_{k'k} \cos \omega t - i\omega \sin \omega t \right] \right\}. \tag{21}$$

The classical polarization of a medium is given by the well known formula

$$P = Np = -Nex, \tag{22}$$

where N is the number of atom in the unite volume of dielectric medium. So we are able to define the quantum analogue form of the polarization as it follows:

$$P = N\bar{p} = -Ne \int \psi_k^*(t) x \psi_k(t) d^3x, \quad (23)$$

or, with

$$\int \psi_k^{0*} x \psi_k^0 d^3x = 0, \quad (24)$$

we have

$$P = \sum_{k'} \left(2\frac{Ne^2 E_0}{\hbar} \right) \cdot \frac{\omega_{k'k} |x_{k'k}|^2}{\omega_{k'k}^2 - \omega^2} \cos \omega t. \quad (25)$$

Using the classical formula for polarization P,

$$\mathbf{P} = \frac{n^2 - 1}{4\pi} \mathbf{E}, \quad (26)$$

we get for the quantum model of polarization

$$\frac{n^2 - 1}{4\pi} = \sum_{k'} \left(2\frac{Ne^2}{\hbar} \right) \cdot \frac{\omega_{k'k} |x_{k'k}|^2}{\omega_{k'k}^2 - \omega^2}. \quad (27)$$

Using the definition of the coefficients $f_{k'k}$ by relation

$$f_{k'k} = \frac{2m}{\hbar} \omega_{k'k} |x_{k'k}|^2, \quad (28)$$

we get the modified equation (27) as follows:

$$\frac{n^2 - 1}{4\pi} = \frac{Ne^2}{m} \sum_{k'} \frac{f_{k'k}}{\omega_{k'k}^2 - \omega^2}. \quad (29)$$

The last formula should be compared with the classical one:

$$\frac{n^2 - 1}{4\pi} = \frac{e^2}{m} \sum_{k} \frac{N_k}{\omega_k^2 - \omega^2}, \quad (30)$$

where N_k is number of electrons moving with frequency ω_k in the unit volume.

3. The classical description of the Čerenkov radiation

In electrodynamics, a fast moving charged particle in a medium when its speed is faster than the speed of light in this medium produces electromagnetic radiation which is called the Čerenkov radiation. This radiation was first observed experimentally by Čerenkov [3, 4] and theoretically interpreted by Tamm and Frank [34], in the framework of the classical electrodynamics [9].

The charge and current density of electron moving with the velocity $\mathbf{v}$ and charge e is as it is well known:

$$\varrho = e\delta(\mathbf{x} - \mathbf{v}t) \tag{31}$$

$$\mathbf{j} = e\mathbf{v}\delta(\mathbf{x} - \mathbf{v}t). \tag{32}$$

The equations for the potentials $\mathbf{A}$, φ are given by equations [17, 18]

$$\Delta\mathbf{A} - \frac{\varepsilon}{c^2}\frac{\partial^2\mathbf{A}}{\partial t^2} = -\frac{4\pi}{c}e\mathbf{v}\delta(\mathbf{x} - \mathbf{v}t) \tag{33}$$

and

$$\Delta' - \frac{\varepsilon}{c^2}\frac{\partial^2\varphi}{\partial t^2} = -4\pi\frac{e}{\varepsilon}\delta(\mathbf{x} - \mathbf{v}t) \tag{34}$$

with the additional Lorentz calibration condition:

$$\mathrm{div}\mathbf{A} + \frac{\varepsilon}{c}\frac{\partial\varphi}{\partial t} = 0, \tag{35}$$

where magnetic permeability $\mu = 1$ and ε is dielectric constant of medium.

After the Fourier transformation the vector potential

$$\frac{1}{(2\pi)^{3/2}}\int \mathbf{A}e^{-i\mathbf{kx}}d^3x = \mathbf{A_k}; \quad \mathbf{A}(\mathbf{x}) = \frac{1}{(2\pi)^{3/2}}\int \mathbf{A_k}e^{i\mathbf{kx}}d^3k, \tag{36}$$

we get

$$\Delta\mathbf{A_k} - \frac{\varepsilon}{c^2}\frac{\partial^2\mathbf{A_k}}{\partial t^2} = \frac{4\pi e}{c(2\pi)^{3/2}}\mathbf{v}\int e^{-i\mathbf{kx}}\delta(\mathbf{x} - \mathbf{v}t)d^3x, \tag{37}$$

or,

$$\Delta \mathbf{A_k} - \frac{\varepsilon}{c^2}\frac{\partial^2 \mathbf{A_k}}{\partial t^2} = -\frac{4\pi e\mathbf{v}}{c(2\pi)^{3/2}}e^{-i\mathbf{kv}t}. \tag{38}$$

On the other hand, we have:

$$\Delta \mathbf{A} = \Delta \int \mathbf{A_k} e^{i\mathbf{kx}} d^3k = -\int k^2 \mathbf{A_k} e^{i\mathbf{kr}} d^3k, \tag{39}$$

from which we have

$$\Delta \mathbf{A_k} = k^2 \mathbf{A_k} \tag{40}$$

and

$$-k^2 \mathbf{A_k} - \frac{\varepsilon}{c^2}\frac{\partial^2 \mathbf{A_k}}{\partial t^2} = -\frac{4\pi e\mathbf{v}}{c(2\pi)^{3/2}}e^{-i\mathbf{kv}t}. \tag{41}$$

Formula (41) shows that the dependence $\mathbf{A_k}$ on time is of the form:

$$\mathbf{A_k} \sim e^{-i\mathbf{kv}t} = e^{-i\omega t}, \tag{42}$$

where

$$\omega = \mathbf{kv}. \tag{43}$$

At the same time

$$\frac{\partial^2 \mathbf{A_k}}{\partial t^2} = -\omega^2 \mathbf{A_k}. \tag{44}$$

We can transcribe eq. (41) in the following form:

$$\mathbf{A_k} = \frac{4\pi e}{c(2\pi)^{3/2}} \frac{\mathbf{v}}{k^2 - \frac{\omega^2 \varepsilon}{c^2}} e^{-i\omega t}. \tag{45}$$

By analogy with the formula (45) we can derive the formula concerning the Fourier transform of φ. Or,

$$\varphi_{\mathbf{k}} = \frac{4\pi e}{\varepsilon(2\pi)^{3/2}} \frac{1}{k^2 - \frac{\omega^2 \varepsilon}{c^2}} e^{-i\omega t}. \tag{46}$$

The intensity of the electric field has the Fourier Component as follows:

$$\mathbf{E}_{\mathbf{k}} = -\frac{1}{c}\frac{\partial \mathbf{A}_{\mathbf{k}}}{\partial t} - \mathrm{grad}\varphi_{\mathbf{k}} = \frac{i\omega}{c}\mathbf{A}_{\mathbf{k}} - i\mathbf{k}\varphi_{\mathbf{k}} =$$

$$\frac{4\pi e}{(2\pi)^{3/2}} \left(\frac{\omega \mathbf{v}}{c^2} - \frac{\mathbf{k}}{\varepsilon}\right) \frac{i}{k^2 - \frac{\omega^2 \varepsilon}{c^2}} e^{-i\omega t}, \tag{47}$$

from which follows that the intensity of the electric field induced in the dielectric medium is:

$$\mathbf{E} = \frac{ie}{2\pi^2} \int \left(\frac{\omega \mathbf{v}}{c^2} - \frac{\mathbf{k}}{\varepsilon}\right) \frac{e^{i(\mathbf{kx} - \omega t)}}{k^2 - \frac{\omega^2 \varepsilon}{c^2}} dk_x dk_y dk_z. \tag{48}$$

The formula (48) gives **E** in the form of the moving plane wave in case that we can write

$$\omega = v k_z = v k \cos\Theta \equiv \frac{kc}{n(\omega)}. \tag{49}$$

From the last equation we have:

$$\cos\Theta = \frac{c}{n(\omega) v}. \tag{50}$$

Now, let us chose the direction of the particle motion along the z-axis and let us introduce the cylindrical coordinates putting

$$k^2 = k_x^2 + k_y^2 + k_z^2 = k_z^2 + q^2. \tag{51}$$

At the same time

$$dk_x dk_y dk_z = q\, dq\, d\varphi\, dk_z. \tag{52}$$

Further

$$dk_z = \frac{d\omega}{v} \tag{53}$$

and therefore

$$dk_x dk_y dk_y = q\, dq\, d\varphi\, \frac{d\omega}{v}. \tag{54}$$

In such a way we have for the intensity of the electrical field:

$$\mathbf{E} = \frac{ie}{\pi} \int q\, dq\, d\omega \left(\frac{\omega \mathbf{v}}{c^2} - \frac{\mathbf{k}}{\varepsilon} \right) \frac{e^{i(\mathbf{kx} - \omega t)}}{v \left[q^2 + \omega^2 \left(\frac{1}{v^2} - \frac{\varepsilon}{c^2} \right) \right]}, \tag{55}$$

where the φ-integration was already performed. The ω-integration involves both positive and negative frequencies.

The quantity, which is experimentally meaningful, is the energy loss of the moving particle per unit length, or, dW/dz in the prescribed frequency interval $d\omega$. This energy loss is in the relation with the work of force which acts on the particle by the induced electromagnetic field. The work is expressed by the formula:

$$dW = -F_z dz = -e(E_z)_{\mathbf{x}=\mathbf{v}t} dz, \tag{56}$$

where $(E_z)_{\mathbf{x}=\mathbf{v}t}$ is the z-component of the electric intensity at the point where the particle is. The sign minus denotes the physical fact that the force acts against the vector of velocity, or, in the negative direction of the z-axis.

Thus we have:

$$\frac{dW}{dz} = -e\left(E_z\right)_{\mathbf{x}=\mathbf{v}t} = -\frac{ie^2}{\pi} \int q\, dq\, d\omega \left(e^{i(\mathbf{kx} - \omega t)} \right)_{\mathbf{x}=\mathbf{v}t} \frac{\left(\frac{\omega \mathbf{v}}{c^2} - \frac{k_z}{\varepsilon} \right)}{v \left[q^2 + \omega^2 \left(\frac{1}{v^2} - \frac{\varepsilon}{c^2} \right) \right]}$$

$$= \frac{ie^2}{\pi} \int q\, dq\, \omega d\omega\, \frac{1}{\varepsilon} \frac{\left(\frac{1}{v^2} - \frac{\varepsilon}{c^2} \right)}{q^2 + \omega^2 \left(\frac{1}{v^2} - \frac{\varepsilon}{c^2} \right)}. \tag{57}$$

The energy loss of particle per unit length and in the frequency interval $\omega, \omega + d\omega$ is obviously given as

$$\frac{d^2W}{dzd\omega} = \frac{ie^2\omega}{\pi}\int qdq\frac{\left(\frac{1}{v^2\varepsilon_+} - \frac{1}{c^2}\right)}{q^2 + \omega^2\left(\frac{1}{v^2} - \frac{\varepsilon_+}{c^2}\right)} - \frac{ie^2|\omega|}{\pi}\int qdq\frac{\left(\frac{1}{v^2\varepsilon_-} - \frac{1}{c^2}\right)}{q^2 + \omega^2\left(\frac{1}{v^2} - \frac{\varepsilon_-}{c^2}\right)}. \quad (58)$$

We introduced in the last formula notation $\varepsilon_+ = \varepsilon(\omega)$ and $\varepsilon_- = \varepsilon(-|\omega|)$ for ε at positive and negative values of ω. We know that in the absorptive dielectric media ε has the imaginary component. This imaginary component of ε is positive for $\omega < 0$ and negative for $\omega > 0$. In fact, the absorption in medium is real effect and it means that $\exp\{-ikx\}$ must correspond to the absorption for $x > 0$ for the arbitrary sign of ω. This experimental requirement determines the sign of the imaginary part of the permitivity of medium.

In formula (58) we can neglect in the numerator the imaginary part of the permitivity and write:

$$\left(\frac{1}{v^2\varepsilon_+} - \frac{1}{c^2}\right) = \left(\frac{1}{v^2\varepsilon_-} - \frac{1}{c^2}\right) = \left(\frac{1}{v^2n^2} - \frac{1}{c^2}\right). \quad (59)$$

On the other hand, such operation cannot be performed in the denominator which follows from the next text. Let us introduce the new complex quantity:

$$u = q^2 + \omega^2\left(\frac{1}{v^2} - \frac{\varepsilon}{c^2}\right). \quad (60)$$

Then we have

$$\frac{d^2W}{dxd\omega} = \frac{ie^2\omega}{2\pi}\left(\frac{1}{v^2n^2} - \frac{1}{c^2}\right)\left\{\int_{C_1}\frac{du}{u} - \int_{C_2}\frac{du}{u}\right\}. \quad (61)$$

In case of the absence of absorption in medium i.e. $\Im\varepsilon = 0$, the formula (61) gives meaningless zero result. However ε and therefore also u has the nonzero imaginary part. It means that the integrals in (61) is considered in the complex plane. The contour of integration is chosen in such a way that it is going in parallel to the real axis above this axis in case of $\Im u > 0$ (it corresponds to $\omega > 0$) and it corresponds to the curve C_1, and under the axis at $\Im u < 0$ (i.e. $\omega < 0$) which corresponds to the curve C_2. The singular point $u = 0$ is avoided along the infinitesimal semi-circles above and under the axis. Thus evidently:

$$\int_{C_1}\frac{du}{u} - \int_{C_2}\frac{du}{u} = \oint\frac{du}{u} = 2\pi i. \quad (62)$$

The integration was performed as a limiting procedure along the infinitesimal circle with the center in the origin of the coordinate system.

Using eq. (62) we can write the energy loss formula formula (61) in the following simple form:

$$\frac{d^2W}{dxd\omega} = \frac{e^2\omega}{c^2}\left(1 - \frac{c^2}{n^2v^2}\right)\omega. \tag{63}$$

This formula was derived for the first time by Tamm and Frank in year 1937. The fundamental features of the Čerenkov radiation are as follows:

1. The radiation arises only for particle velocity greater than the velocity of light in the dielectric medium is.
2. It depends only on the charge and not on mass of the moving particles
3. The radiation is produced in the visible interval of the light frequencies, i. e. in the ultraviolet part of the frequency spectrum. The radiation does not exists for very short waves.
4. The spectral dependency on the frequency is linear for the homogeneous medium.
5. The radiation generated in the given point of the trajectory spreads on the surface of the cone with the vertex in this point and with the axis identical with the direction of motion of the particle. The vertex angle of the cone is given by the relation $\cos\Theta = c/nv$.

Let us remark that the energy loss of a particle caused by the Čerenkov radiation are approximately equal to 1 % of all energy losses in the condensed matter such as the bremsstrahlung and so on. The fundamental importance of the Čerenkov radiation is in its use for the modern detectors of very speed charged particles in the high energy physics. The detection of the Čerenkov radiation enables to detect not only the existence of the particle, however also the direction of motion and its velocity and according to eq. (63) also its charge.

4. The quantum theory of the Čerenkov effect

Let us start with energetic consideration. So, let us suppose that the initial momentum and energy of electron is $\mathbf{p}$ and E and the final momentum and energy of electron is $\mathbf{p}'$ and E'. The momentum of the emitted photon let be $\hbar\mathbf{k}$. Then, after emission of photon the energy conservation laws are as follows:

$$\sqrt{p^2c^2 + m^2c^4} - \hbar\omega = \sqrt{p'^2c^2 + m^2c^4} \tag{64}$$

$$\mathbf{p} - \hbar\mathbf{k} = \mathbf{p}', \tag{65}$$

where $\mathbf{p}'$ is the momentum of electron after emission of photon. Let us make the quadratical operation of both equations and let us eliminate p'. Then we have:

$$2\hbar pkc^2\cos\Theta = c^2\hbar^2k^2 - (\hbar\omega)^2 + 2\hbar\omega\sqrt{p^2c^2+m^2c^4}. \tag{66}$$

where Θ is the angle between the direction of electron motion and the emission of photon. Putting $\omega = ck/n$ and expressing the momentum of electron in dependence of its velocity

$$\mathbf{v} = \frac{\mathbf{p}c^2}{E}, \tag{67}$$

we get with $\beta = v/c$

$$\cos\Theta = \frac{1}{n\beta} + \frac{\hbar k}{2p}\left(1 - \frac{1}{n^2}\right). \tag{68}$$

Now, following Harris [12], we show, using the second quantization method how to derive the Čerenkov effect in a dielectric medium characterized by its dielectric constant $\varepsilon(\omega)$ and its index of refraction n, which is given by relation $\sqrt{\varepsilon\mu}$ where μ is the magnetic permeability.

The relation between frequency and the wave number in a dielectric medium is

$$\omega = \frac{c}{n}k = \frac{c}{\sqrt{\varepsilon\mu}}k. \tag{69}$$

It was shown [17] that in such dielectric medium the energy of the electromagnetic field is given by the relation

$$U = \int d^3x\frac{1}{8\pi}\left\{|\mathbf{E}|^2\frac{\partial}{\partial\omega}\omega\varepsilon(\omega) + |\mathbf{B}|^2\right\}. \tag{70}$$

Since

$$\text{rot }\mathbf{E} = -\frac{1}{c}\frac{\partial\mathbf{B}}{\partial t} \tag{71}$$

$$i\mathbf{k}\times\mathbf{E} = \frac{i\omega}{c}\mathbf{B} \tag{72}$$

$$|\mathbf{B}|^2 = \frac{c^2}{\omega^2}|\mathbf{k}\times\mathbf{E}|^2 = \frac{c^2k^2}{\omega^2}|\mathbf{E}|^2 = \varepsilon|\mathbf{E}|^2, \tag{73}$$

we have:

$$U = \int d^3x \frac{1}{8\pi} |\mathbf{E}|^2 \left[\frac{\partial}{\partial \omega} \omega \varepsilon + \varepsilon \right] = \int d^3x \frac{1}{4\pi} |\mathbf{E}|^2 \frac{1}{2\omega} \frac{\partial \omega^2 \varepsilon}{\partial \omega}. \tag{74}$$

We see that the energy density that a vacuum would have in a vacuum must be corrected by the factor

$$\frac{1}{2\omega} \frac{\partial}{\partial \omega} \omega^2 \varepsilon(\omega) \tag{75}$$

when it moves in a medium of dielectric constant $\varepsilon(\omega)$.

Now, let us consider the Fourier transformation of the electromagnetic potential $\mathbf{A}$:

$$\mathbf{A} = \sum_{\mathbf{k}} \sum_{\sigma=1,2} \left(\frac{2\pi\hbar c^2}{\Omega \omega_k} \right)^{1/2} \mathbf{u}_{\mathbf{k}\sigma} \left\{ a_{\mathbf{k}\sigma}(t) e^{i\mathbf{kx}} + a^+_{\mathbf{k}\sigma}(t) e^{-i\mathbf{kx}} \right\}, \tag{76}$$

where the factor

$$\left(\frac{2\pi\hbar c^2}{\Omega \omega_k} \right)^{1/2} \tag{77}$$

is a normalization factor chosen for later convenience. In other words it is chosen in order the energy of the el. magnetic field to be interpreted as the sum of energies of the free harmonic oscillators, or,

$$H = \sum_{k,\sigma} \hbar \omega_k a^+_{k\sigma} a_{k\sigma}, \tag{78}$$

where a^+, a are creation and annihilation operators fulfilling commutation relations

$$\left[a_{\mathbf{k}\sigma}, a^+_{\mathbf{k}'\sigma'} \right] = \delta_{\mathbf{k},\mathbf{k}'} \delta_{\sigma,\sigma'}. \tag{79}$$

We want the total energy rather than just the el.mag. field energy to have the form of eq. (78). And it means we are forced to replace the normalization factor by

$$\left(\frac{2\pi\hbar c^2}{\Omega [\frac{1}{2\omega} \frac{\partial}{\partial \omega} \omega^2 \varepsilon(\omega)]_{\omega_k}} \right)^{1/2}, \tag{80}$$

http://dx.doi.org/10.5772/50296

and it leads to renormalized $\mathbf{A}(\mathbf{x}, t)$ as follows:

$$\mathbf{A}(\mathbf{x}, t) = \sum_{\mathbf{k},\sigma} \left(\frac{2\pi\hbar c^2}{\Omega[\frac{1}{2\omega}\frac{\partial}{\partial\omega}\omega^2\varepsilon(\omega)]_{\omega_k}} \right)^{1/2} \mathbf{u}_{\mathbf{k}\sigma} \left\{ a_{\mathbf{k}\sigma} e^{i\mathbf{k}\mathbf{x}} + a^{+}_{\mathbf{k}\sigma} e^{-i\mathbf{k}\mathbf{x}} \right\}. \tag{81}$$

The interaction Hamiltonian H' is unchanged except for the change in the normalization factor, or

$$H' = -\frac{e}{mc} \sum_{\mathbf{k},\sigma} \left(\frac{2\pi\hbar c^2}{\Omega[\frac{1}{2\omega}\frac{\partial}{\partial\omega}\omega^2\varepsilon(\omega)]_{\omega_k}} \right)^{1/2} \mathbf{p} \cdot \mathbf{u}_{\mathbf{k}\sigma} \left\{ a_{\mathbf{k}\sigma} e^{i\mathbf{k}\mathbf{x}} + a^{+}_{\mathbf{k}\sigma} e^{-i\mathbf{k}\mathbf{x}} \right\}. \tag{82}$$

Now, let be the initial state $|i\rangle$ and the final state $|f\rangle$. The transition probability per unit time is given by the first order perturbation term, or

$$\left(\frac{\text{trans prob}}{\text{time}} \right) = \frac{2\pi}{\hbar} |\langle f|i\rangle|^2 \delta(E_f - E_i). \tag{83}$$

We use the last formula to calculate the transition probability per unit time for a free electron of momentum $\hbar\mathbf{q}$ to emit a photon of momentum $\hbar\mathbf{k}$ thereby changing its momentum to $\hbar(\mathbf{q} - \mathbf{k})$.

We find:

$$\left(\frac{\text{trans prob}}{\text{time}} \right)_{\mathbf{q}\to\mathbf{q}-\mathbf{k}} = \frac{2\pi}{\hbar} \left(\frac{e}{mc} \right)^2 \left(\frac{2\pi\hbar c^2}{\Omega[\frac{1}{2\omega}\frac{\partial}{\partial\omega}\omega^2\varepsilon(\omega)]_{\omega_k}} \right) \times$$

$$|\langle \mathbf{q} - \mathbf{k}|\mathbf{p} \cdot \mathbf{u}_{\mathbf{k}\sigma} e^{-i\mathbf{k}\mathbf{x}}|\mathbf{q}\rangle|^2 \delta \left[\frac{\hbar^2 q^2}{2m} - \frac{\hbar^2}{2m}|\mathbf{q} - \mathbf{k}|^2 - \hbar\omega_{\mathbf{k}} \right]. \tag{84}$$

The matrix element in (84) is just equal to $\hbar\mathbf{q}u_{\mathbf{k}\sigma}$. Letting Θ to be the angle between $\mathbf{q}$ and $\mathbf{k}$ and writing $\mathbf{v} = \hbar\mathbf{q}/m$ be the particle velocity, we find:

$$\left(\frac{\text{trans prob}}{\text{time}} \right)_{\mathbf{q}\to\mathbf{q}-\mathbf{k}} = \left(\frac{4\pi^2 e^2 \hbar^2 |\mathbf{q} \cdot \mathbf{u}_{\mathbf{k}\sigma}|^2}{m^2 \Omega \hbar v k [\frac{1}{2\omega}\frac{\partial}{\partial\omega}\omega^2\varepsilon(\omega)]_{\omega_k}} \right) \times$$

$$\delta \left[\cos\Theta - \frac{c}{nv} - \frac{\hbar\omega n}{2mcv} \right]. \tag{85}$$

Note that the photon is emitted at an angle to the path of the electron given by

$$\cos\Theta = \frac{c}{nv}\left[1 + \frac{\hbar\omega n^2}{2mc^2}\right]. \tag{86}$$

If the energy of the photon $\hbar\omega$ is much less than the rest mass of the electron mc^2 then $\cos\Theta \approx c/nv$ which gives the classical Čerenkov angle. This can only be satisfied if the velocity of the particle is greater than c/n which is the velocity of the electromagnetic wave in medium. In vacuum where $n = 1$, v can never exceed c and so emission cannot occur.

The quantity of physical interest is the loss of energy per unit length of path of the electron. It is given by the formula:

$$\frac{dW}{dx} = \frac{1}{v}\frac{dW}{dt} = \frac{1}{v}\sum_{\mathbf{k},\sigma}\hbar\omega_{\mathbf{k}}\left(\frac{\text{trans prob}}{\text{time}}\right)_{\mathbf{q}\to\mathbf{q}-\mathbf{k}}. \tag{87}$$

Using

$$\sum_{\sigma}|\mathbf{q}\cdot\mathbf{u}_{\mathbf{k}}\sigma|^2 = q^2(1-\cos^2\Theta) = \frac{m^2v^2}{\hbar^2}(1-\cos^2\Theta) \tag{88}$$

and (for infinite Ω)

$$\lim_{\Omega\to\infty}\sum_{\mathbf{k}} \rightarrow \frac{\Omega}{(2\pi)^3}\int d^3k \tag{89}$$

and introducing spherical coordinates in k-space, we find:

$$\frac{dW}{dx} = e^2\int_0^\infty kdk\int_{-1}^{1} d(\cos\Theta)\frac{(1-\cos^2\Theta)\delta\left[\cos\Theta - (\frac{c}{nv}) - \frac{\hbar\omega n}{2mcv}\right]}{\left[\frac{1}{2\omega}\frac{\partial}{\partial\omega}(\omega^2\varepsilon(\omega))\right]_{nkc}} =$$

$$\frac{e^2}{c^2}\int\frac{\varepsilon(\omega)\omega^2 d\omega}{\left[\frac{1}{2}\frac{\partial}{\partial\omega}(\omega^2\varepsilon(\omega))\right]}\left[1 - \frac{c^2}{n^2v^2}\left(1 + \frac{\hbar\omega n^2}{2mc^2}\right)^2\right]. \tag{90}$$

It is clear from this derivation that the integration over ω is only over those frequencies for that eq. (86) can be satisfied. Since

$$\lim_{\omega\to\infty} n(\omega) \longrightarrow 1, \tag{91}$$

the range of integration does not go to infinity and the integral is convergent.

It is possible to show that the Čerenkov angle relation for the relativistic particles with spin zero particles is given by the relation:

$$\cos\Theta = \frac{c}{nv}\left[1 + \frac{\hbar\omega}{2mc^2}(n^2-1)\sqrt{1-\frac{v^2}{c^2}}\right]. \tag{92}$$

This expression can be also derived using the so called Duffin-Kemmer equation for particles with spin 0 or 1.

4.1. The Dirac electron

We have seen that in the nonrelativistic situation the appropriate Hamiltonian involved the renormalization term

$$\left(\frac{1}{2}\frac{\partial}{\partial\omega}\omega^2\varepsilon(\omega)\right), \tag{93}$$

must be the same also in case of the relativistic situation.

Let us consider the process where an electron of momentum $\hbar(\mathbf{p}+\mathbf{k})$ emits a photon of momentum $\hbar\mathbf{k}$ and polarization σ. The interaction Hamiltonian in case of Dirac electron is

$$H_I = -e\int d^3\psi^+\boldsymbol{\alpha}\cdot\mathbf{A}\psi, \tag{94}$$

where $\boldsymbol{\alpha}$ are the Dirac matrices, $\mathbf{A}$ is the electromagnetic potential [12].

Expanding ψ^+, ψ and $\mathbf{A}$ by the second quantization method, we have for the interacting potential:

$$H_I = \sum_{\mathbf{k},\sigma}\sum_{\mathbf{p},\lambda}\sum_{\lambda'}\left[\frac{2\pi\hbar c^2}{\Omega\frac{1}{2}\frac{\partial}{\partial\omega}(\omega^2\varepsilon)}\right]\left\{u^+_{\mathbf{p}+\mathbf{k},\lambda}(\boldsymbol{\alpha}\cdot\mathbf{u}_{\mathbf{k}\sigma}u_{\mathbf{p}\lambda})b^+_{\mathbf{p}+\mathbf{k},\lambda'}b_{\mathbf{p}\lambda}a_{\mathbf{k},\lambda} + h.c.\right\}, \tag{95}$$

where h.c. denotes operation of Hermite conjugation.

Then, the transition probability per unit time is:

$$\left(\frac{\text{trans prob}}{\text{time}}\right)_{\mathbf{p}+\mathbf{k},\lambda'\to\mathbf{p},\lambda} = \frac{2\pi}{\hbar}e^2\left[\frac{2\pi\hbar c^2}{\Omega\frac{1}{2}\frac{\partial}{\partial\omega}(\omega^2\varepsilon)}\right]|u^+_{\mathbf{p}+\mathbf{k},\lambda}(\boldsymbol{\alpha}\cdot\mathbf{u}_{\mathbf{k}\sigma}u_{\mathbf{p}\lambda})|^2 \quad\times$$

$$\delta\left[\sqrt{\hbar^2c^2|\mathbf{p}+\mathbf{k}|^2+m^2c^4}-\sqrt{\hbar^2c^2p^2+m^2c^4}-\hbar\omega\right] \tag{96}$$

and we may proceed to calculate the energy loss per length as we did in the nonrelativistic case.

There is one modification in this calculation. The sum over final states must include a sum over the final spin states of the electron $\lambda = 1,2$. We also average over the initial spin states. Thus the general formula is of the form:

$$\frac{dW}{dx} = \frac{1}{v}\frac{1}{2}\sum_{\lambda'=1}^{2}\sum_{\lambda=1}^{2}\sum_{\mathbf{k},\sigma}^{2}\hbar\omega_k\left(\frac{\text{trans prob}}{\text{time}}\right). \tag{97}$$

So, we must evaluate

$$\frac{1}{2}\sum_{\lambda'=1}^{2}\sum_{\lambda=1}^{2}(u^{+}_{\mathbf{p}+\mathbf{k},\lambda}\boldsymbol{\alpha}\cdot\mathbf{u}_{\mathbf{k}\sigma}u_{\mathbf{p}\lambda})(u^{+}_{\mathbf{p},\lambda}\boldsymbol{\alpha}\cdot\mathbf{u}_{\mathbf{k}\sigma}u_{\mathbf{p}+\mathbf{k}\lambda'}). \tag{98}$$

Let us demonstrate the easy way of calculation of the sums. The first step is to extend the sums over λ' and λ to include all four values. We can do this by noting that

$$\frac{H_{\mathbf{p}}+|E_{\mathbf{p}}|}{2|E_{\mathbf{p}}|}u_{\mathbf{p},\lambda} = \begin{cases} u_{\mathbf{p},\lambda}, & \lambda = 1,2 \\ 0, & \lambda = 3,4 \end{cases}, \tag{99}$$

where

$$H_{\mathbf{p}} = \boldsymbol{\alpha}\cdot\mathbf{p}+\beta mc^2. \tag{100}$$

We can use the relation (99) and the similar relation with $u_{\mathbf{p}+\mathbf{k},\lambda'}$ to write eq. (98) as follows:

$$\frac{1}{2}\sum_{\lambda'=1}^{2}\sum_{\lambda=1}^{2}(u^{+}_{\mathbf{p}+\mathbf{k},\lambda'}\boldsymbol{\alpha}\cdot\mathbf{u}_{\mathbf{k}\sigma}(H_{\mathbf{p}}+|E_{\mathbf{p}}|)u_{\mathbf{p}\lambda}) \quad \times$$

$$\left[u^{+}_{\mathbf{p},\lambda}\boldsymbol{\alpha}\cdot\mathbf{u}_{\mathbf{k}\sigma}(H_{\mathbf{p}+\mathbf{k}}+|E_{\mathbf{p}+\mathbf{k}}|)u_{\mathbf{p}+\mathbf{k}\lambda'}\right]\frac{1}{4|E_{\mathbf{p}}||E_{\mathbf{p}+\mathbf{k}|}}. \tag{101}$$

Now, let us consider

http://dx.doi.org/10.5772/50296

$$\sum_{\lambda=1}^{4} u_{\mathbf{p},\lambda} u^{+}_{\mathbf{p},\lambda}. \tag{102}$$

Using the relation of completeness, the eq. (102) is just the 4×4 unit matrix. Therefore eq. (101) becomes:

$$\frac{1}{2}\sum_{\lambda'=1}^{4}\left[u^{+}_{\mathbf{p}+\mathbf{k},\lambda'}\boldsymbol{\alpha}\cdot\mathbf{u}_{\mathbf{k}\sigma}(H_{\mathbf{p}}+|E_{\mathbf{p}}|)\boldsymbol{\alpha}\cdot\mathbf{u}_{\mathbf{k}\sigma}(H_{\mathbf{p}+\mathbf{k}}+|E_{\mathbf{p}+\mathbf{k}}|)u_{\mathbf{p}+\mathbf{k}\lambda'}\right]\frac{1}{4|E_{\mathbf{p}}||E_{\mathbf{p}+\mathbf{k}}|} =$$

$$\frac{1}{8|E_{\mathbf{p}}||E_{\mathbf{p}+\mathbf{k}}|}\mathrm{Tr}\left[\boldsymbol{\alpha}\cdot\mathbf{u}_{\mathbf{k}\sigma}(H_{\mathbf{p}}+|E_{\mathbf{p}}|)\boldsymbol{\alpha}\cdot\mathbf{u}_{\mathbf{k}\sigma}(H_{\mathbf{p}+\mathbf{k}}+|E_{\mathbf{p}+\mathbf{k}}|)\right], \tag{103}$$

where Trace can be evaluated with the certain difficulties.

First, we note that

$$\mathrm{Tr}\ \alpha_i = \mathrm{Tr}\ \beta. \tag{104}$$

The trace of a product of any odd number of the matrices α_x, α_y, α_z and β is zero. We may use the following identity:

$$(\boldsymbol{\alpha}\cdot\mathbf{a})(\boldsymbol{\alpha}\cdot\mathbf{b}) = 2(\mathbf{a}\cdot\mathbf{b})1 - (\boldsymbol{\alpha}\cdot\mathbf{b})(\boldsymbol{\alpha}\cdot\mathbf{a}), \tag{105}$$

where **a** and **b** are arbitrary vectors and

$$\mathrm{Tr}\ AB - \mathrm{Tr}\ BA, \tag{106}$$

to show that

$$\mathrm{Tr}\ (\boldsymbol{\alpha}\cdot\mathbf{a})(\boldsymbol{\alpha}\cdot\mathbf{b}) = 4\mathbf{a}\cdot\mathbf{b}. \tag{107}$$

We can show also that

$$\mathrm{Tr}\ (\boldsymbol{\alpha}\cdot\mathbf{a})\beta(\boldsymbol{\alpha}\cdot\mathbf{b})\beta = -4\mathbf{a}\cdot\mathbf{b} \tag{108}$$

and

$$\mathrm{Tr}\ (\boldsymbol{\alpha}\cdot\mathbf{a})(\boldsymbol{\alpha}\cdot\mathbf{b})(\boldsymbol{\alpha}\cdot\mathbf{c})(\boldsymbol{\alpha}\cdot\mathbf{d}) \quad =$$

$$4(\mathbf{a}\cdot\mathbf{b})(\mathbf{c}\cdot\mathbf{d}) - 4(\mathbf{a}\cdot\mathbf{c})4(\mathbf{b}\cdot\mathbf{d}) + 4(\mathbf{a}\cdot\mathbf{d})4(\mathbf{b}\cdot\mathbf{c}) \tag{109}$$

for any three vectors $\mathbf{a}, \mathbf{b}, \mathbf{c}, \mathbf{d}$.

Using the formulas with operation Trace, we can evaluate eq. (103). We find:

$$\frac{1}{2}\left\{1 - \frac{m^2c^4}{|E_{\mathbf{p}}||E_{\mathbf{p}+\mathbf{k}}|} + 2\frac{(\mathbf{u}_{\mathbf{k}\sigma}\cdot\mathbf{v}_1)^2}{c^2} - \frac{\mathbf{v}_1\cdot\mathbf{v}_2}{c^2}\right\}, \tag{110}$$

where we have used

$$\mathbf{v} = \frac{c^2\mathbf{p}}{E}, \tag{111}$$

and where $\mathbf{v}_1$ and $\mathbf{v}_2$ are the velocities before and after emission of photon. The sum over polarizations can be carried out as was done in eq. (102). The result is that eq. (103) summed over polarization is:

$$\frac{v_1^2}{c^2}(1-\cos^2\Theta) + \frac{1}{2}\left\{1 - \sqrt{\left(1-\frac{v_1^2}{c^2}\right)\left(1-\frac{v_2^2}{c^2}\right)} - \frac{\mathbf{v}_1\cdot\mathbf{v}_2}{c^2}\right\}, \tag{112}$$

where again Θ is the angle between $\mathbf{p}$ and $\mathbf{k}$ and it is given by the formula:

$$\cos\Theta = \frac{c}{nv}\left[1 + \frac{\hbar\omega}{2mc^2}(n^2-1)\sqrt{1-\frac{v^2}{c^2}}\right]. \tag{113}$$

We have used

$$E = \frac{mc^2}{\sqrt{1-\frac{v^2}{c^2}}} \tag{114}$$

to obtain eq. (113) from eq. (110). The second term in eq. (113) is a small correction to the result formed in the spin 0 case.

The momentum of photon is negligible in comparison with the momentum of electron. Then $(\mathbf{v}_1 \approx \mathbf{v}_2)$ and the term in braces vanishes. This will be true in both the classical limit $\hbar \to 0$

and the extremal relativistic limit $(\mathbf{v} \to c)$. We neglect this term in the remainder of the calculation. The rest of the calculation is the similar to the case with the spin 0. The only differences is that eq. (113) must be used instead of eq. (112). The result is

$$\frac{dW}{dx} = \frac{e^2}{c^2}\int \frac{\varepsilon(\omega)\omega^2 d\omega}{\frac{1}{2}\left(\frac{\partial}{\partial\omega}\right)\omega^2\varepsilon(\omega)}\left[1 - \frac{c^2}{n^2 v^2}\left(1 + \frac{\hbar\omega}{2mc^2}(n^2-1)\sqrt{1-\frac{v^2}{c^2}}\right)^2\right]. \tag{115}$$

5. The source theory of the Čerenkov effect

Source theory [6, 30–32] is the theoretical construction which uses quantum-mechanical particle language. Initially it was constructed for description of the particle physics situations occurring in the high-energy physics experiments. However, it was found that the original formulation simplifies the calculations in the electrodynamics and gravity where the interactions are mediated by photon or graviton respectively.

The basic formula in the source theory is the vacuum to vacuum amplitude [30]:

$$< 0_+|0_- > = e^{\frac{i}{\hbar}W(S)}, \tag{116}$$

where the minus and plus tags on the vacuum symbol are causal labels, referring to any time before and after space-time region where sources are manipulated. The exponential form is introduced with regard to the existence of the physically independent experimental arrangements which has a simple consequence that the associated probability amplitudes multiply and corresponding W expressions add.

The electromagnetic field is described by the amplitude (116) with the action

$$W(J) = \frac{1}{2c^2}\int (dx)(dx')J^{\mu}(x)D_{+\mu\nu}(x-x')J^{\nu}(x'), \tag{117}$$

where the dimensionality of $W(J)$ is the same as the dimensionality of the Planck constant $\hbar$. J_μ is the charge and current densities. The symbol $D_{+\mu\nu}(x-x')$, is the photon propagator and its explicit form will be determined later.

It may be easy to show that the probability of the persistence of vacuum is given by the following formula [30]:

$$| < 0_+|0_- > |^2 = \exp\{-\frac{2}{\hbar}\mathrm{Im}\,W\} \stackrel{d}{=} \exp\{-\int dt d\omega \frac{P(\omega,t)}{\hbar\omega}\}, \tag{118}$$

where we have introduced the so called power spectral function $P(\omega,t)$. In order to extract this spectral function from $\mathrm{Im}\,W$, it is necessary to know the explicit form of the photon propagator $D_{+\mu\nu}(x-x')$.

The electromagnetic field is described by the four-potentials $A^{\mu}(\varphi, \mathbf{A})$ and it is generated by the four-current $J^{\mu}(c\varrho, \mathbf{J})$ according to the differential equation [30]:

$$(\Delta - \frac{\mu\varepsilon}{c^2}\frac{\partial^2}{\partial t^2})A^{\mu} = \frac{\mu}{c}(g^{\mu\nu} + \frac{n^2-1}{n^2}\eta^{\mu}\eta^{\nu})J_{\nu} \tag{119}$$

with the corresponding Green function $D_{+\mu\nu}$:

$$D_{+}^{\mu\nu} = \frac{\mu}{c}(g^{\mu\nu} + \frac{n^2-1}{n^2}\eta^{\mu}\eta^{\nu})D_{+}(x-x'), \tag{120}$$

where $\eta^{\mu} \equiv (1, \mathbf{0})$, μ is the magnetic permeability of the dielectric medium with the dielectric constant ε, c is the velocity of light in vacuum, n is the index of refraction of this medium, and $D_{+}(x-x')$ was derived by Schwinger et al. [30] in the following form:

$$D_{+}(x-x') = \frac{i}{4\pi^2 c}\int_{0}^{\infty} d\omega \frac{\sin\frac{n\omega}{c}|\mathbf{x}-\mathbf{x}'|}{|\mathbf{x}-\mathbf{x}'|}e^{-i\omega|t-t'|}. \tag{121}$$

Using formulas (117), (118), (120) and (121), we get for the power spectral formula the following expression [30]:

$$P(\omega, t) = -\frac{\omega}{4\pi^2}\frac{\mu}{n^2}\int d\mathbf{x}d\mathbf{x}'dt'\frac{\sin\frac{n\omega}{c}|\mathbf{x}-\mathbf{x}'|}{|\mathbf{x}-\mathbf{x}'|}\cos[\omega(t-t')]\times$$

$$\times\left\{\varrho(\mathbf{x}, t)\varrho(\mathbf{x}', t') - \frac{n^2}{c^2}\mathbf{J}(\mathbf{x}, t)\cdot\mathbf{J}(\mathbf{x}', t')\right\}. \tag{122}$$

Now, we are prepared to apply the last formula to the situations of the charge moving in the dielectric medium.

The charge and current density of electron moving with the velocity $\mathbf{v}$ and charge e is

$$\varrho = e\delta(\mathbf{x} - \mathbf{v}t) \tag{123}$$

$$\mathbf{J} = e\mathbf{v}\delta(\mathbf{x} - \mathbf{v}t). \tag{124}$$

After insertion of eqs. (123) and (124) in equation for spectral density (122), we find:

$$P(\omega, t) = \frac{e^2}{4\pi c^2}\mu\omega v(1 - \frac{1}{n^2\beta^2}); \quad n\beta > 1 \tag{125}$$

$$P(\omega, t) = 0; \quad n\beta < 1, \tag{126}$$

where $\beta = v/c$. Relations (125) and (126) determine the Čerenkov spectrum and the threshold condition for the existence of the Čerenkov effect $n\beta = 1$.

6. The Čerenkov effect in the dielectric 2D hexagonal structure

In case of the two dimension situation, the form of equations (119) and (120) is the same with the difference that $\eta^\mu \equiv (1, \mathbf{0})$ has two space components, or $\eta^\mu \equiv (1, 0, 0)$, and the Green function D_+ as the propagator must be determined by the two-dimensional procedure. I other words, the Fourier form of this propagator is with $(dk) = dk^0 d\mathbf{k} = dk^0 dk^1 dk^2 = dk^0 k dk d\theta$

$$D_+(x - x') = \int \frac{(dk)}{(2\pi)^3} \frac{1}{\mathbf{k}^2 - n^2(k)^2} e^{ik(x-x')}, \tag{127}$$

or, with $R = |\mathbf{x} - \mathbf{x}'|$

$$D_+(x - x') = \frac{1}{(2\pi)^3} \int_0^{2\pi} d\theta \int_0^\infty k dk \int_{-\infty}^{\infty} \frac{d\omega}{c} \frac{e^{ikR\cos\theta - i\omega(t-t')}}{k^2 - \frac{n^2\omega^2}{c^2} - i\varepsilon}. \tag{128}$$

Using $\exp(ikR\cos\theta) = \cos(kR\cos\theta) + i\sin(kR\cos\theta)$ and $(z = kR)$

$$\cos(z\cos\theta) = J_0(z) + 2\sum_{n=1}^{\infty}(-1)^n J_{2n}(z)\cos 2n\theta \tag{129}$$

and

$$\sin(z\cos\theta) = \sum_{n=1}^{\infty}(-1)^n J_{2n-1}(z)\cos(2n-1)\theta, \tag{130}$$

where $J_n(z)$ are the Bessel functions [15], we get after integration over θ:

$$D_+(x - x') = \frac{1}{(2\pi)^2} \int_0^\infty k dk \int_{-\infty}^{\infty} \frac{d\omega}{c} \frac{J_0(kR)}{k^2 - \frac{n^2\omega^2}{c^2} - i\varepsilon} e^{-i\omega(t-t')}, \tag{131}$$

where the Bessel function $J_0(z)$ has the following expansion [15]:

$$J_0(z) = \sum_{s=0}^{\infty} \frac{(-1)^s z^{2s}}{s!s!2^{2s}} \tag{132}$$

The ω-integral in (131) can be performed using the residuum theorem after integration in the complex half ω-plane.

The result of such integration is the propagator D_+ in the following form:

$$D_+(x - x') = \frac{i}{2\pi c} \int_0^{\infty} d\omega J_0 \left(\frac{n\omega}{c} |\mathbf{x} - \mathbf{x}'| \right) e^{-i\omega|t-t'|}. \tag{133}$$

The initial terms in the expansion of the Bessel function with exponent zero is as follows:

$$J_0(z) = 1 - \frac{z^2}{2^2} + \frac{z^4}{2^2 4^2} - \frac{z^6}{2^2 4^2 6^2} + \frac{z^8}{2^2 4^2 6^2 8^2} - \cdots \quad . \tag{134}$$

The spectral formula for the two dimensional Čerenkov radiation is of the analogue of the formula (122), or,

$$P(\omega, t) = -\frac{\omega}{2\pi} \frac{\mu}{n^2} \int d\mathbf{x} d\mathbf{x}' dt' J_0 \left(\frac{n\omega}{c} |\mathbf{x} - \mathbf{x}'| \right) \cos[\omega(t - t')] \times$$

$$\times \left\{ \varrho(\mathbf{x}, t) \varrho(\mathbf{x}', t') - \frac{n^2}{c^2} \mathbf{J}(\mathbf{x}, t) \cdot \mathbf{J}(\mathbf{x}', t') \right\}, \tag{135}$$

where the charge density and current involves only two-dimensional velocities and integration is also only two-dimensional with two-dimensional $d\mathbf{x}, d\mathbf{x}'$.

The difference is in the replacing mathematical formulas as follows:

$$\frac{\sin \frac{n\omega}{c} |\mathbf{x} - \mathbf{x}'|}{|\mathbf{x} - \mathbf{x}'|} \quad \longrightarrow \quad J_0 \left(\frac{n\omega}{c} |\mathbf{x} - \mathbf{x}'| \right). \tag{136}$$

So, After insertion the quantities (123) and (124) into (135), we get:

$$P(\omega, t) = \frac{e^2}{2\pi} \frac{\mu\omega v}{c^2} \left(1 - \frac{1}{n^2 \beta^2} \right) \int dt' J_0 \left(\frac{nv\omega}{c} |t - t'| \right) \cos[\omega(t - t')], \quad \beta = v/c, \tag{137}$$

http://dx.doi.org/10.5772/50296

where the t'-integration must be performed. Putting $\tau = t' - t$, we get the final formula:

$$P(\omega, t) = \frac{e^2}{2\pi}\frac{\mu\omega v}{c^2}\left(1 - \frac{1}{n^2\beta^2}\right)\int_{-\infty}^{\infty} d\tau J_0\left(n\beta\omega\tau\right)\cos(\omega\tau), \quad \beta = v/c. \tag{138}$$

The integral in formula (138) is involved in the tables of integrals [11] on page 745, number 8. Or,

$$J = \int_0^{\infty} dx J_0\left(ax\right)\cos(bx) = \frac{1}{\sqrt{a^2 - b^2}}; \quad 0 < b < a,$$

$$J = \infty; \quad a = b, \quad J = 0; \quad 0 < a < b. \tag{139}$$

In our case we have $a = n\beta\omega$ and $b = \omega$. So, the power spectrum of in eq. (138) is as follows with $J_0(-z) = J_0(z)$:

$$P(\omega, t) = \frac{e^2}{\pi}\frac{\mu\omega v}{c^2}\left(1 - \frac{1}{n^2\beta^2}\right)\frac{2}{\omega\sqrt{n^2\beta^2 - 1}}, \quad n\beta > 1, \quad \beta = v/c. \tag{140}$$

and

$$P(\omega, t) = 0; \quad n\beta < 1, \tag{141}$$

where condition $n\beta = 1$ is the threshold of the existence of the two-dimensional form of the Čerenkov radiation.

7. The Čerenkov radiation in two-dimensional structure generated by a pulse

Let us consider the electron moving perpendicularly to the 2D sheet in the pane $y - z$ with the index of refraction n and the magnetic permeability μ. Then, the charge density and current density for the charge moving along the axis is ($v > 0$)

$$\varrho = e\delta(vt)\delta(\mathbf{x}) = \frac{e}{v}\delta(t)\delta(\mathbf{x}) \tag{142}$$

$$\mathbf{J} = 0. \tag{143}$$

After insertion of the last formulas into the spectral formula for the Čerenkov radiation (135) with regard to (136), we get

$$P(\omega,t) = \frac{e^2}{2\pi}\frac{\mu\omega}{n^2v^2}\int dt'\delta(t)\delta(t')J_0(0)\cos[\omega(t-t')], \tag{144}$$

After performing the t and t′ integration we get

$$\int dtP(\omega,t) = \frac{e^2}{2\pi}\frac{\mu\omega}{n^2v^2}J_0\left(0\right). \tag{145}$$

The derived formula does not involve the Čerenkov radiation threshold. At the same time the formula does not involve the transition radiation which is generated by the charge when it is moving outside of the sheet. Nevertheless, such radiation can be easily determined by the Ginzburg method [10].

8. The Čerenkov effect with massive photons

The massive electrodynamics in medium can be constructed by generalization of massless electrodynamics to the case with massive photon. In our case it means that we replace only eq. (119) by the following one:

$$\left(\Delta - \frac{\mu\epsilon}{c^2}\frac{\partial^2}{\partial t^2} + \frac{m^2c^2}{\hbar^2}\right)A^\mu = \frac{\mu}{c}\left(g^{\mu\nu} + \frac{n^2-1}{n^2}\eta^\mu\eta^\nu\right)J_\nu, \tag{146}$$

where m is mass of photon. The Lorentz gauge of massless photons is conserved also in the massive situation.

In superconductiviy photon is a massive spin 1 particle as a consequence of a broken symmetry of the Landau-Ginzburg Lagrangian. The Meissner effect can be used as a experimental demonstration that photon in a superconductor is a massive particle. In particle physics the situation is analogous to the situation in superconductivity. The masses of particles are also generated by the broken symmetry or in other words by the Higgs mechanism. Massive particles with spin 1 form the analogue of the massive photon.

Kirzhnits and Linde [14] proposed a qualitative analysis wherein they indicated that, as in the Ginzburg-Landau theory of superconductivity, the Meissner effect can also be realized in the Weinberg model. Later, it was shown that the Meissner effect is realizable in renormalizable gauge fields and also in the Weinberg model [35].

We will investigate how the spectrum of the Čerenkov radiation is modified if we suppose the massive photons are generated instead of massless photons. The derived results form an analogue of the situation with massless photons. According to author Pardy [25–27] and Dittrich [6] with the analogy of the massless photon propagator $D(k)$ in the momentum representation

$$D(k) = \frac{1}{|\mathbf{k}|^2 - n^2(k^0)^2 - i\epsilon}, \tag{147}$$

the massive photon propagator is of the form (here we introduce $\hbar$ and c):

$$D(k, m^2) = \frac{1}{|\mathbf{k}|^2 - n^2(k^0)^2 + \frac{m^2c^2}{\hbar^2} - i\epsilon}, \tag{148}$$

where this propagator is derived from an assumption that the photon energetic equation is

$$|\mathbf{k}|^2 - n^2(k^0)^2 = -\frac{m^2c^2}{\hbar^2}, \tag{149}$$

where n is the parameter of the medium and m is mass of photon in this medium.

From eq. (149) the dispersion law for the massive photons follows:

$$\omega = \frac{c}{n}\sqrt{k^2 + \frac{m^2c^2}{\hbar^2}}. \tag{150}$$

Let us remark here that such dispersion law is valid not only for the massive photon but also for electromagnetic field in waveguides and electromagnetic field in ionosphere. It means that the corresponding photons are also massive and the theory of massive photons is physically meaningful. It means that also the Čerenkov radiation of massive photons is physically meaningful and it is meaningful to study it.

The validity of eq. (149) can be verified using very simple idea that for $n = 1$ the Einstein equation for mass and energy has to follow. Putting $\mathbf{p} = \hbar\mathbf{k}, \quad \hbar k^0 = \hbar(\omega/c) = (E/c)$, we get the Einstein energetic equation

$$E^2 = \mathbf{p}^2c^2 + m^2c^4. \tag{151}$$

The propagator for the massive photon is then derived as

$$D_+(x - x', m^2) = \frac{i}{c}\frac{1}{4\pi^2}\int_0^\infty d\omega \frac{\sin[\frac{n^2\omega^2}{c^2} - \frac{m^2c^2}{\hbar^2}]^{1/2}|\mathbf{x} - \mathbf{x}'|}{|\mathbf{x} - \mathbf{x}'|}e^{-i\omega|t-t'|}. \tag{152}$$

The function (152) differs from the the original function D_+ by the factor

$$\left(\frac{\omega^2 n^2}{c^2} - \frac{m^2 c^2}{\hbar^2}\right)^{1/2}. \tag{153}$$

From eq. (152) the potentials generated by the massless or massive photons respectively follow. In case of the massless photon, the potential is according to Schwinger defined by the formula:

$$V(\mathbf{x}-\mathbf{x}') = \int_{-\infty}^{\infty} d\tau D_{+}(\mathbf{x}-\mathbf{x}',\tau) = \int_{-\infty}^{\infty} d\tau \left\{ \frac{i}{c}\frac{1}{4\pi^2}\int_0^{\infty} d\omega \frac{\sin\frac{n\omega}{c}|\mathbf{x}-\mathbf{x}'|}{|\mathbf{x}-\mathbf{x}'|} e^{-i\omega|\tau|} \right\}. \tag{154}$$

The τ-integral can be evaluated using the mathematical formula

$$\int_{-\infty}^{\infty} d\tau\, e^{-i\omega|\tau|} = \frac{2}{i\omega} \tag{155}$$

and the ω-integral can be evaluated using the formula

$$\int_0^{\infty} \frac{\sin ax}{x} dx = \frac{\pi}{2}, \quad \text{for} \quad a > 0. \tag{156}$$

After using eqs. (155) and (156), we get

$$V(\mathbf{x}-\mathbf{x}') = \frac{1}{c}\frac{1}{4\pi}\frac{1}{|\mathbf{x}-\mathbf{x}'|}. \tag{157}$$

In case of the massive photon, the mathematical determination of potential is the analogical to the massless situation only with the difference we use the propagator (152) and the tables of integrals [11]:

$$\int_0^{\infty} \frac{dx}{x} \sin\left(p\sqrt{x^2-u^2}\right) = \frac{\pi}{2} e^{-pu}. \tag{158}$$

Using this integral we get that the potential generated by the massive photons is

$$V(\mathbf{x}-\mathbf{x}', m^2) = \frac{1}{c}\frac{1}{4\pi}\frac{\exp\left\{-\frac{mcn}{\hbar}|\mathbf{x}-\mathbf{x}'|\right\}}{|\mathbf{x}-\mathbf{x}'|}. \tag{159}$$

If we compare the potentials concerning massive and massless photons, we can deduce that also Čerenkov radiation with massive photons can be generated. So, the determination of the Čerenkov effect with massive photons is physically meaningful.

In case of the massive electromagnetic field in the medium, the action W is given by the following formula:

$$W = \frac{1}{2c^2}\int (dx)(dx')J^{\mu}(x)D_{+\mu\nu}(x-x',m^2)J^{\nu}(x'), \tag{160}$$

where

$$D_{+}^{\mu\nu} = \frac{\mu}{c}[g^{\mu\nu} + (1-n^{-2})\eta^{\mu}\eta^{\nu}]D_{+}(x-x',m^2), \tag{161}$$

where $\eta^{\mu} \equiv (1,\mathbf{0})$, $J^{\mu} \equiv (c\varrho,\mathbf{J})$ is the conserved current, μ is the magnetic permeability of the medium, ϵ is the dielectric constant od the medium and $n = \sqrt{\epsilon\mu}$ is the index of refraction of the medium.

The probability of the persistence of vacuum is of the following form:

$$|\langle 0_{+}|0_{-}\rangle|^2 = e^{-\frac{2}{\hbar}\mathrm{Im}W}, \tag{162}$$

where Im W is the basis for the definition of the spectral function $P(\omega,t)$ as follows:

$$-\frac{2}{\hbar}\mathrm{Im}W \stackrel{d}{=} -\int dtd\omega\frac{P(\omega,t)}{\hbar\omega}. \tag{163}$$

Now, if we insert eq. (161) into eq. (160), we get after extracting $P(\omega,t)$ the following general expression for this spectral function:

$$P(\omega,t) = -\frac{\omega}{4\pi^2}\frac{\mu}{n^2}\int d\mathbf{x}d\mathbf{x}'dt'\left[\frac{\sin[\frac{n^2\omega^2}{c^2} - \frac{m^2c^2}{\hbar^2}]^{1/2}]|\mathbf{x}-\mathbf{x}'|}{|\mathbf{x}-\mathbf{x}'|}\right]\times$$

$$\cos[\omega(t-t')][\varrho(\mathbf{x},t)\varrho(\mathbf{x}',t') - \frac{n^2}{c^2}\mathbf{J}(\mathbf{x},t)\cdot\mathbf{J}(\mathbf{x}',t')]. \tag{164}$$

Now, let us apply the formula (164) in order to get the Čerenkov distribution of massive photons. let as consider a particle of charge Q moving at a constant velocity $\mathbf{v}$. In such a way we can write for the charge density and for the current density:

$$\varrho = Q\delta(\mathbf{x} - \mathbf{v}t), \qquad \mathbf{J} = Q\mathbf{v}\delta(\mathbf{x} - \mathbf{v}t). \tag{165}$$

After insertion of eq. (165) into eq. (164), we get ($v = |\mathbf{v}|$).

$$P(\omega, t) = \frac{Q^2}{4\pi^2}\frac{v\mu\omega}{c^2}\left(1 - \frac{1}{n^2\beta^2}\right)\int_{\infty}^{\infty}\frac{d\tau}{\tau}\sin\left(\left[\frac{n^2\omega^2}{c^2} - \frac{m^2c^2}{\hbar^2}\right]^{1/2} v\tau\right)\cos\omega\tau, \tag{166}$$

where we have put $\tau = t' - t, \beta = v/c$.

For $P(\omega, t)$, the situation leads to evaluation of the τ-integral. For this integral we have:

$$\int_{-\infty}^{\infty}\frac{d\tau}{\tau}\sin\left(\left[\frac{n^2\omega^2}{c^2} - \frac{c^2}{m^2}\right]^{1/2} v\tau\right)\cos\omega\tau = \begin{cases} \pi, \; 0 < m^2 < \frac{\omega^2}{c^2v^2}(n^2\beta^2 - 1) \\ 0, \; m^2 > \frac{\omega^2}{c^2v^2}(n^2\beta^2 - 1). \end{cases} \tag{167}$$

From eq. (167) immediately follows that $m^2 > 0$ implies the Čerenkov threshold $n\beta > 1$. From eq. (166) and (167) we get the spectral formula of the Čerenkov radiation of massive photons in the form:

$$P(\omega, t) = \frac{Q^2}{4\pi}\frac{v\omega\mu}{c^2}\left(1 - \frac{1}{n^2\beta^2}\right) \tag{168}$$

for

$$\omega > \frac{mcv}{\hbar}\frac{1}{\sqrt{n^2\beta^2 - 1}} > 0, \tag{169}$$

and $P(\omega, t) = 0$ for

$$\omega < \frac{mcv}{\hbar}\frac{1}{\sqrt{n^2\beta^2 - 1}}. \tag{170}$$

Using the dispersion law (150) we can write the power spectrum $P(\omega)$ as a function dependent on k^2. Then,

$$P(k^2) = \frac{Q^2}{4\pi}\frac{v\mu}{nc}\sqrt{k^2 + \frac{m^2c^2}{\hbar^2}}\left(1 - \frac{1}{n^2\beta^2}\right); \quad k^2 > \frac{m^2c^2}{\hbar^2}\frac{1}{n^2\beta^2 - 1} \tag{171}$$

and $P(\omega, t) = 0$ for $k^2 < (m^2c^2/\hbar^2)(n^2\beta^2 - 1)^{-1}$.

http://dx.doi.org/10.5772/50296

The most simple way how to get the angle Θ between vectors $\mathbf{k}$ and $\mathbf{p}$ is the use the conservation laws for an energy and momentum.

$$E - \hbar\omega = E' \tag{172}$$

$$\mathbf{p} - \hbar\mathbf{k} = \mathbf{p}', \tag{173}$$

where E and E' are energies of a moving particle before and after act of emission of a photon with energy $\hbar\omega$ and momentum $\hbar\mathbf{k}$, and $\mathbf{p}$ and $\mathbf{p}'$ are momenta of the particle before and after emission of the same photon.

If we raise the equations (172) and (173) to the second power and take the difference of these quadratic equations, we can extract the $\cos\Theta$ in the form:

$$\cos\Theta = \frac{1}{n\beta}\left(1 + \frac{m^2c^2}{\hbar^2k^2}\right)^{1/2} + \frac{\hbar k}{2p}\left(1 - \frac{1}{n^2}\right) - \frac{m^2c^2}{2n^2p\hbar k}, \tag{174}$$

which has the correct massless limit. The massless limit also gives the sense of the parameter n which is introduced in the massive situation. We also observe that while in the massless situation the angle of emission depends only on $n\beta$, in case of massive situation it depends also on the wave vector k. It means that the emission of the massive photons are emitted by the Čerenkov mechanism in all space directions.

So, in experiment the Čerenkov production of massive photons can be strictly distinguished from the Čerenkov production of massless photons, or, from the hard production of spin 1 massive particles.

9. Perspective

The article is in some sense the preamble to the any conferences of ideas related to the Čerenkov effect in the graphene-like dielectric structures. At present time, the most attention is devoted in graphene physics with a goal to construct the computers with the artificial intelligence. However, we do not know, a priori, how many discoveries are involved in the investigation of the Čerenkov effect in graphene-like structures.

The information on the Čerenkov effect in graphene-like structures and also the elementary particle interaction with graphene-like structures is necessary not only in the solid state physics, but also in the elementary particle physics in the big laboratories where graphene can form the substantial components of the particle detectors. We hope that these possibilities will be consider in the physical laboratories.

The monolithic structures can be also built into graphene-like structures by addition and re-arrangement of deposit atoms [20]. The repeating patterns can be created to form new carbon allotropes called haeckelites. The introducing such architectonic defects modifies mechanical, electrical, optical and chemical properties of graphene-like structures and it is not excluded that special haeckelites are superconductive at high temperatures. The unconventional graphene-like materials can be prepared by special technique in order to do revolution in the solid state physics.

While the last century economy growth was based on the inventions in the Edison-Tesla electricity, the economy growth in this century will be obviously based on the graphene-like structures physics. We hope that these perspective ideas will be considered at the universities and in the physical laboratories.

Author details

Miroslav Pardy

* Address all correspondence to: pamir@physics.muni.cz

Department of Physical Electronics, Masaryk University, Brno, Czech Republic

10. References

[1] Berestetzkii, V. B.; Lifshitz, E. M. and Pitaevskii, L. P. (1989). Quantum electrodynamics (Moscow, Nauka). (in Russian).

[2] Yu. V. Bludov, N. M. R. and Peres, M. I. Vasilevskiy, Graphene-based polaritonic crystal Yu. V.,[cond-mat.mes-hall] 17 Apr. 2012

[3] Čerenkov. P. A. (1934), The visible radiation of pure liquids caused by γ-rays, *Comptes Rendus Hebdomaclaires des Scances de l' Academic des Sciences* USSR 2, 451.

[4] Čerenkov, P. A., *C. R. Acad. Sci.* (USSR) 3 (1936), 413.

[5] Dato, A.; Radmilovic, V.; Lee, Z.; Philips, J. and Frenklach, M. (2008). Substrate-free gas phase synthesis of graphene sheets, *Nano Letters*, 8 (7), 2012.

[6] Dittrich, W.(1978). Source methods in quantum field theory *Fortschritte der Physik* 26, 289.

[7] Drummond, N. D.; Zťolyomi, V. and Falko, V. I. (2011). Electrically Tunable band gap in silicene, arXiv:1112.4792v1 [cond-mat.mes-hall] 20 Dec 2011.

[8] Ezawa, M. (2012). Topological insulator and helical zero mode in silicene under inhomogeneous electric field, arXiv:1201.3687v1 [cond-mat.mes-hall] 18 Jan 2012.

[9] Frank, I. M. (1988). The Vavilov-Čerenkov radiation, (Nauka). (in Russian).

[10] Ginzburg, V. L. and Tsytovich, V. N. (1984). Transition radiation and transition scattering, (Moscow, Nauka). (in Russian).

[11] Gradshteyn, S. and Ryzhik, I. M.(1962). Tables of integrals, sums, series and products, (Moscow). (in Russian).

[12] Harris, E. G. (1972). A pedestrian approach to quantum field theory, (John Willey and Sons, Inc., New York, London).

[13] Kane, Ch. L. (2005). Erasing electron mass, *Nature*, 438, November 205, 168.

[14] Kirzhnits, A. D. and Linde, A. D. (1972). Macroscopic consequences of the Weinberg model, *Physics Letters* 42 B, No. 4, 471.

[15] Kuznetsov, D. S. (1962). The special functions, (Moscow). (in Russian).

[16] Landau, L. D. (1937). Zur Teorie der Phasenumwandlungen II., *Phys. Z. Sowjetunion* 11, 26.

[17] Landau, L. D. and Lifshitz, E. M. (1984). Electrodynamics of continuous media, Second revised edition, (Pergamon Press, Oxford, ..).

[18] Levich, V. G.; Vdovin, Yu. A. and Miamlin, V. A. (1962). Course of theoretical Physics, II., (GIFML, Moscow). (in Russsian).

[19] Lozovik, Yu. E.; Merkulov, S. P. and Sokolik, A. A. (2008). Collective electron phenomena in graphene, *Uspekhi Fiz. Nauk*, 178, (7), 758.

[20] Lusk, M. T.; Carr, L. D. (2008). Creation of graphene allotropes using patterned defects, cond-mat.mtrl-sci/0809.3160v2.

[21] Mermin, N. D. (1968). Crystalline order in two dimensions, *Phys. Rev.*, 176.

[22] Novoselov, K. S.; Geim, A.K.; Morozov, S. V.; Jiang, D.; Zhang. Y.; Dubonos, S. V.; Grigorieva, I.V. and Firsov, A.A. (2004). Electric field effect in atomically thin Carbon films, *Science*, 306, 666-669.

[23] Novoselov, K.S.; Geim, A. K.; Morozov, S.V. et al. (2005). Two-dimensional gas of massless Dirac fermions in graphene, *Nature*, 438, 197.

[24] Pardy, M. (1983). Particle production by the Čerenkov mechanism, *Phys. Lett.* 94 A , 30.

[25] Pardy, M. (1989). Finite temperature Čerenkov radiation, *Phys. Lett.* A 134, No.6, 357.

[26] Pardy, M. (1994). The Čerenkov effect with radiative corrections, *Phys. Lett.* B 325, 517.

[27] Pardy, M. (2002). Čerenkov effect with massive photons, *Int. Journal of Theoretical Physics* 41 No. 5, 887.

[28] Pardy, M. (2004). Massive photons and the Volkov solution, *International Journal of Theoretical Physics*, 43(1), 127.

[29] Peierls, R. E. (1934). Bemerkungen über Umwandlungstemperaturen. *Helvetica Physica Acta* 7, 81; ibid. (1935). Quelques proprietes typiques des corpses solides. *Ann. Inst. Henri Poincaré* 5, 177.

[30] Schwinger, J., Tsai, W. Y. and Erber, T. (1976). Classical and quantum theory of synergic synchrotron-Čerenkov radiation, *Annals of Physics* (NY) 96, 303.

[31] Schwinger, J. (1970).Particles, Sources and Fields Vol. I (Addison-Wesley, Reading, Mass.).

[32] Schwinger, J. (1988). Particles, sources and fields, (Addison-Wesley Publishing Company, Inc.).

[33] Sokolov, A. A.; Loskutov, Yu. M. and Ternov, I.M. (1962). Quantum mechanics, (Moscow). (in Russian).

[34] Tamm, I. E. and Frank, I. M. (1937). The coherent radiation of a fast electron in a medium, *Dokl. Akad. Nauk SSSR* 14, 109.

[35] Yildiz, A. (1977). Meissner effect in gauge fields *Phys. Rev. D* 16, No. 12 , 3450.

Chapter 5

Electronic and Vibrational Properties of Adsorbed and Embedded Graphene and Bigraphene with Defects

Alexander Feher, Eugen Syrkin, Sergey Feodosyev,
Igor Gospodarev, Elena Manzhelii,
Alexander Kotlar and Kirill Kravchenko

Additional information is available at the end of the chapter

http://dx.doi.org/10.5772/50562

1. Introduction

The perennial interest in studying the physical properties of nanofilms has increased substantially over the last few years due to the development of nanotechnologies and the synthesis of new compounds – especially those based on carbon, which are extremely interesting for both fundamental research and potential applications.

An important feature of carbon nanofilms (including those with defects) is a close relation between the electronic and phonon properties, which is exhibited, for example, in the graphene-based systems with superconducting properties [1,2].

It is well known that graphene monolayers cannot exist as planar objects in the free state, because in flat 2D-crystals the mean-square amplitudes of the atoms in the direction normal to the layer plane diverge even at $T=0$ (see, e.g., [3]). So we can study and practically apply only such graphene, which is deposited on a certain substrate providing the stability of the plane carbon nanofilms (see, e.g., [4-6]). Only small flakes can be detached from the substrate and these flakes immediately acquire a corrugated shape [7]. When studying the electronic properties of graphene a dielectric substrate is often used. The presence of the substrate greatly increases the occurrence of various defects in graphene and carbon nanofilms. Our investigations make it possible to predict the general properties of phonon and electron spectra for graphene and bigraphene containing different defects.

This chapter consists of three sections: first section is devoted to the calculation of local discrete levels in the electron spectra of graphene with different defects. In the second section we describe the electronic properties of bilayered graphene and, finally, the third section deals with the influence of defects on electron spectra of bigraphene.

2. Impurity levels in the electron spectra of graphene

The exceptionality of graphene is manifested in the phonon and electron properties. Graphene is a semimetal whose valence and conduction bands touch at the points K and K′ of the Brillouin zone [8,9]. In the pure graphene unique electronic properties are manifested by the charge carriers behaving as massless relativistic particles - the dependence of energy on the momentum is linear rather than - as in ordinary solids - quadratic. Thus, the lower-dimensionality affects the formation of phonon localized states [10] and also the formation of localized states in the electronic spectrum. Absent gap between the valence and conduction bands is a consequence of the symmetry between two equivalent sublattices in graphene [11]. Presence of impurities lowers the symmetry. The influence of vacancies placed into one of the graphene sublattices was investigated in [12], where it was shown that the equivalence of the sublattices is broken.

In this section we describe the characteristics of localized and local states present in graphene due to the impurities of nitrogen and boron, respectively. The presence of the substrate greatly increases the possibility to introduce various defects into graphene. For example, in the graphene deposited on silicon, vacancies can occur [13, 14], whereas in graphite (a set of weakly interacting graphene monolayers) vacancies heal and form a stacking fault with local fivefold symmetry axis [15]. Impurity atoms embedded in graphene may lead to the appearance of impurity states outside the band of quasi-continuous spectrum. At low impurity concentrations (when impurity is considered as an isolated defect) these states appear in the form of local discrete levels (LDL).

Although such levels in various quasiparticle spectra have been known and studied over 60 years, an adequate description is still absent, even in the harmonic approximation for sufficiently realistic models of the crystal lattice. The dependence of the appearance conditions and characteristics of LDL on the parameters of a perfect lattice and defect was identified only in the most general terms. However, LDL may be used as an important source of information about the defect structure and force interactions in real crystals. To extract such information it is useful to have analytical expressions that relate main characteristics of LDL to the parameters of both the defect and the host lattice.

Here we present the results of our calculations and analyses of the characteristics of the electronic local discrete levels for substitutional impurities in graphene, especially for a boron substitutional impurity, using an analytical approximation based on the Jacobi matrices method [16, 17].

The fact that the charge carriers in graphene are formally described by the Dirac equation and not by the Schrödinger equation is due to the symmetry of the crystal lattice of graphene, which consists of two equivalent carbon sublattices. Electronic subbands formed by the symmetric and antisymmetric combinations of wave functions in the two sublattices intersect at the edge of the Brillouin zone, which leads to a cone-shaped energy spectrum near the K and K' points of the first Brillouin zone. The electrons obey the linear dispersion law (in ordinary metals and semiconductors the dispersion law is parabolic).

The electronic spectrum of graphene can be described by a strong coupling approximation, and it is sufficient to consider the interaction between nearest neighbors only (see, e.g., [5,6,18-20]). The corresponding Hamiltonian is

$$\hat{H}=\sum_i \varepsilon_i \mid ii \mid -\sum_{i,j} J_{ij} \mid i j \mid \tag{1}$$

where i and j are the labels of the nodes of the two-dimensional lattice, ε_i is the energy of electron at node i, and J_{ij} is the so-called overlap integral.

Curve 1 in Figure 1 shows the density of electronic states of graphene as calculated using the method of Jacobi matrices [16, 17]. In a perfect graphene the local Green's function $G(\varepsilon, i)=i \mid (\varepsilon \hat{I}-\hat{H})^{-1} \mid i$ coincides with the total Green's function $G(\varepsilon)=\lim_{N\to\infty}\frac{1}{N}\sum_{i=1}^{N} i \mid (\varepsilon \hat{I}-\hat{H})^{-1} \mid i$ because of the physical equivalence of the atoms of both sublattices. Peculiarity of the density of states at $\varepsilon=\varepsilon(K)$ (the value $\varepsilon(K)$ corresponds to the Fermi energy ε_F in graphene) determines the behavior of the real part of the Green's function near ε_F . For a wide class of perturbations caused by defects we can find, using the Lifshitz equation [21], quasilocalized states in the interval $[-\varepsilon(M), \varepsilon(M)]$ (in this model $\varepsilon(M)=J$). This equation, which determines the energy of these states, can be written as (see, e.g., [3,17])

$$\mathrm{Re}G(\varepsilon)=S(\varepsilon, \Lambda_{ik}) \tag{2}$$

where the $S(\varepsilon, \Lambda_{ik})$ function is determined by the perturbation operator $\hat{\Lambda}$ (Λ_{ik} are matrix elements of this operator on defined basis).

The local spectral densities $\rho(\varepsilon, i)\equiv\frac{1}{\pi}\lim_{\gamma\downarrow 0}\mathrm{Im}G(\varepsilon+i\gamma, i)$ of impurity atoms are calculated in [6]. For an isolated substitutional impurity with the energy $\varepsilon_0=\tilde{\varepsilon}$ of the impurity node $i=0$ and with the overlap integral $J_{i0}=(1+\eta)J$, the function $S(\varepsilon, \tilde{\varepsilon}, \eta)$ has the form

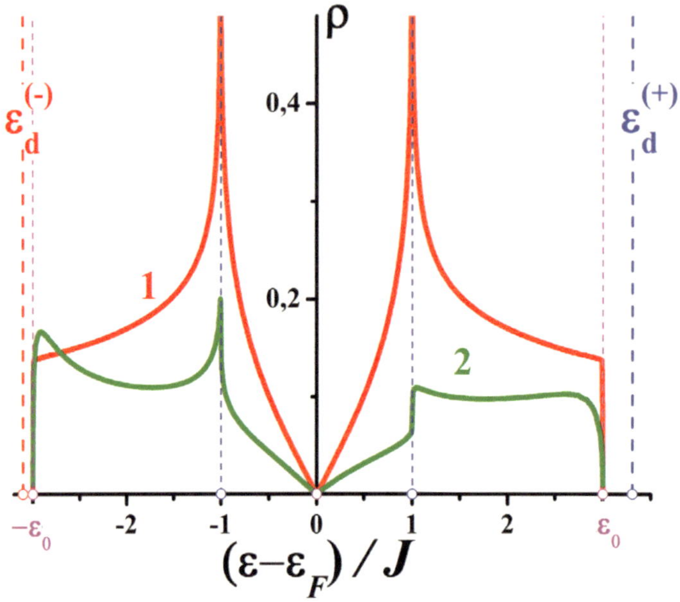

Figure 1. Electronic density of states of perfect graphene (curve 1) and the local density of states for an isolated boron substitutional impurity (curve 2).

$$S(\varepsilon, \tilde{\varepsilon}, \eta) = \frac{(1+\eta)^2}{\tilde{\varepsilon} + \varepsilon\eta(2+\eta)} \tag{3}$$

For a nitrogen impurity $\tilde{\varepsilon} - \varepsilon(K) \approx -0.525J$ and $\eta \approx -0.5$ (according to [4]). As shown in [22], equation (2) has a solution for both interval $[-\varepsilon(M), \varepsilon(K)]$ and interval $[\varepsilon(K), \varepsilon(M)]$.

The local density of states of the nitrogen substitutional impurity calculated in [11] has quasi-local maxima in both intervals. For an boron substitutional impurity ($\eta \approx 0.5$) [6], quasi-localized states are absent in the $[-\varepsilon(M), \varepsilon(M)]$ interval [12]. Figure 2 shows the graphical solution of the Lifshitz equation (2) for a given impurity atom. In this case the Lifshitz equation has no solutions in interval $[-\varepsilon(M), \varepsilon(M)]$ (corresponding dependences $S(\varepsilon)$ are shown as curves 3 in Figure 2). The local Green's function of the boron impurity (curve 2 in Figure 1) has two poles outside the band of quasi-continuous spectrum, which are called local discrete levels and which are also solutions of equation (2). As is clearly seen in Figure 1 the area under the curve 2 is smaller (by the sum of the residues at these poles) than the area under the curve 1.

Local discrete levels can be an important source of information about defective structure and force interactions in real crystals. To extract this useful information we should have analytical expressions that relate the main characteristics of LDL (primarily their energy) to the parameters of the defect and the host lattice.

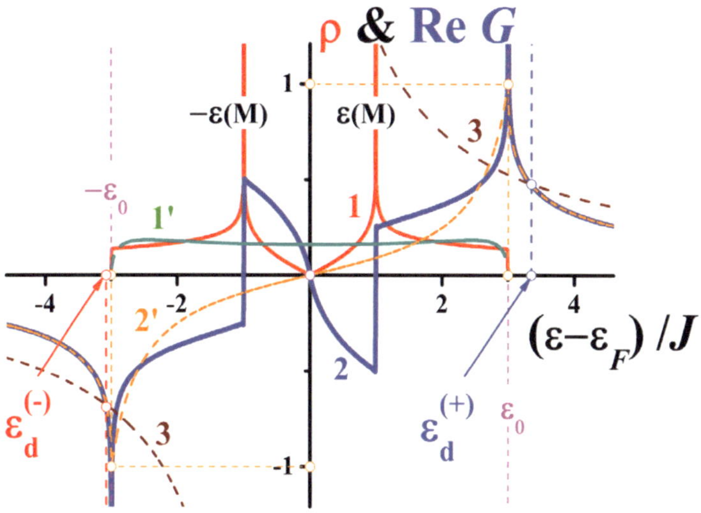

Figure 2. Graphical solution of equation (2) for boron substitutional impurity in graphene. Curve 1 is the electronic density of states of ideal graphene, curve 2 is the corresponding real part of the Green's function. Curves 1′ and 2′ are the "approximations of two moments" of these functions. Curve 3 represents the function $S(\varepsilon)$.

Such expressions were obtained in [23] for localized vibrations in the phonon spectrum of a three-dimensional crystal. Authors proposed an analytical approximation of the basic characteristics of local vibrations based on the rapid convergence of the real part of the Green's function outside the band of quasi-continuous spectrum using the method of Jacobi matrices [16,17].

Let us, briefly, to the extent necessary to understand the use of the classification of the eigenfunctions of Hamiltonian (1), to present the basics of the method of Jacobi matrices. This method allows, without finding the dispersion laws, to calculate directly the local partial Green's functions of the system, corresponding to the perturbation of one or more atoms. This perturbation is described by the so-called generating vector $\vec{h}_0 \in H$, where H is the space of electronic excitations of atoms. Its dimension is qN , where N is the number of atoms in the system, and q is the dimension of the displacement of a single atom ($q = 1, 2, 3$). Vectors of the space H are denoted by an arrow above the symbol, and "ordinary" q-dimensional vectors are in bold italics.

If, using the generating vector $\vec{h}_0 = p^{-1}\sum_{j=1}^{p}|j\rangle$ (p is the number of excited atoms) and the Hamiltonian (1), we construct the sequence $\{\hat{H}^n \vec{h}_0\}_{n=0}^{\infty}$, then the linear envelope covering the vectors of this sequence forms, in the H space, a cyclic subspace invariant to the operator $\hat{H}$. This subspace contains, within itself, all the atomic displacements generated by the vector $\vec{h}_0$. The corresponding partial Green's function is determined as a matrix element $G_{00}(\lambda) \equiv (\vec{h}_0, [\varepsilon\hat{I} - \hat{H}]^{-1}\vec{h}_0)$, where $\lambda \equiv \omega^2$ is the eigenvalue. Quantity $\rho(\lambda) \equiv \pi^{-1}\mathrm{Im}G_{00}(\lambda)$ is called the spectral density generated by the initial displacement $\vec{h}_0$. In the basis $\{\vec{h}_n\}_{n=0}^{\infty}$ which is obtained by the orthonormalization of the sequence $\{\hat{H}^n \vec{h}_0\}_{n=0}^{\infty}$, the operator (1) is represented in the form of a tridiagonal Jacobi matrix (or J-matrix). This matrix has a simple spectrum, what greatly simplifies finding the partial Green's functions and spectral densities. As can be seen, this method does not use explicitly the translational symmetry of the crystal, making it extremely effective for treating systems in which such symmetry is broken. The method of Jacobi matrices is particularly effective for treating systems with a simply connected quasi-continuous band of spectrum *D*. In this case, with increasing rank of the J-matrix ($n \to \infty$), its diagonal elements a_n converge to *a* corresponding to the middle of the bandwidth *D*, and nondiagonal elements b_n converge to *b* corresponding to the one-quarter of the bandwidth *D*.

For the local Green's function (LGF), corresponding to the excitations of one or more atoms, which are determined by the generating vector $\vec{h}_0$, we get following expression using the *J*-matrix method

$$G(\varepsilon, \vec{h}_0) = \left(\vec{h}_0, [\varepsilon \hat{I} - \hat{H}]^{-1} \vec{h}_0\right) = \lim_{n \to \infty} \frac{Q_n(\varepsilon) - b_{n-1} Q_{n-1}(\varepsilon) K_\infty(\varepsilon)}{P_n(\varepsilon) - b_{n-1} P_{n-1}(\varepsilon) K_\infty(\varepsilon)} \tag{4}$$

where $\hat{I}$ is the unit operator and polynomials $P_n(\varepsilon)$ and $Q_n(\varepsilon)$ are determined by the following recurrence relations

$$b_n \{P, Q\}_{n+1}(\varepsilon) = (\varepsilon - a_n)\{P, Q\}_n(\varepsilon) - b_{n-1}\{P, Q\}_{n-1}(\varepsilon) \tag{5}$$

The initial conditions are $P_{-1}(\varepsilon) = Q_0(\varepsilon) \equiv 0$, $P_0(\varepsilon) \equiv 1$, $Q_0(\varepsilon) \equiv 0$, and $Q_1(\varepsilon) \equiv b_0^{-1}$, function $K_\infty(\varepsilon)$ corresponds to the LGF operator, with all elements of its J-matrix being equal to their limit values *a* and *b*:

$$K_\infty(\varepsilon) = 2b^{-2}\left[\varepsilon - a - Z(\varepsilon) \cdot \sqrt{(\varepsilon - a + 2b)(\varepsilon - a - 2b)}\right] \tag{6}$$

$$Z(\varepsilon) \equiv \Theta(a - 2b - \varepsilon) + i\Theta(\varepsilon - a + 2b) \cdot \Theta(a + 2b - \varepsilon) - \Theta(\varepsilon - a - 2b) \tag{7}$$

The method of Jacobi matrices can treat as a regular singular perturbation a much larger number of perturbations of the phonon spectrum due to the presence of various crystal defects than the traditional methods [19, 20]. In addition, perturbations do not change the bandwidth of the quasi-continuous spectrum, and consequently, the asymptotic values of the elements of the J-matrix can be regarded as an asymptotically degenerated regular perturbation [23]. This type of perturbations covers virtually all perturbations of the phonon spectrum caused by local defects. The calculation of vibration characteristics of such systems is performed, using the method of J-matrices, with the same accuracy as for the initial ideal system.

In practice, it is usually possible to calculate the Jacobi matrix of the Hamiltonian of a finite rank. The expression

$$G(\varepsilon,\vec{h}_0)\approx\frac{Q_n(\varepsilon)-b_{n-1}Q_{n-1}(\varepsilon)K_\infty(\varepsilon)}{P_n(\varepsilon)-b_{n-1}P_{n-1}(\varepsilon)K_\infty(\varepsilon)} \qquad (8)$$

is called analytical approximation of LGF. All dependences in Figure 1 and curves 1 and 2 in Figure 2 were calculated by the formula (8), using the Jacobi matrix of the Hamiltonian (1) with rank $n=600$. If we count the energy from the Fermi energy level, then all diagonal elements of Jacobi matrices are zero ($a_n=a=0;b=\varepsilon_0/2$, where $\varepsilon_0=3J$ is the half-width of the quasi-continuous spectrum). A good accuracy of the approximations shown in figures is confirmed also by the fact that they show the nonanalyticity effects corresponding to the densities of states of systems with the dimension larger than unity (so-called van Hove singularities). In the vicinity of these singularities the expression (8) slowly converges to the true values of the really and imaginary parts of LGF.

Curves $1^{'}$ and $2^{'}$ in Figure 2 show local density of states and their corresponding real parts of the LGF calculated by formula (8) with n=1. As can be seen in the band of the quasi-continuous spectrum these relationships have very little in common with curves 1 and 2. Thus, the curve $1^{'}$ does not even hint at the *V*-shaped "Dirac" singularity at $\varepsilon=\varepsilon(K)=\varepsilon_F$, and on the curve $2^{'}$ in the interval $[-\varepsilon(M),\ \varepsilon(M)]$ both non-monotonous parts and the logarithmic singularities at the edges of the band of quasi-continuous spectrum, characteristic for the 2*D* systems are absent. However, outside the band of quasi-continuous spectrum (also in the area of intersection of the real part of LGF with curve (3)) curves 2 and $2^{'}$ practically coincide, and if we put in the Lifshitz equation (2) instead of LGF its approximation (8) for n=1, the obtained solutions give the energies of LDL with quite high accuracy. Moreover, these solutions can be easily found analytically. The LGF approximation by formula (8) for n=1 was named the approximation of two moments in [23]. Indeed, it follows from the orthonormality of the polynomials defined in [16,17] that

$$\int_{-\varepsilon_0}^{-\varepsilon_0}P_1(\varepsilon)\rho(\varepsilon)d\varepsilon=0\Rightarrow a_0=\int_{-\varepsilon_0}^{-\varepsilon_0}\varepsilon\rho(\varepsilon)d\varepsilon=M_1,$$

$$\int_{-\varepsilon_0}^{-\varepsilon_0}P_1^2(\varepsilon)\rho(\varepsilon)d\varepsilon=\int_{-\varepsilon_0}^{-\varepsilon_0}\frac{\left(\varepsilon^2-a_0^2\right)}{b_0^2}\rho(\varepsilon)d\varepsilon=1\Rightarrow b_0=\sqrt{M_2-M_1^2}$$

Finding the characteristics of LDL is more convenient without using equation (2), looking for them as the poles of the LGF perturbed Hamiltonian $\tilde{G}(\varepsilon,\vec{h}_0)=\left(\vec{h}_0,[\varepsilon\hat{I}-\hat{H}-\hat{\Lambda}]^{-1}\vec{h}_0\right)$. In the approximation of two moments for the subspace generated by the excitation of an impurity atom, we get

$$\widetilde{G}(\varepsilon)=\frac{1}{\varepsilon-a_0-b_0^2K_\infty(\varepsilon)} \tag{9}$$

For the case of an isolated substitutional impurity

$$a_0=\tilde{\varepsilon};b_0=\sqrt{3}(1+\eta)J=\frac{1+\eta}{\sqrt{3}}\varepsilon_0 \tag{10}$$

from where we get

$$\widetilde{G}(\varepsilon)=\frac{(1-\gamma)\varepsilon-a_0+Z(\varepsilon)\gamma\sqrt{|\varepsilon^2-\varepsilon_0^2|}}{R(\varepsilon)} \tag{11}$$

where $R(\varepsilon)=(1-2\gamma)\varepsilon^2-2(1-\gamma)a_0\varepsilon+a_0^2+\gamma^2\varepsilon_0^2$ and

$$\gamma\equiv b_0^2/2b^2=2(1+\eta)^2/3 \tag{12}$$

Local discrete levels are poles (11), i.e. the roots of $R(\varepsilon)$ are

$$\varepsilon_d^{(\pm)}=\frac{(\gamma-1)a_0\pm\gamma\sqrt{a_0^2+(2\gamma-1)\varepsilon_0^2}}{2\gamma-1} \tag{13}$$

Residues at these poles $\mu_0^{(\pm)}=\underset{\varepsilon=\varepsilon_d^{(\pm)}}{r\grave{e}s}\widetilde{G}(\varepsilon)$ are called intensities of LDL and they determine the relative LDL "amplitude" on the impurity atom: $\mu_0^{(\pm)}=1-\pi^{-1}\int_{-\varepsilon_0}^{\varepsilon_0}\mathrm{Im}\widetilde{G}(\varepsilon)d\varepsilon$. The condition that the intensity differs from zero defines the existence region of LDL. In this case

$$\begin{aligned}\mu_0^{(+)}&=\frac{\gamma a_0+(\gamma-1)\sqrt{a_0^2+(2\gamma-1)\varepsilon_0^2}}{(2\gamma-1)\sqrt{a_0^2+(2\gamma-1)\varepsilon_0^2}}\cdot\Theta\left(\frac{a_0}{\varepsilon_0}-1+\gamma\right);\\ \mu_0^{(-)}&=\frac{-\gamma a_0+(\gamma-1)\sqrt{a_0^2+(2\gamma-1)\varepsilon_0^2}}{(2\gamma-1)\sqrt{a_0^2+(2\gamma-1)\varepsilon_0^2}}\cdot\Theta\left(\gamma-1-\frac{a_0}{\varepsilon_0}\right).\end{aligned} \tag{14}$$

It was shown in [23] that

$$G_{mn}(\varepsilon, \vec{h}_0)=\left(\vec{h}_{m'} [\varepsilon\hat{I}-\hat{H}]^{-1}\vec{h}_n\right)=-P_m(\varepsilon)Q_n(\varepsilon)+P_m(\varepsilon)P_m(\varepsilon)G(\varepsilon, \vec{h}_0) \qquad (m<n).$$

This implies that the damping of LDL, i.e. the decay of its intensity with the increasing distance from the impurity atom (i.e. with the increase of n) follows the equation $\mu_n^{(\pm)}=P_n^2\left(\varepsilon_d^{(\pm)}\right)\cdot \mu_0^{(\pm)}$. Using the method of mathematical induction we can prove that

$$P_n\left(\varepsilon_d^{(\pm)}\right)=\sqrt{2\gamma}\cdot\left[\pm\frac{\sqrt{a_0^2+(2\gamma-1)\varepsilon_0^2}\mp a_0}{\varepsilon_0(2\gamma-1)}\right]^n \tag{15}$$

The intensities $\mu_n^{(\pm)}$ decay with increasing n according to $\mu_{n0}^{(\pm)}=2\gamma\cdot q^n\cdot \mu_0^{(\pm)}$, that is, starting from n=1 they form an infinitely decreasing geometric progression whose denominator

$$q^{(\pm)}=\left[\frac{\sqrt{a_0^2+(2\gamma-1)\varepsilon_0^2}\mp a_0}{\varepsilon_0(2\gamma-1)}\right]^2 \tag{16}$$

Summing these progressions we see that $\sum_{n=0}^{\infty}\mu_n^{(+)}=\sum_{n=0}^{\infty}\mu_n^{(-)}=1$, that is the formation of each LDL is the formation of one quasi-particle outside the band of the quasi-continuous electron spectrum.

Formulas (10, 12-14, 16) give simple analytical expressions of the local conditions of the existence of discrete levels due to the presence of a substitutional impurity in graphene.

Regions of the LDL existence for $\varepsilon_d^{(+)}$ (in this case $\tilde{\varepsilon}\varepsilon_0(1-\gamma)-\varepsilon_0[1-2(1+\eta)^2/3]$) and for $\varepsilon_d^{(\ )}$ (in this case $\tilde{\varepsilon}\varepsilon_0(\gamma-1)=\varepsilon_0[2(1+\eta)^2/3-1]$) lie above and under curves in Figure 3, as indicated by arrows.

It is seen that such levels exist in a very wide range of variables $\tilde{\varepsilon}$ and η. The absence of LDL is possible only in a narrow range of values $\tilde{\varepsilon}$ at $\eta\sqrt{3/2}-1$. If $\eta\sqrt{3/2}-1$ at least one local discrete level exists. In fact, the lines delineating the area of existence of LDL in graphene in the plane $\{\tilde{\varepsilon}, \eta\}$ must pass through the origin of coordinates, since $\mathrm{Re}G(\varepsilon)\to\infty$ for $\varepsilon\to\pm\varepsilon_0$. However, since this divergence is logarithmic, for any appreciable splitting of LDL from the boundary of quasi-continuous spectrum there is a certain threshold. Curves 2 and 2$'$ in Figure 2 merge at $|\varepsilon|>\varepsilon_0$.

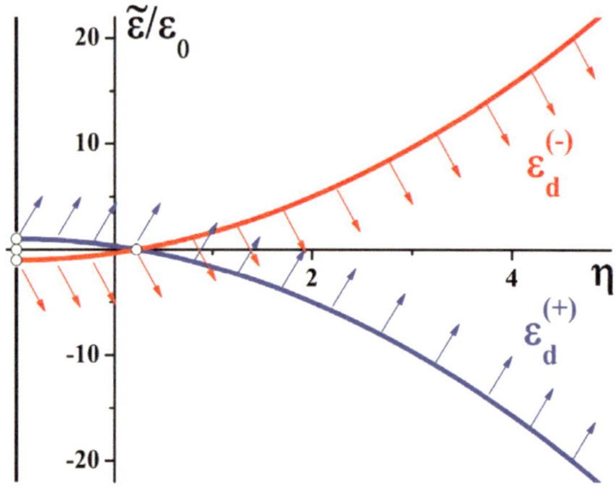

Figure 3. Regions of the existence of discrete levels for substitutional impurities in graphene.

Figure 4 shows, for $\tilde{\varepsilon} \approx 0.525J$ (for boron, from [6]), the dependences of energies, the LDL intensities at the boron impurity and the damping parameters of the value η that characterizes the change in the overlap integral of the boron impurity atom (10). Solid lines show the characteristics of LDL, calculated using the approximation of the Green's function (9), i.e. according to the analytical formulas (13), (14) and (16). Open circles show the results of numerical calculations of dependences $\varepsilon_d^{(\pm)}(\eta)$ and $\mu_0^{(\pm)}(\eta)$, using the Green's function in the form of (8), calculated by the Jacobi matrices of the $n \geq 100$ rank.

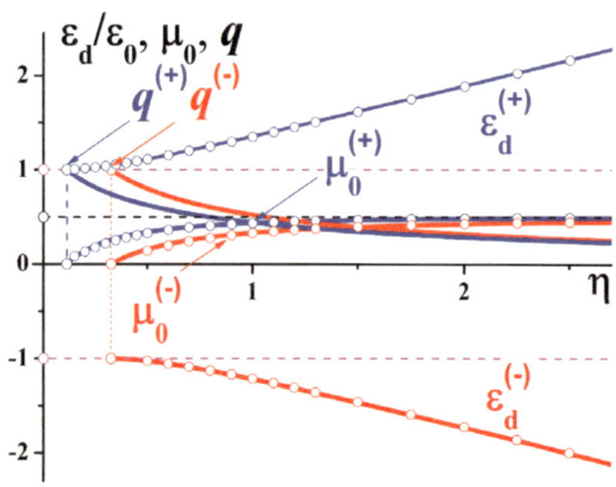

Figure 4. The basic characteristics of LDL in the presence of a boron substitutional impurity in graphene.

It is seen that at the threshold values $\varepsilon_d^{(\pm)} = \pm\,\varepsilon_0$ of the LDL formation, the intensities of LDL equal zero and the parameters of damping are equal to unit. Further increase of $|\,\varepsilon_d^{(\pm)}\,|$ is accompanied by the increase of $\mu_0^{(\pm)}$ and by the strengthening of the containment level.

So, a good agreement between the results of numerical calculation of the LDL characteristics using Jacobi matrices of high rank, and their analytical description by the Green's function (9), which relates these characteristics to the parameters of the defect (13) (14), makes it relatively easy to extract the defect parameters from the known characteristics of LDL. Experimental measurement (e.g. by scanning tunneling microscopy) of values $\varepsilon_d^{(\pm)}$ should lead, using (10), (12) and (13), to the determination of the parameters $\tilde{\varepsilon}$ and η and this might represent a significant advance in creating nanomaterials with predetermined spectral characteristics. As can be seen from Figure 4, with increasing η the intensity of LDL $\mu_0^{(+)} \to \mu_0^{(-)} \to 1/2$. That is, the impurity levels can not be completely localized on the impurities, but they also appear in the spectra of surrounding carbon atoms. This greatly increases the probability of experimental detection of such levels, even at low concentrations of impurities.

3. The electronic spectrum of bilayer graphene

Bilayer graphene is a carbon film consisting of two graphene monolayers, linked together by (as in bulk graphite) van der Waals forces. Since the distance between the layers (film thickness) $h \sim 3.5$ Å the bilayer graphene can be considered not a nanofilm but a subnanofilm. The constants of the interatomic interaction of bilayer graphene were determined and its phonon density of states and partial contributions to this quantity from the atomic displacements along different crystallographic directions were calculated [7,24]. On the basis of the analyses of the mean-square amplitudes of atomic displacements calculated using data from the spectral densities, we have shown that the flat shape of a free bilayer graphene remains stable up to the temperatures much higher than the room temperature, which makes this compound promising for nanoelectronics. In this section we calculate and analyze the electronic spectrum of a defect-free bilayer graphene. Naturally of greatest interest is its behavior in the energy range close to ε_F where there are characteristic Dirac points on the spectrum of graphene monolayer (whose plane shape is unstable).

The unit cell of graphene contains two physically equivalent atoms and therefore local Green's function and the local density of states (LDOS) of the atoms of different sublattices are identical. On the other hand, bilayer graphene unit cell contains four atoms, and atoms of different sublattices of a single graphene layer interact differently with the atoms of the other layer and their physical equivalence is disrupted (Figure 5a).

The electronic spectrum of bilayer graphene, as well as the electronic spectrum of graphene can be described in the strong coupling approximation. Corresponding Hamiltonian has form (1). For graphene and bigraphene we assume that the electron hopping within the layer is possible only between nearest neighbors $J_{ij} = J \approx 2.8$ eV (see for example [25]). Electron hopping between layers is also assumed to be possible only between nearest neighbors from different layers, that is, between those which lie at a distance h from each other. Denote the corresponding hopping integral J' . Note that only half of the bilayer graphene atoms have such neighbors (sublattice AI and AII, see Figure 1). In sublattices BI and BII no such neighbors exist, since nearest neighbors from different layers are at a distance $\sqrt{h^2 + a^2}$ ($a \approx 1.415$ Å is the distance between nearest neighbors in the layer plane). Since this distance is only by less than 10% greater than h , we can neglect the interaction with the atoms of the sublattices *BI* and *BII* and this does not lead to qualitative changes in the behavior of the spectra near ε_F (see, e.g. [6]). Then the dispersion relation of each of the four branches of the electronic spectrum of bilayer graphene can be written as

$$\begin{aligned} \varepsilon_{1,2}(k) &= \pm\sqrt{\varepsilon_0^2(k) + \frac{J'^2}{2} - J'\sqrt{\varepsilon_0^2(k) + \frac{J'^2}{4}}}; \\ \varepsilon_{3,4}(k) &= \pm\sqrt{\varepsilon_0^2(k) + \frac{J'^2}{2} + J'\sqrt{\varepsilon_0^2(k) + \frac{J'^2}{4}}}, \end{aligned} \tag{17}$$

where $\varepsilon_0(k)$ is the electronic spectrum of graphene, calculated in the strong coupling approximation:

$$\varepsilon_0(k) = \pm J\sqrt{1 + 4\cos\left(k \cdot \frac{a_1 - a_2}{2}\right)\left[\cos\left(k \cdot \frac{a_1 + a_2}{2}\right) + \cos\left(k \cdot \frac{a_1 - a_2}{2}\right)\right]} \tag{18}$$

where $a_1 = \left(\frac{a}{2}, \frac{a\sqrt{3}}{2}, 0\right)$ and $a_2 = \left(\frac{a}{2}, -\frac{a\sqrt{3}}{2}, 0\right)$ are two-dimensional Bravais lattice vectors (see Figure 5b).

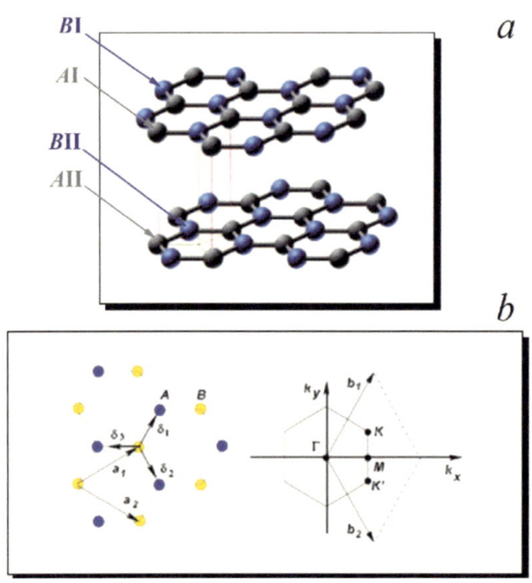

Figure 5. The structure of bilayer graphene (a); Bravais lattice and the first Brillouin zone (b).

http://dx.doi.org/10.5772/50562

Dispersion curves along highly symmetric directions ΓK, ΓM and KM for $J' = 0.1J$ are shown in Figure 6. Region near the K-point $K=\left(0,\pm 4\pi / 3a\sqrt{3},0\right)$ and $K=\left(\pm 2\pi / 3a,\ \pm 2\pi / 3a\sqrt{3},0\right)$ is shown in the inset. The same inset shows also the dispersion curves for graphene (18). We clearly see the quasi-relativistic nature of the electronic spectrum of graphene as well as ordinary quadratic dispersion curves $\varepsilon_{1,2}(k)$ for bigraphene. Spectral branches $\varepsilon_{3,4}(k) \notin (-J',\ J')$ are determined in the $(-J',\ J')$ interval. Indeed, if k takes value along ΓK, then $\varepsilon_0(k) = \pm\left(1+2\cos\frac{ak\sqrt{3}}{2}\right)$ and putting $k = K + \kappa$ ($\kappa 1$) we find $\varepsilon_0(K+\kappa) \approx \mp\frac{3a\kappa}{2}$, i.e. a linear (relativistic) dispersion relation. For electronic modes of bilayer graphene $\varepsilon_{1,2}(k)$ near the K point, we can write

$$\varepsilon_{1,2}^2(K+\kappa)\approx\frac{J_1^2}{2}+\varepsilon_0^2(K+\kappa)-\sqrt{\left(\frac{J_1^2}{2}+\varepsilon_0^2(K+\kappa)\right)^2-\varepsilon_0^4(K+\kappa)}\approx\frac{J_1^2}{2}+\varepsilon_0^2(K+\kappa)- -\left[\frac{J_1^2}{2}+\varepsilon_0^2(K+\kappa)\right]\cdot\left\{1-\frac{\varepsilon_0^4(K+\kappa)}{2\left[J_1^2/2+\varepsilon_0^2(K+\kappa)\right]^2}\right\}=\frac{\varepsilon_0^4(K+\kappa)}{J_1^2+2\varepsilon_0^2(K+\kappa)}\approx\frac{81J^4a^4\kappa^4}{16J_1^2}, \tag{19}$$

that is, we get an ordinary quadratic dispersion law

$$\varepsilon_{1,2}(K+\kappa)\approx\pm\frac{9J^2a^2\kappa^2}{4J_1} \tag{20}$$

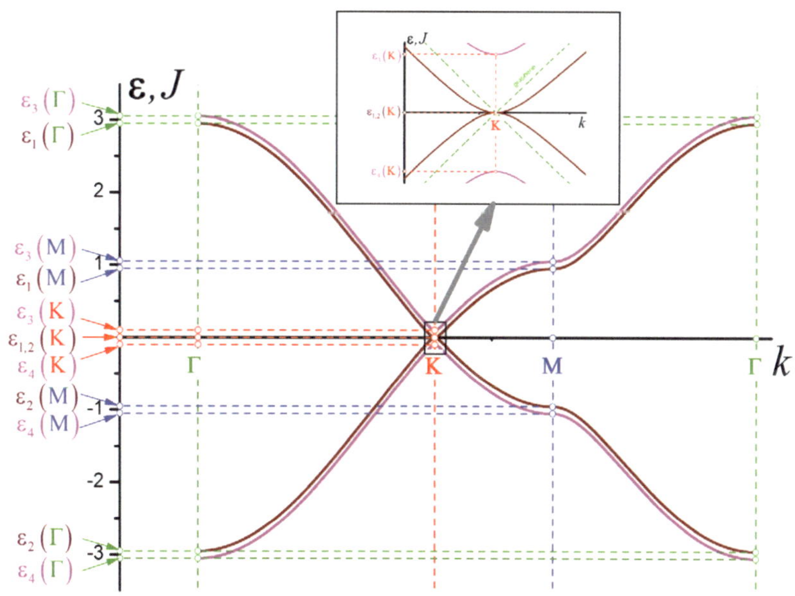

Figure 6. Dispersion curves along high-symmetry directions of bilayer graphene.

The electron effective mass in considered branches, determined from the relation $\varepsilon=(\hbar\kappa)^2/2m^*$, is equal to

$$m^*=\frac{2\hbar^2 J'}{9J^2a^2} \tag{21}$$

In the case considered above $J'=0.1J$ the effective mass is $m^*\approx 2.75\cdot 10^{-32}$ kg, and if $J'\to J$ it tends to a value close to the free electron mass. Since $m^*\sim J'$, then by changing the interlayer hopping integral we can change the effective mass of charge carriers.

It should be noted that for $J'\to 0$, that is for the transition from bilayer graphene to two non-interacting graphene monolayers the effective mass, $m^*\to 0$, and formulas (19 - 21) cannot correctly describe this transition, since they were obtained under the assumption $\varepsilon_0(k)<<J'$.

Electron density of states for values of energy near ε_F is determined by branches of ε_1 and ε_2 only, and it follows from (3) that $g_1(\varepsilon)=g_2(-\varepsilon)$. Then

$$g(\varepsilon)=\frac{\Sigma_0}{(2\pi)^2}\oint\limits_{\varepsilon(\kappa)=\varepsilon}\frac{dl_{1,2}}{|\partial\varepsilon_{1,2}/\partial\kappa|} \tag{22}$$

where $\Sigma_0=3a^2\sqrt{3}/2$ is the area of the two-dimensional Bravais cell, and integration is done along a closed isoenergetic line $\varepsilon_{1,2}(k)=\varepsilon$. At $\varepsilon=0$ (Fermi level) the line contracts into a point and near ε_F the contour of integration is a circle. Taking into account (20) we may write

$$\begin{gathered}g_{1,2}(\varepsilon_F)\approx\frac{S_0}{(2\pi)^2}\oint\limits_{\varepsilon(\delta)=\varepsilon}\frac{\delta d\varphi}{|\partial\varepsilon_{1,2}/\partial\kappa|}=\frac{3a^2\sqrt{3}}{2(2\pi)^2}\cdot 2\pi\frac{\delta}{\frac{18J^2a^2\kappa}{4J'}}=\frac{J'\sqrt{3}}{6\pi J^2}=const\\ g_{3,4}(\varepsilon_F)=0;\Rightarrow g(\varepsilon_F)\approx\frac{J'}{J^2\sqrt{3}}.\end{gathered} \tag{23}$$

This means that the electron density of states (DOS) is constant and different from zero. As follows from (20), DOS is an analytical function and has minimum at $\varepsilon=0$ and near ε_F the function $g(\varepsilon)\sim\varepsilon^2$.

Total electron DOS can be represented as a mean-arithmetic function of the two LDOS corresponding to atoms of sublattices A and B: $\rho_{AI}(\varepsilon)=\rho_{AII}(\varepsilon)\equiv\rho_A(\varepsilon)$; $\rho_{BI}(\varepsilon)=\rho_{BII}(\varepsilon)\equiv\rho_B(\varepsilon)$ and $g(\varepsilon)=[\rho_A(\varepsilon)+\rho_B(\varepsilon)]/2$. For each sublattice with perfect structure the LDOS may be written as

$$\rho_s(\varepsilon_F)\approx\frac{S_0}{(2\pi)^2}\sum_{\alpha=1}^{4}\oint_{\varepsilon(\delta)=\varepsilon}\frac{\delta d\varphi\,|\,\psi_s(\alpha,\,\kappa)\,|^{\,2}}{|\,\partial\varepsilon_{1,2}/\partial\kappa\,|} \tag{24}$$

where index s is the designation of sublattice, index α is the designation of branch, and $\psi_s(\alpha,\,\kappa)$ are the eigenwave functions corresponding to atoms from sublattices. LDOS are calculated by the method of Jacobi matrix [16, 17] and are shown in Figure 7.

In this figure we clearly see two-dimensional van-Hove peculiarities at energy values corresponding to points Γ and M of the first Brillouin zone (see Figure 5b). These eigenvalues are given in the top inset in Figure 7. The bottom inset shows LDOS, in enlarged scale, near ε_F . It is seen that near the Fermi level the local density of states as well the total density of states are analytical and their dependences on energy are essentially nonlinear (for comparison the DOS of graphene is also shown in the same inset). Besides, $\rho_A(\varepsilon)$ differs from both $\rho_B(\varepsilon)$ and the total density of states and it approaches to zero for $\varepsilon\rightarrow 0$.

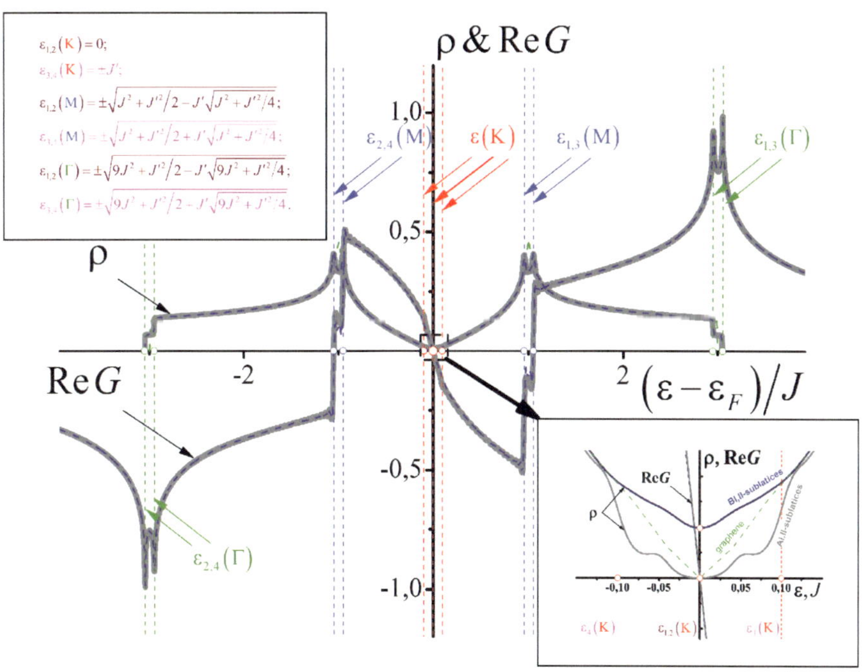

Figure 7. LDOS for atoms of different sublattices of bigraphene.

Indeed, putting zero eigenvalue in the equation of eigenfunctions of Hamiltonian (1) we get the values $\psi_{AI} \sim \psi_{AII} = O(\kappa^2)$, $\psi_{BI} \sim \psi_{BII} \sim 1$. Therefore, near the Fermi level $\rho_A(\varepsilon) \sim \varepsilon^2 \rho_B(\varepsilon)$.

Peculiarities at both ends, i.e. at $\varepsilon \approx \pm 0.1J = \pm J'$, originate from the contribution of modes $\varepsilon_{3,4}$ which are not represented on interval $\varepsilon \in [-J', J']$. Peculiarities at $\varepsilon \approx \pm 0.05J$ are due to the fact that beginning from these energies the anisotropy of isofrequency lines becomes essential.

So we can conclude that, in contrary to graphene, the bigraphene has an ordinary non-relativistic form of the electronic dispersion law. The effective mass of electron in the bigraphene strongly depends on the value of integral describing the hopping between two layers, and this value may be changed by external conditions (for example by pressure). Near the Fermi level the LDOS of atoms of different sublattices qualitively differ from each other. If the LDOS for the atoms of sublattice *A* at the Fermi level equals to zero and it slowly increases near this level ("a quasi-gap" appears), then the LDOS of sublattice *B* for the same energy values differs from zero and increases very quickly.

4. The influence of defects on the electron spectrum of bigraphene

Some peculiarities in the behavior of the bigraphene electron spectrum near $\varepsilon = \varepsilon(K)$, that is near the Fermi level, indicate the possibility of a strong influence of various defects [26-28].

Let us first consider point defects [26]. Figure 8 shows the real parts of the local Green's functions $\mathrm{Re}G_A(\varepsilon)$ and $\mathrm{Re}G_B(\varepsilon)$ of atoms of the sublattices *A* and *B*.

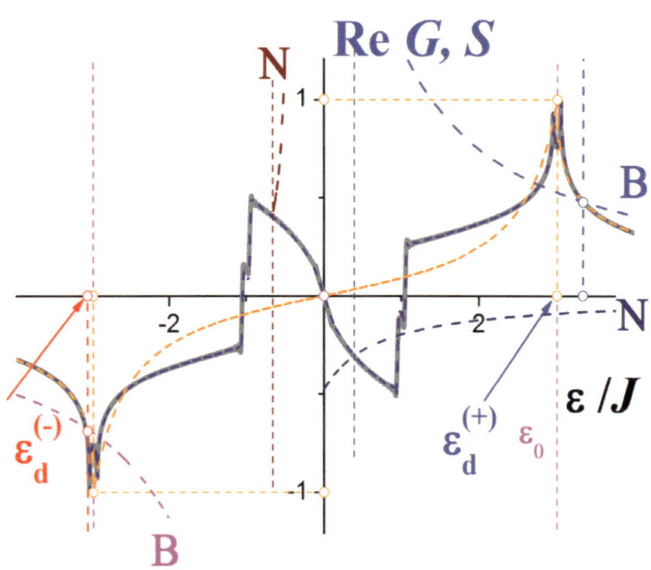

Figure 8. Real parts of the local Green's functions for the atoms of the two sublattices of bigraphene. The dashed lines are $S(\varepsilon)$ functions correspond to substitutional impurities of Boron (B-curves) and Nitrogen (N-curves).

http://dx.doi.org/10.5772/50562

This figure also shows the dependences S(ε) appearing in the Lifshitz equation (2) and corresponding correspond to the presence of substitutional impurities of boron and nitrogen in bigraphene.

For an isolated substitutional impurity that differs from the atom of the basic lattice by the values of energies at the impurity site and also by the overlap integral, function $S(\varepsilon)$ has for each sublattice form (3). Also, as in the case of graphene, for the nitrogen impurity in the interval $[\varepsilon_2(M),\ \varepsilon_1(M)]$ equation (2) has a solution and this impurity forms quasilocalized states in this interval. For the boron impurity, equation (2) has two solutions outside the band of quasi-continuous spectrum $[\varepsilon_4(M),\ \varepsilon_3(M)]$, corresponding to local discrete levels.

LDOS of isolated impurity atom of nitrogen in sublattice *A* or *B* are shown in Figure 9 (we remind that atoms *A*I and *A*II as well as *B*I and *B*II are physically equivalent).

As the inset shows, the nitrogen impurity does not forms a quasilocal maximum on LDOS and substantially changes it near the Fermi level. Figure 10 shows a LDOS of the boron impurity in sublattices *A* and *B*. As in the case of boron impurity in graphene (see Figure 1), the area under this curve is less than unity. Outside the band of the quasi-continuous spectrum local discrete energy levels are formed.

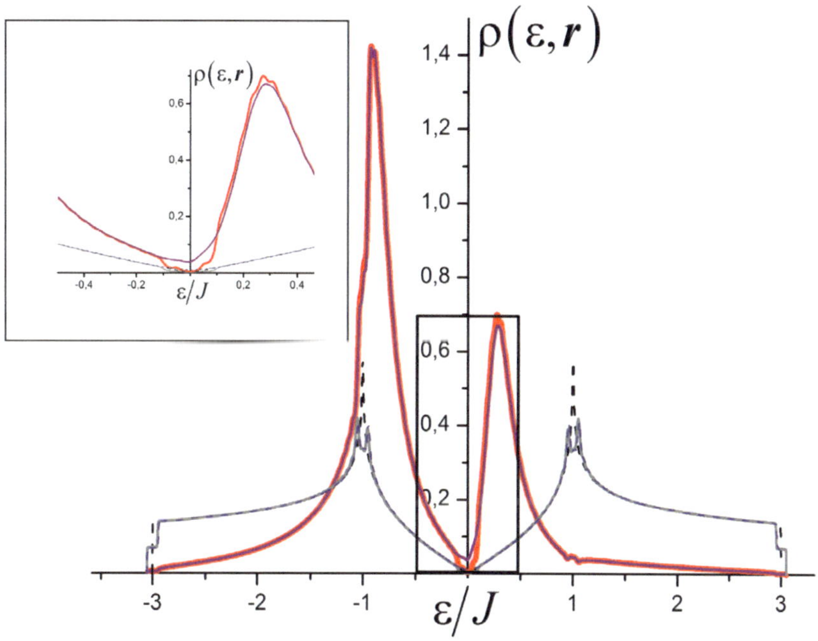

Figure 9. LDOS of nitrogen impurity in the sublattices of bigraphene (red curve is for *A* sublattice, purple is for *B* sublattice). For comparison the figure shows the DOS of graphene (black dashed line) and LDOS (thin solid gray and dashed gray, respectively).

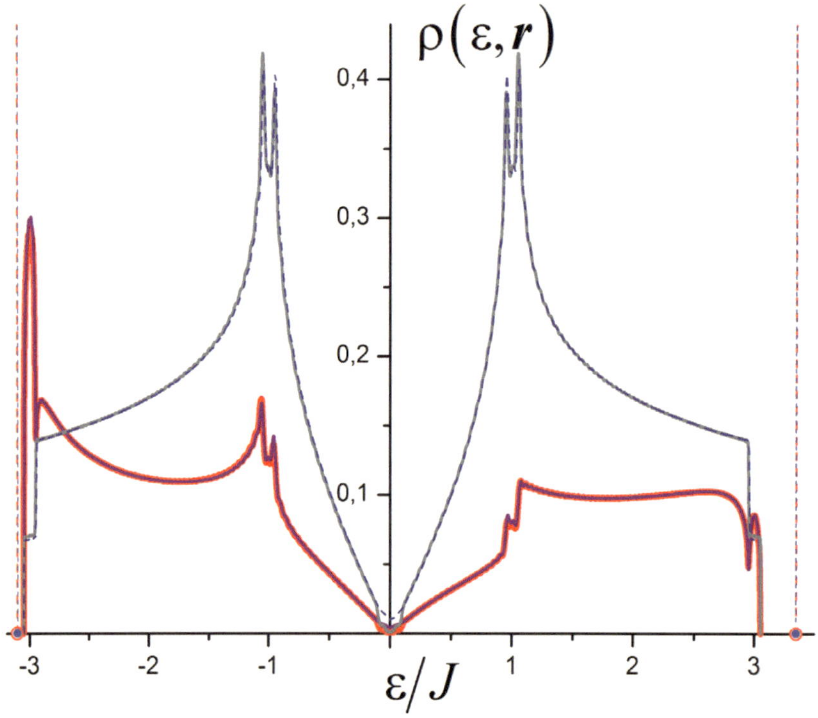

Figure 10. LDOS of the boron impurity in different sublattices of bigraphene (designations are the same as in Figure 9).

The energies of these levels for the considered ratio between the overlap integrals J and J' are slightly different. As can be seen from Figure 8, the energies of the local discrete levels for the substituted boron atom in sublattice *A* or sublattice *B* can be calculated using the two-moment approximation (13). Near the Fermi level the LDOS of the boron atom, substituting an atom of sublattice *B*, is considerably lower than the LDOS of the carbon atom of this sublattice. That is the boron impurity lowers the conductivity of bigraphene.

Because in bigraphene atoms of sublattices *A* and *B* are physically inequivalent, the influence on their electron spectra by various defects is different. In the first part of this section we have described the influence of substitutional impurities on the electron density of states. However, the influence of vacancies in sublattices on the electron DOS is even more profound. Figures 11 and 12 show the LDOS of neighbors of vacancies in sublattices *A* (Figure 11) and *B* (Figure 12).

Neighboring atoms are in the same layer as a vacancy, either in sublattice *B* (top) or in sublattice *A* (bottom). Insets show the arrangement of atoms. Atoms are shown in the same color as the corresponding LDOS.

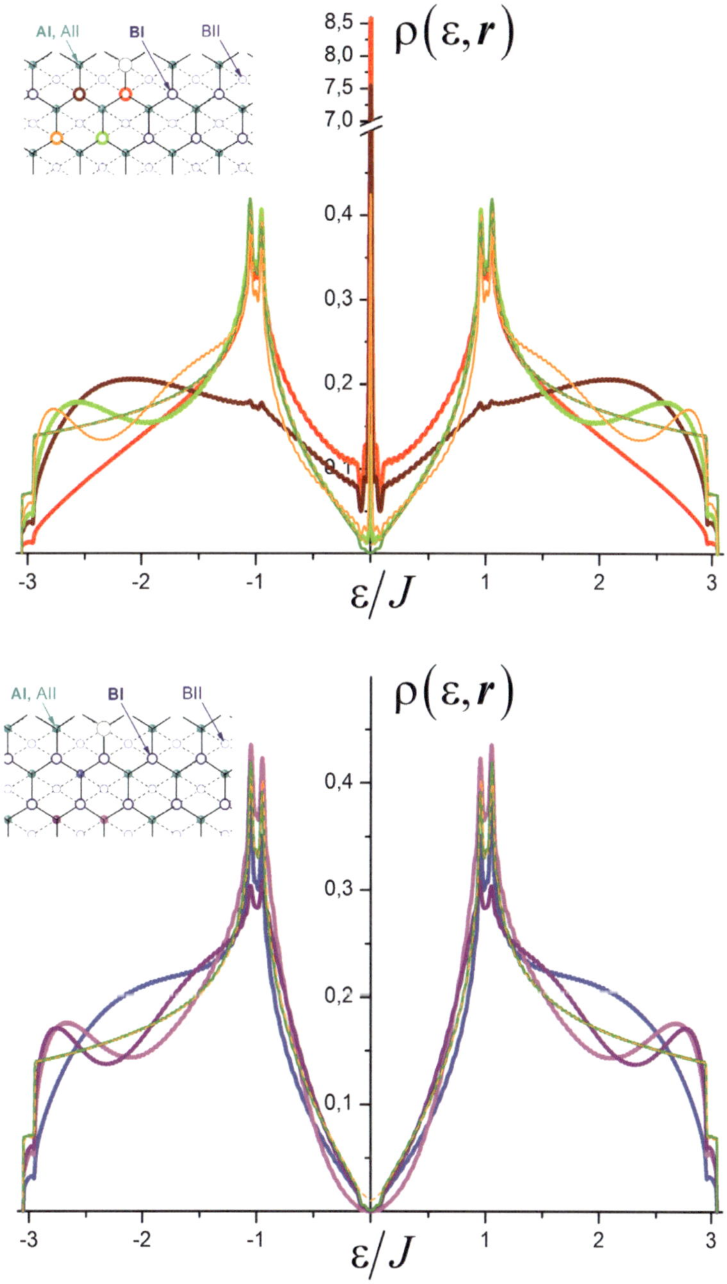

Figure 11. LDOS of bigraphene atoms which are neighbors of a vacancy in the sublattice *A*. Neighboring atoms are in the same layer as a vacancy, either in sublattice *B* (top) or in sublattice *A* (bottom). Insets show the arrangement of atoms. Atoms are shown in the same color as the corresponding LDOS.

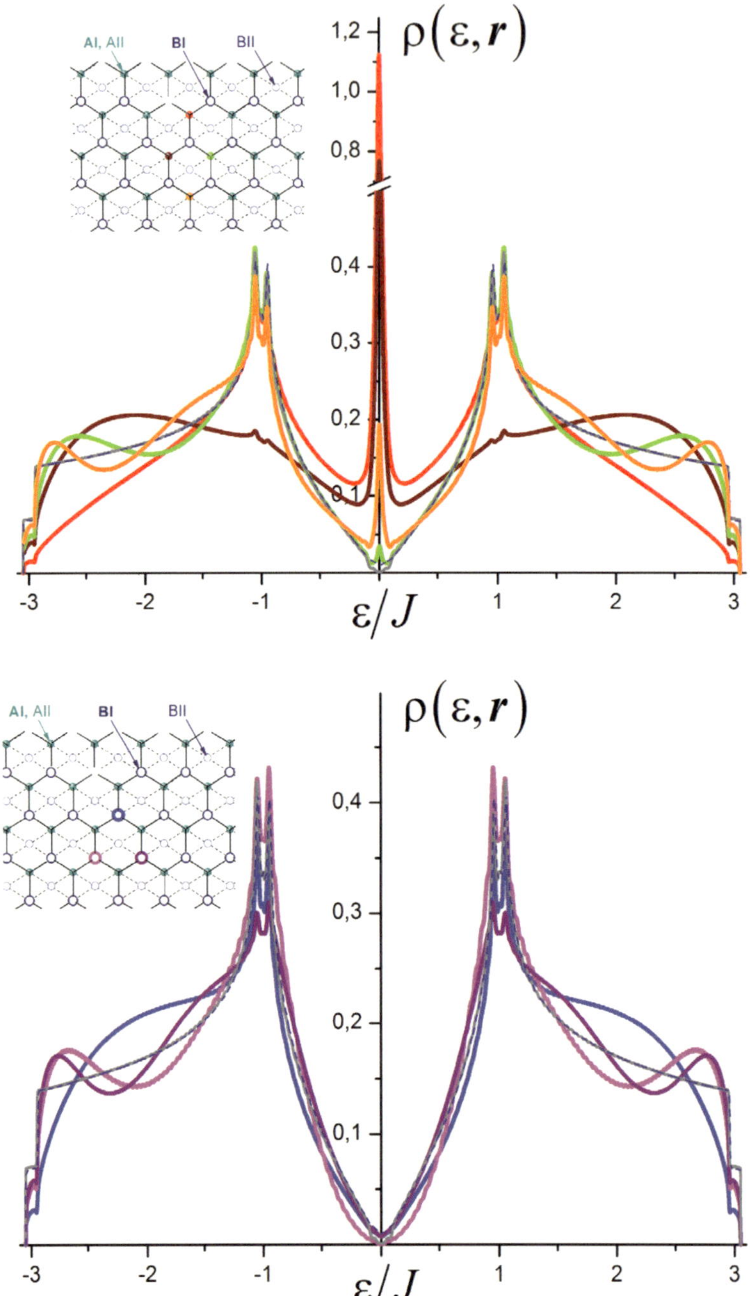

Figure 12. LDOS of bigrafene atoms which are neighbors of a vacancy in the sublattice *B*.

The reason for the specific evolution of the LDOS with increasing distance from the vacancy has been explained in [22]. In bigraphene there is an analogous situation. Electronic spectra of vacancy neighbors belonging to other sublattice also have sharp resonance peaks at $\varepsilon = \varepsilon_F$ (see upper parts Figure 11 and 12). So, if the vacancy is in the sublattice *A*, then the maximum of the LDOS of the sublattice *B* atoms is sharp, and the maximum on the spectrum the sublattice *A* atoms is blurred. Therefore we can conclude that the vacancy in bigraphene should have a more pronounced effect.

5. Conclusion

Unique properties of both graphene and bigraphene are caused, above all, by an unusual symmetry leading to the absence of a gap between the valence and conduction bands. The quantum states of quasiparticle in these bands are described by the same wave function, i.e. quasiparticle in graphene and bigraphene have the so-called chiral symmetry. In graphene this leads to the fact that dispersion relation of electron spectrum is linear and is described by the Dirac equation, characteristic for massless ultrarelativistic quasiparticle in quantum electrodynamics. On other hand, in bigraphene the presence of chiral symmetry leads at low energies to an ordinary parabolic dispersion relation, i.e. quiasiparticle of a new type appear – massive chiral fermions having no analogy in quantum electrodynamics. At symmetry breaking a gap between the valence and conduction band appears, allowing to tune the conducting properties of these materials.

In this chapter, using the method of Jacobi matrices, we analyzed how different impurities affect the energy gap width and the local density of states. The method of Jacobi matrices enables to investigate heterogeneous systems and to calculate the densities of states for each atom in different sublattices. Such analysis is necessary for correct comparison with experimental results. As defects we considered the vacancies and the impurity of nitrogen or boron. The presence of nitrogen leads to the formation of sharp resonance peaks (quasilocal states) inside the continuous spectrum; on the other hand, the boron impurity leads to states outside the continuous spectrum (local states). Both quasilocal and local states can be investigated experimentally by, for example, a scanning tunneling microscope. In the presence of vacancies we have analyzed how the density of states in each of the sublattices A or B changes. Different situations were analyzed. For example, if the vacancy in this sublattice vanishes at Fermi level, whereas in sublattice B the density has maximum at this point. We also investigated the conditions for opening the gap and changing its width. Main attention was paid to the analysis of electronic properties of considered systems. Moreover, computational method we used has also been successfully applied to the analysis of vibrational states. For example, increasing the overlap integral between the boron impurity and carbon atom leads to the strengthening of force interaction contacts between them. In addition, boron atom is 16% lighter than carbon atom, i.e. all necessary conditions are sent for the appearance of local vibrations in the phonon spectrum of both graphene and bigraphene with boron impurity. We hope that predicted peculiarities in electronic and vibrational spectra of perfect graphene and bigraphene as well as of graphene and bigraphene with defects will be detected experimentally.

Acknowledgements

This work is the result of the project implementation: Research and Education at UPJŠ – Heading towards Excellent European Universities, ITMS project code: 26110230056, supported by the Operational Program Education funded by the European Social Fund (ESF). This work was also supported by the grant of the Ukrainian Academy of Sciences under the contract No 4/10-H and by the grant of the Scientific Grant Agency of the Ministry of Education, Science, Research and Sport of the Slovak Republic and Slovak Academy of Sciences under No. 1/0159/09.

Author details

Alexander Feher[1*], Eugen Syrkin[2], Sergey Feodosyev[2], Igor Gospodarev[2], Elena Manzhelii[2], Alexander Kotlar[2] and Kirill Kravchenko[2]

*Address all correspondence to: alexander.feher@upjs.sk

1 Institute of Physics, Faculty of Science, P. J. Šafárik University in Kosice, Slovakia

2 B.I.Verkin Institute for Low Temperature Physics and Engineering NASU, Ukraine

References

[1] Uchoa, B., & Castro Neto, A. H. (2007). Superconducting States of Pure and Doped Graphene. *Physical Review Letters*, 98, 14681-4.

[2] McChesney, J. L., Bostwick, A., Ohta, T., Seyller, T., Horn, K., Gonza´lez, J., & Rotenberg, E. (2010). Extended van Hove Singularity and Superconducting Instability in Doped Graphene. *Physical Review Letters*, 104, 13683-4.

[3] Kosevich, A. M. (1999). *The Crystal Lattice (Phonons, Solitons, Dislocations)*, Berlin, Wiley-VCH, 10.1002/3527603085.

[4] Novoselov, K. S., Gein, A. K., Morozov, S. V., Diang, D., Katsnelson, M. I., Grigorieva, I. V., Dubonos, S. V., & Firsov, A. A. (2005). Two-Dimensional Gas of Massless Dirac Fermions in Graphene. *Nature*, 438, 197-201.

[5] Castro Neto, A. H., Guinea, F., Peres, N. M. R., Novoselov, K. S., & Geim, A. K. (2009). The Electronic Properties of Graphene. *Review of Modern Physics*, 81, 109-155.

[6] Peres, N. M. R., Klironomos, F. D., Tsai, S-W., Santos, J. R., Lopes dos Santos, J. M. B., & Castro Neto, A. H. (2007). Electron Waves in Chemistry Substituted Graphene. *EPL*, 80, 67007-67017.

[7] Meyer, J. C., Geim, A. K., Katsnelson, M. I., Novoselov, K. S., Booth, T. J., & Roth, S. (2007). The Structure of Suspended Graphene Sheets. *Nature*, 446, 60-63.

[8] Wallace, P. (1947). The Band Theory of Graphite. *Physical Review*, 71(9), 622-634.

[9] McClure, J. W. (1956). Diamagnetizm of Graphite. *Physical Review*, 104(3), 666-671.

[10] Syrkin, E. S., & Feodosyev, S. B. (1979). The Phonon Spectrum and Local Vibrations in Laminar Crystals. *Fizika Nizkih Temperatur*, 5(9), 1069-1073.

[11] Slonchechevski, J. C., & Weiss, P. R. (1958). Band Structure of Graphite. *Physical Review*, 109(2), 272-279.

[12] Feher, A., Syrkin, E. S., Feodos'ev, S. B., Gospodarev, I. A., & Kravchenko, K. V. (2011). Quasi-Particle Spectra on Substrate and Embedded Graphene Monolayers. In: Mikhailov S. (ed), *Physics and Applications of Graphene Theory*, Rijeka, InTech, 93-112.

[13] Wu, S., Jing, L., Li, Q., Shi, Q. W., Chen, J., Su, H., Wang, X., & Yang, J. (2008). Average Density of States in Disordered Graphene Systems. *Physical Review B*, 77(19), 195411.

[14] Feher, A., Gospodarev, I. A., Grishaev, V. I., Kravchenko, K. V., Manzhelii, E. V., Syrkin, E. S., & Feodos'ev, S. B. (2009). Effect of Defects on Quasi-particle Spectra of Graphite and Graphene. *Low Temperature Physics*, 35(7), 679-686.

[15] Chen, L., Zhang, Y., & Shen, Y. (2007). Self-healing in Defective Carbon Nanotubes: a Molecular Dynamic Study. *Journal of Physics: Cond. Matter*, 9, 38612-6.

[16] Peresada, V. I. (1968). New Calculation Method in the Theory of Crystal Lattice. *Condensed Matter Physics*, 4, ed B.I. Verkin (Kharkov: FTINT AN Ukr. SSR), 172, (in Russian).

[17] Peresada, V. I., Afanas'ev, V. N., & Borovikov, V. S. (1975). On Calculation of Density of States of Single-Magnon Perturbations in Ferromagnetics. *Soviet Low Temperature Physics*, 1(4), 227-232.

[18] Skrypnyk, Yu. V., & Loktev, V. M. (2006). Impurity Effects in a Two-Dimensional Systems with the Dirac Spectrum. *Physical Review B*, 73, 241402-6.

[19] Skrypnyk, Yu. V., & Loktev, V. M. (2008). Spectral Function of Graphene with Short-Range Imourity Centers. *Low Temperature Physics*, 34(9), 818-825.

[20] Bena, C., & Kivelson, S. A. (2005). Qusiparticle Scattering and Local Density of States in Graphite. *Physical Review B*, 72(12), 125432-7.

[21] Lifshits, I. M. (1945). On the Theory of regular Perturbations Report of AS SSSR (in Russian). 48, 83-86.

[22] Feher, A., Syrkin, E. S., Feodosyev, S. B., Gospodarev, I. A., Manzhelii, E. V., Kotlyar, A. V., & Kravchenko, K. V. (2011). The Features of Low Frequency Atomic Vibrations and Properties of Acoustic Waves in Heterogeneous Systems. In: Vila R. (ed.), *Waves in Fluids and Solids*, Rijeka, InTech, 103-126.

[23] Kotlyar, A. V., & Feodosyev, S. B. (2006). Local Vibrational Modes in Crystal Lattices with a Simply Connected Region of the Quasi-continuous Phonon Spectrum. *Low Temperature Physics*, 32(3), 256-269.

[24] Gospodarev, I. A., Eremenko, V. V., Kravchenko, K. V., Sirenko, V. A., Syrkin, E. S., & Feodos'ev, S. B. (2010). Vibrational Charakteristics of Niobium Diselenide and Graphite Nanofilms. *Low Temperature Physics*, 36(4), 344-350.

[25] Novoselov, K. S. (2010). Grafen: Materialy Flatlandii. *Uspehi Fizicheskih Nauk (in Russian)*, 181(12), 1299-1311.

[26] Dahal Hari, P., Balatsky, A. V., & Zhu, J-X. (2008). Tuning Impurity States in Bilayer Graphene. *Physical Review*, 77(11), B 115114, 1-10.

[27] Wang, Z. F., Li, Q., Su, H., Wang, H., Shi, Q. W., Chen, J., Yang, J., & Hou, J. G. (2007). Electronic Sructure of Bilayer Graphene; Areal-Space Green's Function Study. *Physical Review*, 78(8), B 085424, 1-8.

[28] Castro, E. V., Peres, N. M., Lopes dos Santos, J. M. B., Castro Neto, A. H., & Guinea, F. (2008). Localized States at Zigzag Edges of Bilayer Graphene. *Physical Review Letters*, 100(26), 026802, 1-4.

Section 2

Experimental Aspect

Chapter 6

Quantum Transport in Graphene Quantum Dots

Hai-Ou Li, Tao Tu, Gang Cao, Lin-Jun Wang,
Guang-Can Guo and Guo-Ping Guo

Additional information is available at the end of the chapter

http://dx.doi.org/10.5772/52873

1. Introduction

Graphene exhibits unique electrical properties and offers substantial potential as building blocks of nanodevices owing to its unique two-dimensional structure (Geim et al., 2007; Geim et al., 2009; Ihn et al., 2010). Besides being a promising candidate for high performance electronic devices, graphene may also be used in the field of quantum computation, which involves exploration of the extra degrees of freedom provided by electron spin, in addition to those due to electron charge. During the past few years, significant progress has been achieved in implementation of electron spin qubits in semiconductor quantum dots (Hanson et al., 2007; Hanson et al., 2008). To realize quantum computation, the effects of interactions between qubits and their environment must be minimized (Fischer et al., 2009). Because of the weak spin-orbit coupling and largely eliminated hyperfine interaction in graphene, it is highly desirable to coherently control the spin degree of freedom in graphene nanostructures for quantum computation (Trauzettel et al., 2007; Guo et al., 2009). However, the low energy quasiparticles in single layer grapheme behave as massless Dirac fermions (Geim et al., 2007; Geim et al., 2009), and the relativistic Klein tunneling effect leads to the fact that it is hard to confine electrons within a small region to form quantum dot in graphene using traditional electrostatical gates (Ihn et al., 2010; Trauzettel et al., 2007). It is now possible to etch a grapheme flake into nano-constrictions in size, which can obtain electron bound states and thus act as quantum dots. As a result, usually a diamond-like characteristic of suppressed conductance consisting of a number of sub-diamonds is clearly seen (Stampfer et al., 2009; Gallagher et al., 2010), indicating that charge transport in the single graphene quantum dot device may be described by the model of multiple graphene quantum dots in series along the nanoribbon. The formation of multiple quantum dot structures in the nanoribbons may be attributed to edge roughness or local potential. The rough edges also lift the valley degeneracy, which could suppress the exchange coupling between spins in the grapheme quantum dots (Trauzettel et al., 2007; Ponomarenko et al., 2008). Recently, there was a striking advance on experimental production of graphene single (Ponomarenko et al., 2008; Stampfer et al., 2008a; Stampfer et al., 2008 b; Schenz et al., 2009; Wang et al., 2010; Guttinger et al., 2011) or double quantum dots (Molitor et al., 2009; Molitor et al., 2010; Liu et al., 2010; Wang et al., 2011a; Volk et al., 2011; Wang et al., 2012;) which is an important first step towards such promise.

In this chapter, we introduce the design and fabrication of etched gate tunable single and double quantum dots in single-layer and bilayer graphene and present several important quantum transport measurements in these systems. A quantum dot with an integrated charge sensor is becoming a common architecture for a spin or charge based solid state qubit. To implement such a structure in graphene, we have fabricated a twin-dot structure in which the larger QD serves as a single electron transistor (SET) to read out the charge state of the nearby gate controlled small QD. A high SET sensitivity allowed us to probe Coulomb charging as well as excited state spectra of the QD, even in the regime where the current through the QD is too small to be measured by conventional transport means (Wang et al., 2010; Wang et al., 2011b). We also have measured quantum transport properties of gates controlled parallel-coupled double quantum dots (PDQD) and series-coupled double quantum dots (SDQD) device on both single layer and bilayer graphene (Wang et al., 2011a; Wang et al., 2012). The inter-dot coupling strength can be effectively tuned from weak to strong by in-plane plunger gates. All the relevant energy scales and parameters can be extracted from the honeycomb charge stability diagrams. We precisely extract a large inter-dot tunnel coupling strength for the series-coupled quantum dots (SDQD) allowing for the observation of tunnel-coupled molecular states extending over the whole double dot. The present method of designing and fabricating graphene QD is demonstrated to be general and reliable and will enhance the realization of graphene nanodevice and desirable study of rich QD physical phenomena in grapheme. These results demonstrate that both single and double quantum dots in single-layer and bilayer graphene bode well for future quantum transport study and quantum computing applications. The clean, highly controllable systems serves as an essential building block for quantum devices in a nuclear-spin-free world.

2. A graphene quantum dot with a single electron transistor as an integrated charge sensor

The measurement of individual electrons or its spins in GaAs quantum dots (QDs) has been realized by so-called charge detection via a nearby quantum point contact (QPC) or single electron transistor (SET) (Lu et al., 2003; Elzerman et al., 2004a). In particular, the combination of high speed and high charge sensitivity has made SET useful in studying a wide range of physical phenomena such as discrete electron transport (Lu et al., 2003; Bylander et al., 2005; Gotz et al., 2008), qubit read out (Lehnert et al., 2003; Duty et al., 2004; Vijay et al., 2009) and nanomechanical oscillators (Knobel et al., 2003; Lahaye et al., 2004). So far, most SETs have been using $Al/AlO_x/Al$ tunnel junctions. However, the graphene SET reported here is technologically simple and reliable, making it an attractive substitute for use in various charge detector applications.

In this section, we realize an all graphene nanocircuit integration with a SET as charge read out for a QD. In conventional semiconductor systems, the gate-defined structure limits the distance between the QD and the detector. However, in our device reported here, the QD and the SET in the same material are defined in a single etching step, and the distance between the graphene nanostructures is determined by the etched area, which enables optimized coupling and sensing ability. The SET is placed in close proximity to the QD giving rise to a strong capacitive coupling between the two systems. Once an additional electron occupies the QD, the potential in the neighboring SET is modified by capacitive interaction that gives rise to a measurable conductance change. Even if charge transport through the QD is too small to be measured by conventional transport means, the SET charge sensor also allows measurements. These devices demonstrated here provide robust building blocks in a practical quantum information processor.

The graphene flakes were produced by mechanical cleaving of graphite crystallites by Scotch tape and then were transferred to a highly doped Si substrate with a 100 nm SiO_2 top layer. Thin flakes were found by optical microscopy, and single layer graphene flakes were selected by the Raman spectroscopy measurement. We used the standard electron beam lithography and lift off technique to make the Ohmic contact (Ti/Au) on the present graphene devices. Next, a new layer of polymethyl methacrylate is exposed by electron beam to form a designed pattern. Then, the unprotected areas are removed by oxygen reactive ion etching. One of our defined sample structures with a quantum dot and proximity SET is shown in Fig. 1. The quantum dot is an isolated central island of diameter 90 nm, connected by 30nm wide tunneling barriers to source and drain contacts. Here, the Si wafer was used as the back gate and there is also a graphene side gate near the small dot. The SET has a similar pattern while the conducting island has a much larger diameter (180nm). Electronic transport through both the devices exhibits Coulomb blockade (CB) characteristics with back/side gate voltage. The distance between the CB peaks is determined by the sum of charging and quantum confinement energies, and the former contribution becomes dominant for our devices with diameter >100nm (Kouwenhoven et al., 1997). Accordingly, we refer to it as a SET rather than a QD. The device was first immersed into a liquid helium storage Dewar at 4.2K to test the functionality of the gates. The experiment was carried out in a top-loading dilution refrigerator equipped with filtered wiring and low-noise electronics at the base temperature of 10 mK. In the measurement, we employed the standard ac lock-in technique.

Fig. 2(a) shows the conductance through the dot Al/AlO_x/Al for applied side gate voltage V_{sg}. Clear CB peaks are observed related to charging of the tunable dot on the graphene. The dashed green lines in the range of 0.2-0.7V for side gate voltages show that the current through the dot becomes too small to be seen clearly. Fig. 2(b) shows the conductance through the SET versus side gate voltage V_{sg}. The SET is as close as possible to the QD and in this way charging signals of the dot were detected by tracking the change in the SET current. The addition of one electron to the QD leads to a pronounced change of the conductance of the charge detector by typically 30%. The slope of the SET conductance is the steepest at both sides of its CB resonances giving the best charge read-out signal. To offset the large current background, we used a lock-in detection method developed earlier for GaAs dot (Elzerman et al., 2004b). A square shaped pulse was superimposed on the dc bias on side gate voltage V_{sg}. A lock-in detector in sync with the pulse frequency measured the change of SET current due to the pulse modulation. Fig. 2(c) shows a typical trace of the lock-in signal of the transconductance through the SET dI_{SET}/dV_{sg}. These sharp spikes or dips originate from the change of the charge on the dot by one electron. It shows essentially the same features as Fig. 2(a), but is much richer, especially in the regime where the direct dot current is too small to be seen clearly. The vertical dashed lines in Fig. 2 illustrate the SET sensor signals correspond to the QD transport measurements perfectly and indicate the SET is reliable. We also note that the individual charge events measurement has been demonstrated in a graphene QD with a QPC detector based on graphene nanoribbon (Güttinger et al., 2008).

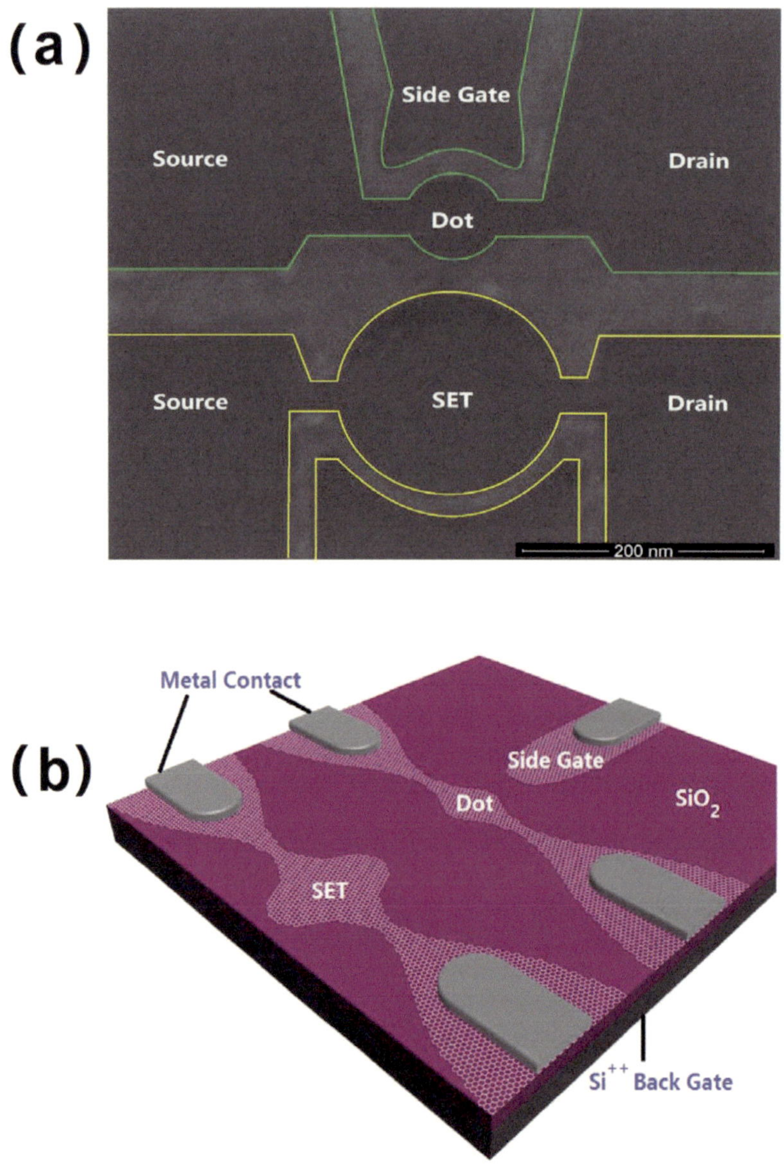

Figure 1. a) Scanning electron microscope image of the etched sample structure. The bar has a length of 200 nm. The upper small quantum dot as the main device has a diameter of 90 nm while the bottom single electron transistor as charge sensor has a diameter of 180 nm. The bright lines define barriers and the graphene side gate. (b) Schematic of a representative device.

http://dx.doi.org/10.5772/52873

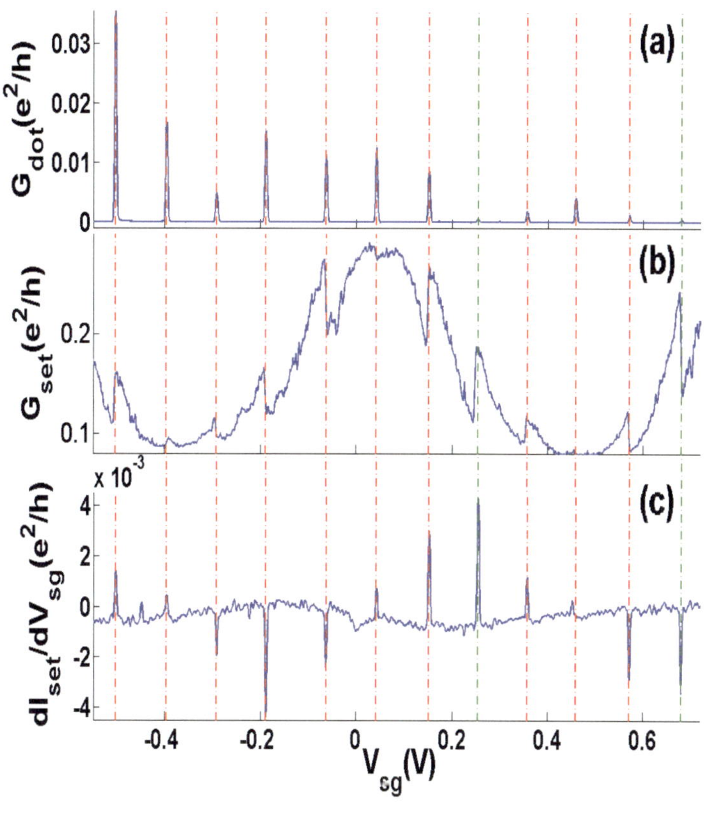

Figure 2. a) Conductance through the quantum dot vs the side gate voltage. (b) The example of conductance through the single electron transistor for the same parameter ranges as in panel (a). The steps in conductance have about 30% change of the total signal and are well aligned with the CB in panel (a). (c) Transconductance of the single electron transistor for the same parameters as in panel (a). The spikes and dips indicate the transitions in the charge states by addition of single electron in quantum dot. In particular, the dashed green lines show that the charge detection can allow measurement in the regime where the current through the dot is too small to be seen clearly by direct means. The vertical dashed red lines are a guide for the eyes to relate features in these graphs.

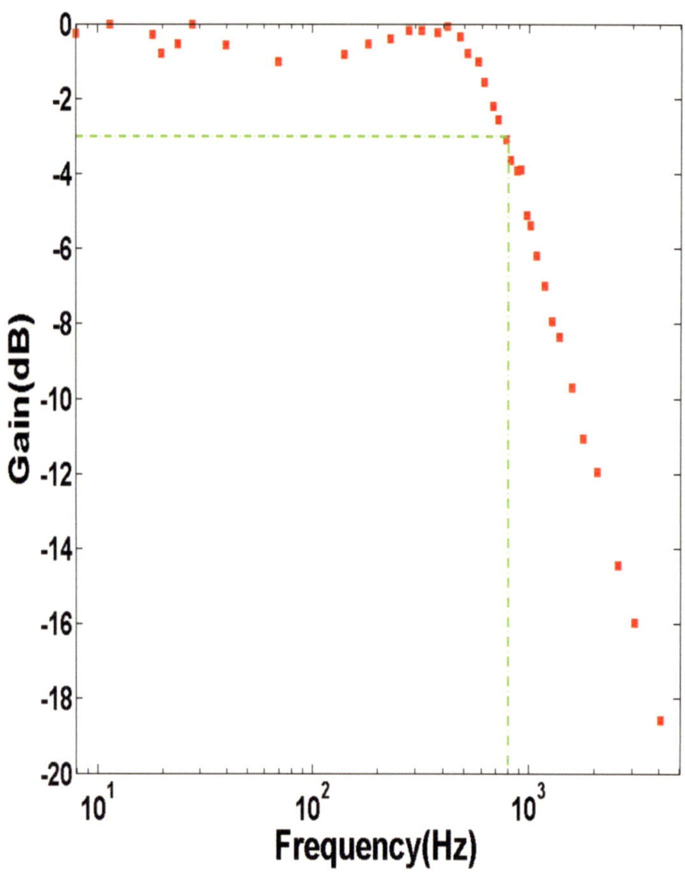

Figure 3. The magnitude of the SET signal dI_{SET}/dV_{sg} as a function of the modulating pulse frequency. The dashed green line illustrates that the bandwidth of the SET device is about 800 Hz corresponding to a gain of 0.707(-3 dB). Due to the stray capacitances, the response decreases rapidly after 800 Hz.

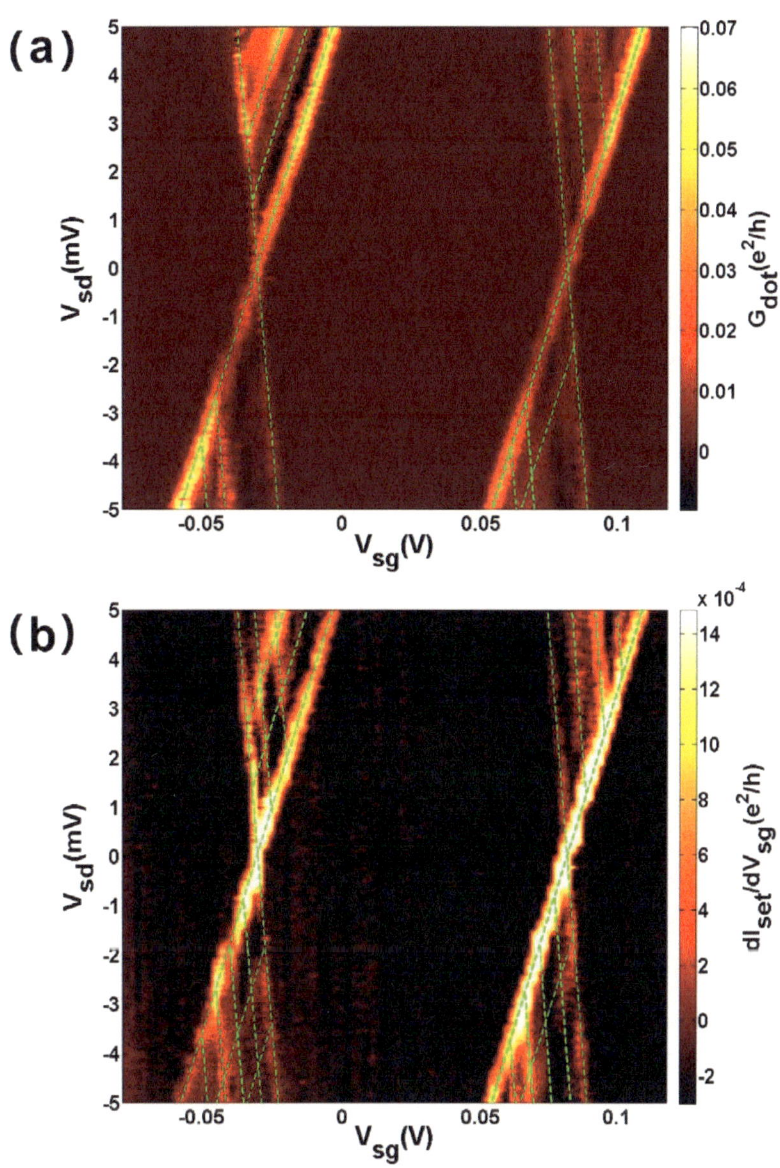

Figure 4. a) Plot of the differential conductance of the quantum dot as a function of the bias voltage and the side gate voltage applied on the dot. From the lines parallel to the edges of Coulomb diamonds, we can identify the excited states. (b) Transconductance of the single electron transistor with the same parameters as in panel (a). Perfect matching with panel (a) and resolving more excited states spectra indicate that the single electron transistor can be used as a highly sensitive charge detector. Data in panels (a) and (b) were recorded simultaneously during a single sweep. Dashed green lines are the guide for identifying the excited states.

More quantitative information on the system can be obtained from the measurement of the height response of the peak at 0.152 V in Fig. 2(c) as a function of the modulating pulse frequency on the side gate. The resulting diagram for the SET dI_{SET}/dV_{sg} gain magnitude is shown in Figure 3.The dashed green line indicates the gain of 0.707 (-3 dB), corresponding approximately to the bandwidth of 800 Hz of the SET device. By applying a signal of 5×10^{-2} electrons on back gate of the SET and measuring the signal with a signal-to-noise ratio of 1, we achieved a charge sensitivity $10^{-3}e/\sqrt{Hz}$of which is similar to that obtained previously in a GaAs QD and superconducting Al SET detector system (Berman et al., 1999). The system can be simply considered as a resistor-capacitor circuit (RC circuit), and the bandwidth is limited by the resistor and capacitance of the cable connecting the SET and the room temperature equipments. As a result, we would expect the bandwidth can be greatly improved by adding a cold amplifier (Vink et al., 2007). It is also expected that adding a side gate near the SET to independently set the SET operating point to about 25 kΩ can obviously enhance the bandwidth.

The information contained in the signal goes beyond simple charge counting. For instance, the stability diagram measurement can reveal excited states, which is crucial to get information of the spin state of electrons on a quantum dot (Hanson et al., 2003). Fig. 4(a) shows Coulomb diamonds for the conductance through the dot GQD versus bias voltage V_{sd} and side gate voltage V_{sg}. For comparison, Fig. 4(b) shows the transconductance of the SET dI_{SET}/dV_{sg} as a function of the same parameters. A perfect match between the QD transport measurements and the detector signal is observed. Moreover, the discrete energy spectra of the graphene quantum dot are revealed by the presence of additional lines parallel to the diamond edges. These lines indicate the quantum dot is in the high bias regime where the source-drain bias is high enough that the excited states can participate in electron tunneling (Hanson et al., 2007). The excited states become much more visible in the SET charge detector signal than the direct measurement. All of these features have been seen in GaAs QD with QPC (Hanson et al., 2007), but here we achieve the goal with an all graphene nanocircuit of QD with SET. In the previous reports, the QD and QPC detector are separated by typically 100 nm in width. In the present case, the SET detector is 50 nm from the edge of the QD. Therefore it is expected that the capacitance coupling between QD and SET is enhanced compared to the conventional case realized in semiconductor QD and QPC. This enhanced coupling leads to a larger signal-to-noise ratio of the SET detector signal that can be exploited for time resolved charge measurement or charge/spin qubit read out on the QD.

In summary, we have presented a simple fabrication process that produces a quantum dot and highly sensitive single electron transistor charge detector with the same material, graphene. Typically the addition of a single electron in QD would result in a change in the SET conductance of about 30%. The charging events measured by both the charge detector and direct transport through the dot perfectly match and more excited states information beyond the conventional transport means is also obtained. The devices demonstrated here represent a fascinating avenue towards realizing a more complex and highly controllable electronic nanostructure formed from molecular conductors such as graphene.

3. Controllable tunnel coupling and molecular states in a graphene double quantum dot

Previously, the charge stability diagram in coupled quantum-dot systems has been studied by the classical capacitance model (van der Wiel et al., 2003). However the quantum effect should also manifest itself (Yang et al., 2011). In particular, the tunnel coupling t between the two dots in a double dot is an important quantity, because it can affect the geometry of the overall charge stability diagram. Furthermore, several different spin qubit operations can be performed by controlling this tunnel rate as a function of time. For approaches based on single electron spin qubit, utilizing t enables the$\sqrt{\mathrm{SWAP}}$gate operations between two qubits (Petta et al., 2005). In an architecture in which each qubit is composed of two-electron single-triplet states, control of t in the presence of a non-uniform magnetic field enables universal single qubit rotations (Foletti et al., 2009).

In this Section, we report an experimental demonstration and electrical transport measurement in a tunable graphene double quantum dot device. Depending on the strength of the inter-dot coupling, the device can form atomic like states on the individual dots (weak tunnel coupling) or molecular like states of the two dots (strong tunnel coupling). We also extract the inter-dot tunnel coupling t by identifying and characterizing the molecule states with wave functions extending over the whole graphene double dot. The result implies that this artificial grapheme device may be useful for implementing two-electron spin manipulation.

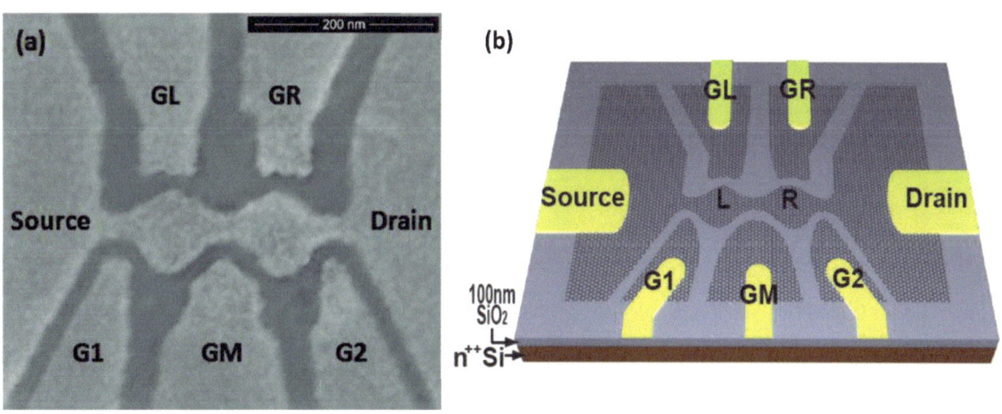

Figure 5. a) Scanning electron microscope image of the structure of the designed multiple gated sample studied in this work. The double quantum dot have two isolated central islands of diameter 100 nm in series, connected by 20×20 nm tunneling barriers to source and drain contacts (S and D) and 30×20 nm tunneling barrier with each other. These gates are labeled by G1, GL, GM, GR, G2, in which gate GM, G1 and G2, are used to control the coupling barriers between the dots as well as the leads. Gates GL and GR are used to control and adjust the energy level of each dot. (b) Schematic of a representative device.

A scanning electron microscope image of our defined sample structure with double quantum dot is shown in Fig. 5(a) and Fig. 5(b). The double quantum dot has two isolated central island of diameter 100 nm in series, connected by 20× 20 nm narrow constriction to source and drain contacts (S and D electrodes) and 30× 20 nm narrow constriction with each other. These constrictions are expected to act as tunnel barriers due to the quantum size effect. In addition, the highly P-doped Si substrate is used as a back gate and five lateral side gates, labeled the left gate G1, right gate G2, center gate GM and GL(R), which are expected for local control. All of side gates are effective, in which gates GL, GR and G2 have very good effect on two dots and middle barrier, while gates G1 and GM have weak effect on those. The device was first immersed into a liquid helium storage dewar at 4.2K to test the functionality of the gates. The experiment was carried out in a He3 cryostat equipped with filtered wiring and low-noise electronics at the base temperature of 300mK. In the measurement, we employed the standard AC lock-in technique with an excitation voltage 20 μVat 11.3 Hz.

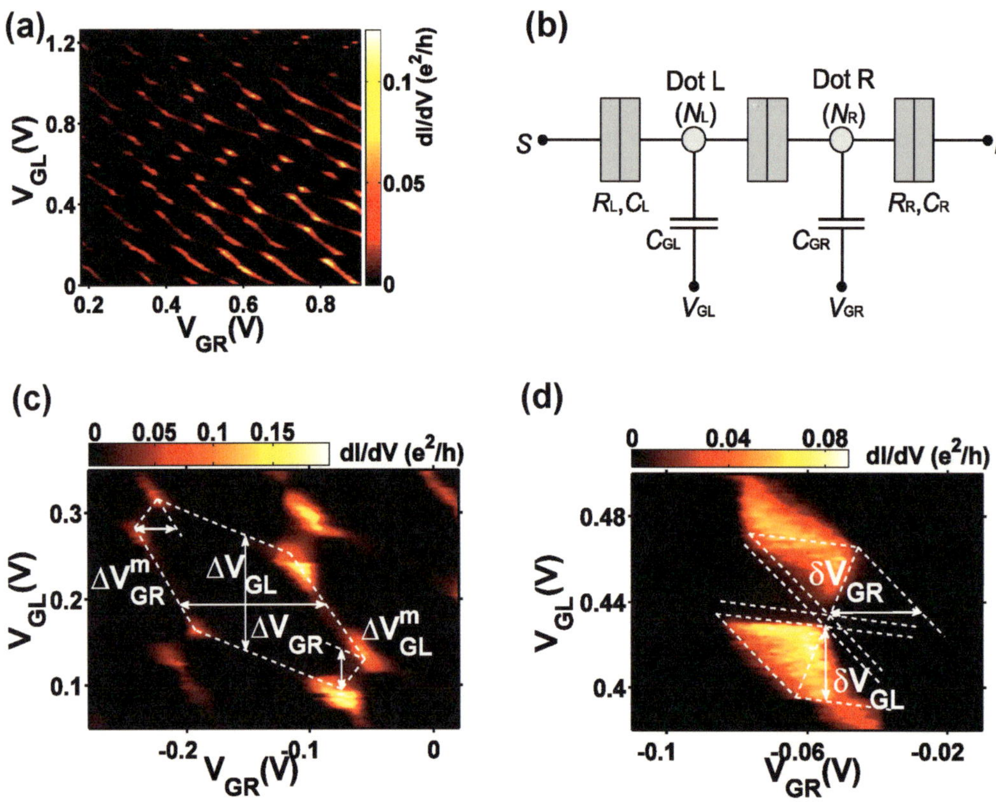

Figure 6. a) Color scale plot of the differential conductance versus voltage applied on gate GL (V_{GL}) and gate GR (V_{GR}) at V_{sd}= 20μV, V_{G1} = 0V, V_{GM} = 0V, V_{G2} = 0V and V_{bg} = 2.5 V. The honeycomb pattern we got stands for the typical charge stability diagram of coupled double quantum dots. (b) Pure capacitance model of a graphene double dot system. Zoom in of a honeycomb structure (c) and a vertex pair (d) at V_{sd} = 900μV.

Fig. 6(a) displays the differential conductance through the graphene double quantum dot circuit as a function of gate voltages V_{GL} and V_{GR}. Here the measurement was recorded at

V_{sd}= 20μV, V_{G1} = 0V, V_{GM} = 0V, V_{G2} = 0V and V_{bg} = 2.5 V. The honeycomb pattern is clearly visible and uniforms over many times. Each cell of the honeycomb corresponds to a well-defined charge configuration (NL,NR) in the nearly independent dots, where NL and NR denote the number of electrons on the left and right dot, respectively. The conductance is large at the vertices, where the electrochemical potentials in both dots are aligned with each other and the Fermi energy in the leads and resonant sequential tunneling is available. These vertices are connected by faint lines of much smaller conductance along the edges of the honeycomb cells. At these lines, the energy level in one dot is aligned with the electrochemical potential in the corresponding lead and inelastic cotunneling processes occur. The observed honeycomb pattern resembles the charge stability diagram found for weakly coupled GaAs double quantum dot (van der Wiel et al., 2003). Such similarities indicate that graphene quantum dot devices will continue to share features with well-studied semiconductor quantum dot systems. The energy-level statistics of single graphene quantum dot was probed and shown to agree well with the theory of chaotic Dirac billiards (Ponomarenko et al., 2008). It is interesting and important to know whether these Dirac fermions' behaviors can be realized and observed in grapheme double quantum dot. Nevertheless, it will be studied in the future work.

More quantitative information such as double dot capacitances can be extracted using a electrostatic model as shown in Fig. 6(b) (van der Wiel et al., 2003). First, the capacitance of the dot to the side gate can be determined from measuring the size of the honeycomb in Fig. 6(c) as $C_{GL} = e / \Delta V_{GL} \approx 1.27$aF and$C_{GR} = e / \Delta V_{GR} \approx 1.49$aF. Next, the capacitance ratios can be determined from measuring the size of the vertices in Fig. 6(d) at finite bias $V_{sd} = 900\mu$V as $\alpha_L = | V_{sd} | / \delta V_{GL} = 0.029$ and$\alpha_R = | V_{sd} | / \delta V_{GR} = 0.035$. Using the relation $C_{GL} / C_L = \alpha_L$ and $C_{GR} / C_R = \alpha_R$, we can obtain the typical values of dot capacitances as $C_L = 44.8$aFand $C_R = 44.1$aF, respectively. The amount of interdot coupling can be achieved by measuring the vertices splitting in Fig. 6(c). Assuming the capacitively coupling is dominant in the weakly coupled dots regime (van der Wiel et al., 2003; Mason et al., 2004), the mutual capacitance between dots is calculated as

$$C_M = \Delta V_{GL}^{m} C_{GL} C_R / e = \Delta V_{GR}^{m} C_{GR} C_L / e = 9.2\text{aF}$$

It has been expected that opening the interdot constriction by gate voltage will cause the tunnel coupling to increase exponentially faster than the capacitive coupling (Kouwenhoven et al., 1997). Fig. 7(a)-(c) represent a selection of such measurements by holding the same V_{GR} and V_{bg} and scanning different ranges of V_{GL} between -0.5 V to 0.35 V. An evolution of conductance pattern indicates that the stability diagram changes from weak to strong tunneling regimes (van der Wiel et al., 2003; Mason et al., 2004). The conductance near the vertices depends on the relative contributions of the capacitive coupling and tunnel coupling. For the former, the vertices become a sharpened point, while for the latter, the vertices become blurred along the edges of the honeycomb cell (Graber et al., 2007). In Fig. 7(b), the vertices is not obvious as those in Fig. 7(a), which indicates a stronger tunnel coupling. The results suggest that two graphene dots are interacting with each other through the large quantum mechanical tunnel coupling, which is analogous to covalent bonding. We will analyze it in details below. An increase in inter-dot coupling also leads to much larger separation of vertices in Fig. 7(b) (Mason et al., 2004), and finally, to a smearing of honeycomb features in Fig. 7(c). In this case, the double dots behave like a single dot, as illustrated in Fig. 7(g). We note that a similar evolution is observed for four different values of V_{bg} from 2.5 V to 2.0 V at the same V_{GL} and V_{GR} regimes as shown in Fig. 7(d)-(f). Thus the inter-dot tunnel coupling could also be changed by V_{GL} or V_{bg}. This can be explained by the fact that the side gates and back gate may influence the central barrier through the existing capacitances between the gates and the central barrier.

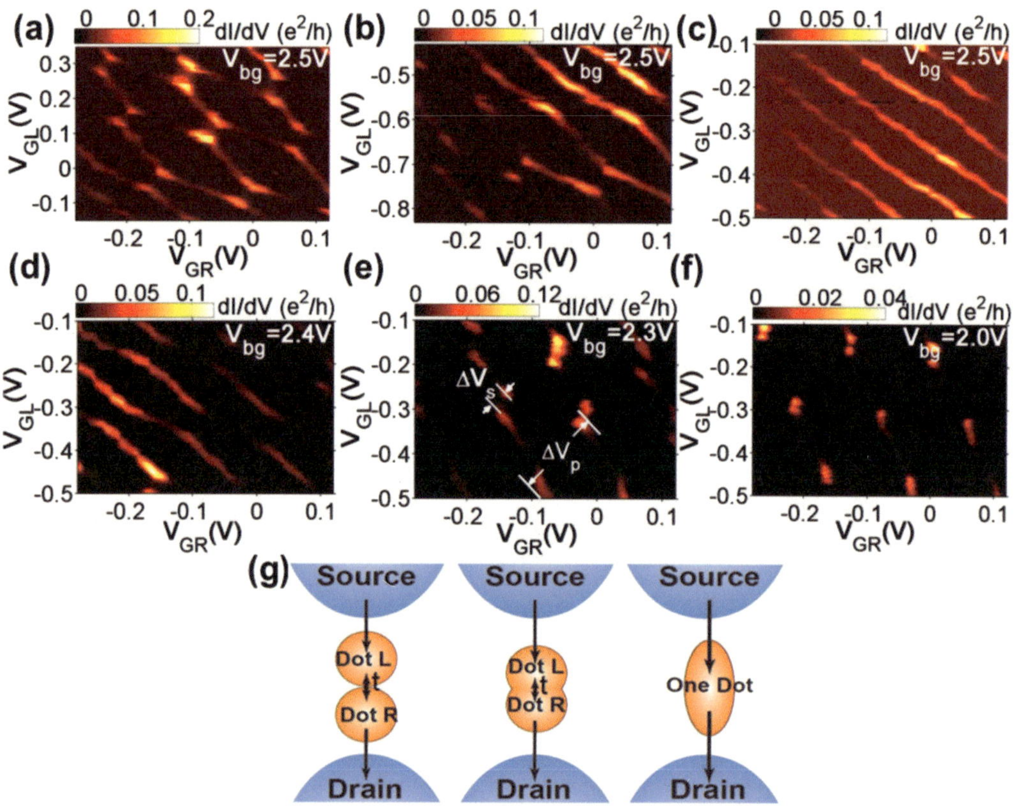

Figure 7. a)-(c) Colorscale plot of the differential conductance versus voltage applied on gate GL (V_{GL}) and gate GR (V_{GR}) at V_{bg} = 2.5 V for different V_{GL} regimes. (d)-(f) Color scale plot of the differential conductance versus voltage applied on gate L (V_{GL}) and gate R(V_{GR}) for different back gate voltage V_{bg}. The trend of interdot tunnel coupling changing from weak to strong can be seen clearly. (g) Sketches of the characteristic electronic configurations with interdot tunnel coupling *t*.

Similar to the definitions in Ref. (Livermore et al., 1996), we define $f = 2\Delta V_S / \Delta V_P$ with ΔV_S representing the splitting between vertices in the diagonal direction and ΔV_P the vertex pairs distance (Fig. 7(e)). Thus, the case $f = 1$ stands for strong coupling limit where the double dots behave like a single dot, while the case $f = 0$ represents weak coupling limit where the double dots behave like two isolated dots. This way, f should have a certain relationship with tunnel couplings which offers us a method to measure the contribution of the interdot tunneling to the splitting of the vertex. In our double-dot sample, a clear evolution of f is obtained through scanning different regimes of V_{GL} with fixed V_{GR} (Fig. 7(a)-(c)). Through extracting ΔV_S and ΔV_P, we get $f = 0.5$ for (a) and $f = 0.65$ for (b) and $f = 1$ for (c) respectively. These values indicate that control of tunnel coupling as a function of such a gate voltage is conceivable.

Having understood the qualitative behavior of the graphene device in the strong coupling regime, we extract the quantitative properties based on a quantum model of graphene artificial molecule states (Yang et al., 2003; Graber et al., 2007; Hatano et al., 2005). Here we only take into account the topmost occupied state in each dot and treat the other electrons as an inert core (van der Wiel et al., 2003; Golovach et al., 2004). In the case of neglected tunnel coupling, the nonzero conductance can only occur right at the vertices which are energy degenerate points as $E(N_L + 1, N_R) = E(N_L, N_R + 1)$. When an electron can tunnel coherently between the two dots, the eigenstates of the double dot system become the superposed states of two well-separated dot states with the form

$$|\psi_B\rangle = -\sin\frac{\theta}{2}e^{-\frac{i\varphi}{2}}|N_L+1, N_R\rangle + \cos\frac{\theta}{2}e^{\frac{i\varphi}{2}}|N_L, N_R+1\rangle \tag{1}$$

$$|\psi_A\rangle = \cos\frac{\theta}{2}e^{-\frac{i\varphi}{2}}|N_L+1, N_R\rangle + \sin\frac{\theta}{2}e^{\frac{i\varphi}{2}}|N_L, N_R+1\rangle \tag{2}$$

Where $\theta = \arctan(\frac{2t}{\varepsilon})$, $\varepsilon = E_L - E_R$, E_L and E_R are the energies of state $|N_L + 1, N_R\rangle$ and $|N_L, N_R + 1\rangle$, respectively. Thus $|\psi_B\rangle$ and $|\psi_A\rangle$ are the bonding and anti-bonding state in terms of the uncoupled dot, and the energy difference between these two states can be expressed by

$$E_\Delta = U' + \sqrt{\varepsilon^2 + (2t^2)} \tag{3}$$

Here $U' = \frac{2e^2 C_m}{C_L C_R - C_m^2}$ is the contribution from electrostatic coupling between dots (Ziegler et al., 2000).

Provided that the graphene double-dot molecule eigenstate $|\psi\rangle$ participates in the transport process, sequential tunneling is also possible along the honeycomb edges. In Fig. 8(a) and Fig. 8(b), a colorscale plot of the differential conductance is shown at $V_{sd} = 20\ \mu$V in the vicinity of a vertex. As expected the visible conductance is observed at both the position of the vertex and the honeycomb edges extending from the vertex. Fig. 8(c) shows a fit of the energy difference E_Δ from the measured mount of splitting of the positions of the differential conductance resonance peak in the ε-direction. Here we use $\varepsilon = E_L - E_R = e\alpha_L V_{GL} - e\alpha_R V_{GR}$ to translate the gate voltage detuning $V_{GL} - V_{GR}$ with the conversion factors α_L and $\alpha_R$determined above. The fitting with Eq. (3) yields the values of tunnel coupling strength $t \approx 727\mu$eVandU$' \approx 209\mu$eV. Similar measurements have been performed in a carbon nanotube double dots with $t \approx 358\mu$eV and U$' \approx 16\mu$eV (Graber et al., 2007) and semiconductor double dots with $t \approx 80\mu$eV and U$' \approx 175\mu$eV (Hatano et al., 2005). The fact that the tunnel coupling t is dominant than capacitive coupling U$'$implies the inter-dot tunnel barrier in the etched grapheme double dot is much more transparent than those gated carbon nanotube or semiconductor double dot.

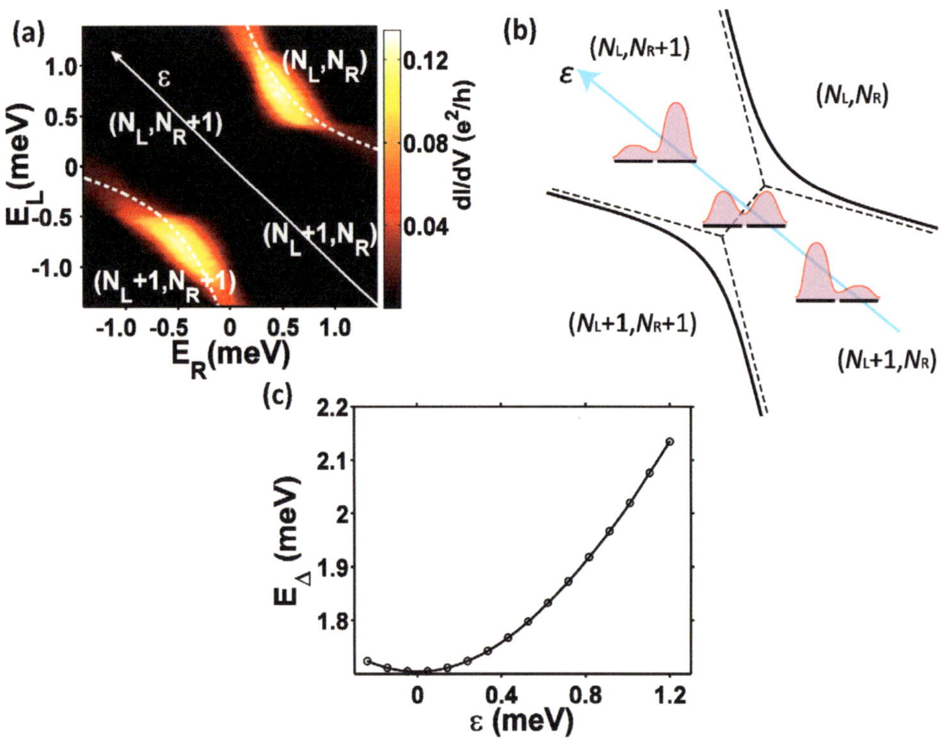

Figure 8. a) Colorscale plot of the differential conductance versus the energies of each dotE_L and E_R at V_{sd} = 20 μV near the selected two vertices with dashed lines as guides to the eye. (b) Schematic of a single anticrossing and the evolution from the state localized in each dot to a molecule state extending across both dots (Hatano et al., 2005). (c) $E_\Delta$dependence of the detuning $\varepsilon = E_L - E_R$ E_Δ(circles) is measured from the separation of the two high conductance wings in Fig. 4(a). The line illustrates a fit of the data to Eq. (3).

Finally, we discuss the relevance of graphene double dot device for implementing a quantum gate and quantum entanglement of coupled electron spins. A $\sqrt{\text{SWAP}}$ operation has already been demonstrated in a semiconductor double dot system using the fast control of exchange coupling J(Petta et al., 2005). The operation time$\tau$is about 180 ps for $J \approx 0.4$meV corresponding to$t \approx 0.16$meV. In the present graphene device, we have obtained much larger $t \approx 0.72$meV and the estimated $\tau \approx 50$ps is much shorter than the predicted decoherence time

(μs) (Fischer et al., 2009). The results indicate the ability to carry out two-electron spin operations in nanosecond timescales on a graphene device, four times faster than perviously shown for semiconductor double dot.

In summary, we have measured a graphene double quantum dot with multiple lectrostatic gates and observed the transport pattern evolution in different gate configurations. This way offers us a method to identify the molecular states as a quantum-mechanical superposition of double dot and measure the contribution of the interdot tunneling to the splitting of the differential conductance vertex. The precisely extracted values of inter-dot tunnel coupling t for this system is much larger than those in previously reported semiconductor device. These short operation times due to large tunneling strength together with the predicted very long coherence times suggest that the requirements for implementing quantum information processing in graphene nanodevice are within reach.

4. Gates controlled parallel-coupled double quantum dot on both single layer and bilayer grapheme

In contrast to DQD in series, where the applied current passes through the double dot in serial, the parallel-coupled double quantum dot (PDQD) requires two sets of entrances and exits, one for each dot. In addition, the source and drain must maintain coherence of the electron waves through both dots, in a manner analogous to a Young's double slit. Thus PDQD is an ideal artificial system for investigating the interaction and interference. Rich physical phenomena, such as Aharonov- Bohm (AB) effect, Kondo regimes and Fano effect, have been predicted to be observed in parallel DQD (Holleitner et al., 2001; Lo´pez et al., 2002; Ladro´n de Guevara et al., 2003; Orellana et al., 2004; Chen et al., 2004). Particularly excitement is the prospect of accessing theoretically predicted quantum critical points in quantum phase transitions (Dias da Silva et al., 2008). The grapheme PDQD is an attractive system for investigating the quantum phase transitions due to its intrinsically large energy separation between on-dot quantum levels, thus offering a significant advantage over conventional systems as GaAs or silicon based quantum dots.

In this section, we present the design, fabrication, and quantum transport measurement of double dot structure coupled in parallel, on both bilayer and single layer grapheme flakes, which may open a door to study the rich PDQD physical phenomena in this material the parallel graphene structure can be tuned from a strong-coupling resulted artificial molecule state to a weak-coupling resulted two-dot state by adjusting in plane plunger gates. The tuning is found to be very reliable and reproducible, with good long-term stability on the order of days.

Graphene flakes are produced by mechanical cleaving of bulk graphite crystallites by Scotch tape (Novoselov et al., 2004). For this kind of exfoliated graphene flakes on SiO_2 substrate, the mobility is normally about 15000 $cm^2/(Vs)$ (Geim et al., 2007). By using heavily doped Si substrate with 100 nm thick SiO_2 on top, we can identify monolayer, bilayer, and few layer graphenes through optical microscope. Monolayer and bilayer graphenes were further checked by Raman spectrum. Firstly, graphene flakes are transferred to the substrate with gold markers. Then, a layer of 50nm thick polymethyl methacrylate (PMMA) is spun on the substrate for electron beam lithography (EBL) to form a designed pattern. After that, O_2/Ar (50:50) plasma is used to remove unprotected parts of graphene. Next, an area of over exposed PMMA is used to separate a bridge plunger gate from the drain part of graphene (Chen et al., 2004; Huard et al., 2007). The final step is to make the metal contacts, which are defined by the standardized EBL process, followed by the E-beam evaporation of Ti/Au (2 nm/50 nm).

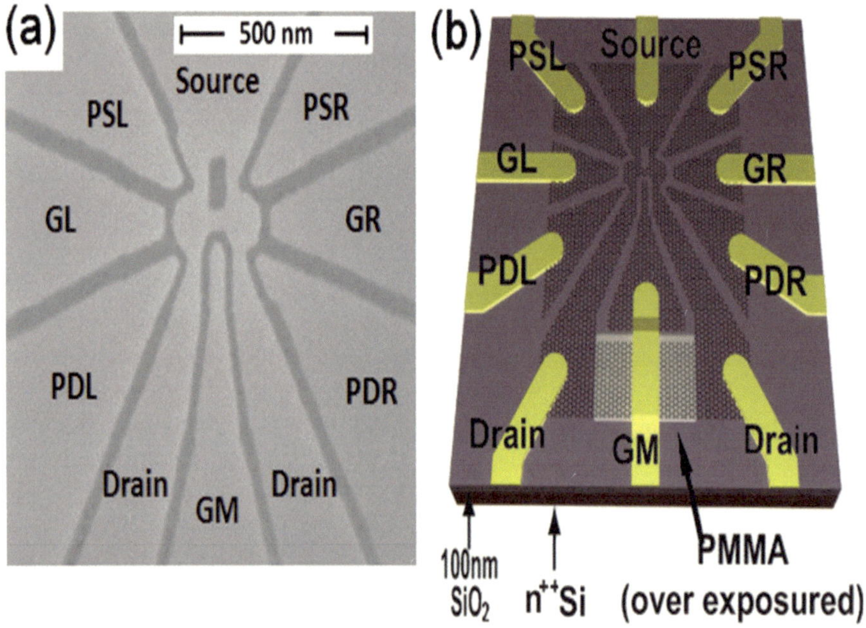

Figure 9. Color online) (a) Scanning electron microscope image of the etched parallel coupled graphene double dot sample structure. The bar has a length of 500 nm. The diameters of the two dots are both 100 nm, constriction between the two dots is 35 nm in width and length. The four narrow parts connecting the dot to source and drain parts have a width of 30 nm. Seven in-plane plunger gates labeled as GL, GR, GM, PSL, PDL, PSR, and PDR are integrated around the dot for fine tuning. (b) Schematic picture of the device. N-type heavily doped silicon substrate is used as a global back gate. A layer of overexposed PMMA is used as a bridge to make gate GM separated from the drain part of graphene.

Fig. 9(a) shows a scanning electron microscope (SEM) image of one sample with the same structure as the bilayer device we measured. Two central islands with diameter of 100 nm connect through 30 nm wide narrow constrictions to the source and the drain regions. Another narrow constriction (35nm in both width and length) connects the two central islands. Seven in-plane plunger gates labeled as GL, GR, GM, PSL, PDL, PSR, and PDR are integrated in close proximity to the dots. GL, GR, and GM are, respectively, designed to adjust the energy level of left dot, right dot, and inter-dot coupling strength. And PSL, PDL (PSR, PDR) are used for the tuning of the coupling of the left (right) dot to source and drain. The

http://dx.doi.org/10.5772/52873

n-type heavily doped silicon substrate is used as a global back gate. The bridge plunger gate GM is separated from the drain part of graphene by a layer of over exposed PMMA. All the devices were primarily tested to check the functionality of all the gates in a liquid helium storage dewar at 4.2 K. Then the samples were mounted on a dilution refrigerator equipped with filtering wirings and low-noise electronics at the base temperature of 10 mK. To maintain consistency, we will use the data from one sample only in the following.

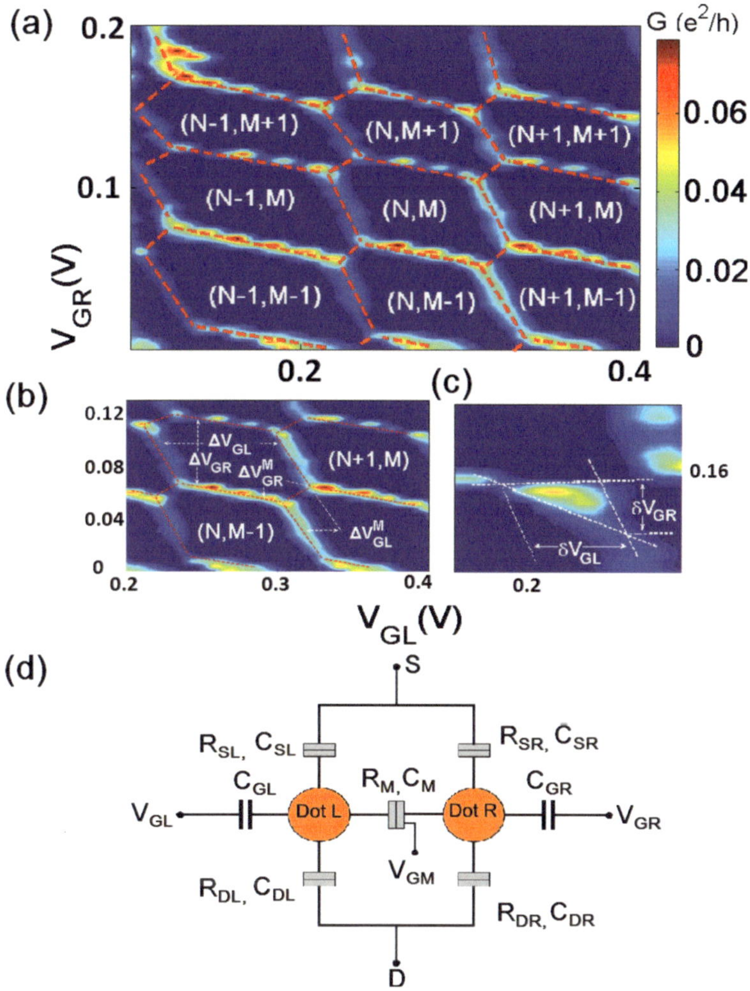

Figure 10. Color online) (a), (b), and (c) PDQD differential conductance as a function of plunger gate voltage V_{GL} and V_{GR}. The red dash lines are guides to the eyes showing the honeycomb pattern. (N,M) represents the carriers in the left and right dot, respectively. (b) Zoom-in of the area (N,M) of the honeycomb pattern. (c) Zoom-in of a vertex pair with white dash lines. (d) Capacitance model for the analysis of the double dot system. Graphene nanoconstrictions behave as tunneling barriers, which are presented, for example, as RSL, CSL (a capacitance and a resistance coupled in parallel). Gate GL and GR are capacitively coupled to the dots; C_{GL} and C_{GR} represent the capacitance.

Fig. 10(a) shows color scale plot of the measured differential conductance of the double dot as a function of V_{GL} and V_{GR} detected in standard ac lock-in technique with anexcitation ac voltage $20\mu V$ at frequency of 11.3 Hz. A dc bias of 0.3 mV is applied, the back gate voltage V_{bg} is fixed at 5 V and the middle plunger gate V_{GM} is -0.45 V. The hexagon pattern characteristic for double dot coupled in parallel is clearly visible. Figure 2(b) Zoom-in of the area (N, M) of the honeycomb pattern, Figure 2(c) Zoom-in of a vertex pair with white dashed lines. From the model of purely capacitively coupled dots as illuminated by Figure 2(d), the energy scales of the system can be extracted (van der Wiel et al., 2003; Molitor et al., 2010; Moriyama et al., 2009). The capacitance of the dot to the side gate can be determined from measuring the size of the honeycomb as shown in Fig. 10(a) and 10(b), ΔV_{GL} =0.087V, ΔV_{GR}=0.053V ,ΔV_{GL}^{m} =0.0261V ΔV_{GR}^{m}=0.0133V, therefore, $C_{GL} = e / \Delta V_{GL}$ =1.84aF,$C_{GR} = e / \Delta V_{GR}$=3.0aF. With a large DC bias of 0.3 mV, we can get δV_{GL} =0.013Vand δV_{GR}=0.01V as shown in Figure 2(c). The lever arm between the left (right) gate $V_{GL}$and the left (right) dot can be calculated as $\alpha_{GL} = V_{bias} / \delta V_{GL}$ =0.023($\alpha_{GR} = V_{bias} / \delta V_{GR}$=0.03). The total capacitances of the dots can then be calculated as $C_L = C_{GL} / \alpha_L$ =79.8aFand$C_R = C_{GR} / \alpha_R$=100.4aF, the corresponding charging energy$E_{CL} = \alpha_{GL} . \Delta V_{GL}$ =2.0meVand$E_{CR} = \alpha_{GR} . \Delta V_{GR}$=1.6meV, the coupling energy between the two dots$E_{CM} = \alpha_{GL} . \Delta V_{GL}^{m}$ =0.3meV. It is also noted that the lever arms between the left gate and the right dot and vice versa can be determined from the slope of the co-tunneling lines delimiting the hexagons. These crossing couplings only modify the results slightly and are neglected usually (Molitor et al., 2010; Moriyama et al., 2009; Liu et al., 2010). Here, by calculating dots area and carrier density (related to V_{BG}), or from the Coulomb charging period, we estimate each dot contains more than 20 electrons when V_{BG}=5V.

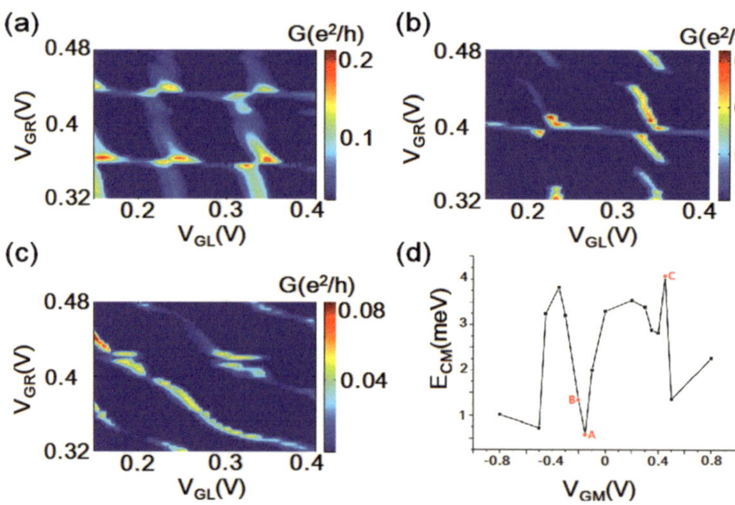

Figure 11. Color online) Interdot coupling vs middle gate voltage V_{GM}. Conductance as a function of gate voltage V_{GL} and V_{GR} at V_{BG}=3 V, V_{bias}=- 1 mV, the scan regions of GL and GR are the same. (a), (b), and (c) represent three different coupling regimes of the two dots. (a) weak coupling regime, V_{GM}= -0.15 V, (b) medium coupling regime, V_{GM}= -0.2 V (c) strong coupling regime, V_{GM}= 0.45 V. (d) shows coupling energy E_{CM} (V) as a non-monotonic function of the middle gate voltage V_{GM}. A, B, C point here represent the corresponding coupling energy in (a), (b), and (c).

By applying voltage to the middle plunger gate GM, the interdot coupling can be tuned efficiently. Fig.11(a), 11(b),and 11(c) show the charge stability diagrams of the PDQD in three different coupling regimes. [(a) weak, (b) medium, and (c) strong]. In these measurements, back gate voltage V_{BG} =3V, Source-Drain DC bias V_{bias} is set to-1.0mV, the scan regions of GL and GR are the same. Only the voltage applied to the gate GM is adjusted as (a) V_{GM}= -0.15 V, (b) V_{GM}=0.2 V, and (c) V_{GM}=0.45 V. By using the same model as in Figure 2, we can calculate the corresponding coupling energy between the dots: (a) E_{CM}=0.58 meV, (b) E_{CM}=1.34 meV, and (c) E_{CM}=4.07 meV. The honeycomb diagrams of the parallel and serial DQD look similar except for the weak coupling regime, as shown in Figure3(a). In this case, the lines delimiting the hexagons are more visible in comparison with serial DQD, because the leads have two parallel accesses to the dots in parallel DQD, which also enables correlated tunneling of two valence electrons simultaneously (Holleitner et al., 2002). Fig. 11(d) indicates the coupling energy changes with the gate voltage VGM. As in the previous reports of graphene DQD in series (Molitor et al., 2010; Liu et al., 2010), the inter-dot coupling is non-monotonically depended on the applied gate voltage. Although the detailed reasons for this non-monotony are undetermined, we assumed that one key factor will be the disorders in graphene introduced by either fabrication steps or substrate (Todd et al., 2009). Many more efforts are still needed to address this issue for the realization of practical graphene based nanodevices.

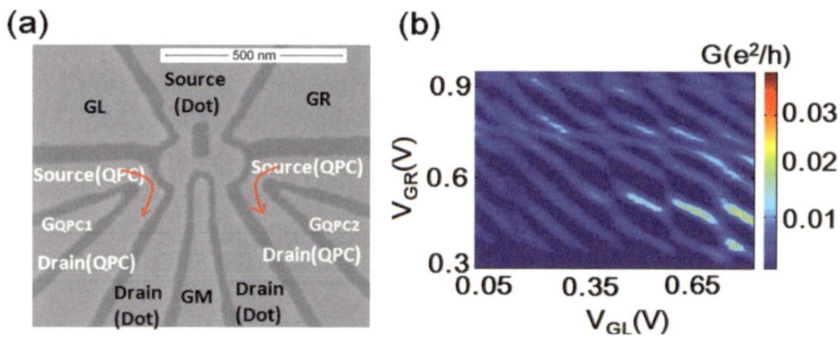

Figure 12. Color online) (a) SEM image of single layer graphene PDQD integrated with two QPCs. The bar has a length of 500 nm. (b) Characteristic honeycomb structure of the conductance through the PDQD as a function of two in-plane plunger gates voltage V_{GL} and V_{GR}, revealed by direct transport measurement of the PDQD at 4.2 K.

We have designed and fabricated an alternative structure of a PDQD integrated with two quantum point contact sensors (QPCs) in single layer graphene, as shown in Figure 4(a). The integrated QPCs can be used as a non-invasive charge detector which may have various applications (van der Wiel et al., 2003; Hanson et al., 2007; Guo et al., 2001; Zhang et al., 2007). As primary tests of the present structure, we can get similar charge stability diagram of the PDQD as in Fig.12 (b) by the direct quantum transport tests at 4.2 K. Although the non-invasive measurements by QPC are still under processing, no remarkable difference is founded between PDQD in bilayer and monolayer graphenes from direct transport measurement. Making tunable coupling double dot is the first step towards the quantum dot based quantum computation bits, the architectonics with integrated charge detector around double quantum dot demonstrated here offers the chance to achieve the charge or spin reading out, which is essential for the quantum computation device. Therefore, a lot of extended and follow-up works can be done on this basis in the future. Both bilayer and single layer graphenes can be exploited in this application.

In summary, we have discussed low temperature quantum transport measurement of gate-controlled parallel coupled double quantum dot on both bilayer and single layer graphenes. The inter-dot coupling strength can be largely tuned by graphene in-plane gates. With the quantum transport honeycomb charge stability diagrams, a common model of purely capacitively coupled double dot is used to extract all the relevant energy scales and parameters of grapheme PDQD. Although many more effects are still needed to further upgrade and exploit the present designed grapheme quantum dot system, the results have intensively demonstrated the promise of the realization of graphene nanodevice and desirable study of rich PDQD physical phenomena in graphene.

5. Conclusion

To conclude, we have discussed the design and fabrication of etched gate tunable single and double quantum dots in single-layer and bilayer graphene and present several important quantum transport measurements in these systems. A quantum dot with an integrated charge sensor is becoming a common architecture for a spin or charge based solid state qubit. To implement such a structure in graphene, we have fabricated a twin-dot structure in which the larger QD serves as a single electron transistor (SET) to read out the charge state of the nearby gate controlled small QD. A high SET sensitivity of $10^{-3}e/\sqrt{Hz}$allowed us to probe Coulomb charging as well as excited state spectra of the QD, even in the regime where the current through the QD is too small to be measured by conventional transport means. We also have measured quantum transport properties of gates controlled parallel-coupled double quantum dots (PDQD) and series-coupled double quantum dots (SDQD) device on both single layer and bilayer graphene with multiple electrostatic gates that are used to enhance control to investigate it. At low temperatures, the transport measurements reveal honeycomb charge stability diagrams which can be tuned from weak to strong inter-dot tunnel coupling regimes. We precisely extract a large inter-dot tunnel coupling strength for this system allowing for the observation of tunnel-coupled molecular states extending over the whole series-coupled double dot. The inter-dot coupling strength also can be effectively tuned from weak to strong by in-plane plunger gates for parallel-coupled double quantum dots. All the relevant energy scales

and parameters can be extracted from the honeycomb charge stability diagrams. The present method of designing and fabricating graphene DQD is demonstrated to be general and reliable and will enhance the realization of graphene nanodevice and desirable study of rich DQD physical phenomena in graphene, and highly controllable system serves as an essential building block for quantum devices in a nuclear-spin-free world.

Acknowledgments

This work was supported by the National Basic Research Program of China (Grants No. 2011CBA00200 and 2011CB921200), and the National Natural Science Foundation of China (Grants No. 10934006, 11074243, 11174267, 91121014, and 60921091)

Author details

Hai-Ou Li, Tao Tu, Gang Cao, Lin-Jun Wang, Guang-Can Guo and Guo-Ping Guo

Key Laboratory of Quantum Information, University of Science and Technology of China, Chinese Academy of Sciences, Hefei, P.R. China

References

[1] Berman D, Zhitenev N. B, Ashoori R. C, and Shayegan M, Phys. Rev. Lett. 82, 161 (1999).

[2] Bylander J, Duty T, and Delsing P, Nature London 434, 361 (2005).

[3] Chen J. C, Chang A. M, and Melloch M. R., Phys. Rev. Lett. 92, 176801 (2004).

[4] Dias da Silva L. G. G. V, Ingersent K, Sandler N, and Ulloa S. E, Phy. Rev. B 78, 153304 (2008).

[5] Duty T, Gunnarsson D, Bladh K, and Delsing P, Phys. Rev. B 69, 40503(R) (2004).

[6] Elzerman J. M, Hanson R, L. van Beveren H. W, L. Vandersypen M. K, and Kouwenhoven L. P, Appl. Phys. Lett. 84, 4617 (2004b).

[7] Elzerman J. M, Hanson R, van Beveren L. H. Witkamp W, B, Vandersypen L. M. K, and L. P. Kouwenhoven, Nature London 430, 431 (2004a).

[8] Fischer J, Trauzettel B, and Loss D, Phys. Rev. B 80, 155401 (2009).

[9] Fischer J. and Loss D, Science 324, 1277 (2009).

[10] Foletti S, Bluhm H, Mahalu D, Umansky V, and Yacoby A, Nat. Phys. 5, 903 (2009).

[11] Gallagher P., Todd K., and Goldhaber-Gordon D., Phys. Rev. B 81, 115409 (2010).

[12] Geim A. K and Novoselov K. S, Nature Mater. 6, 183 (2007).

[13] Geim A. K., Science 324, 1530 (2009).

[14] Golovach V. N, and Loss D, Phys. Rev. B 69, 245327 (2004).

[15] Gotz G, Steele G. A, Vos W. J, and Kouwenhoven L. P, Nano Lett. 8, 4039 (2008).

[16] Graber M. R, Coish W. A, Hoffmann C, Weiss M, Furer J, Oberholzer S, Loss D, and Scho°nenberger C, Phys. Rev. B 74, 075427 (2007).

[17] Guo G. P, Li C. F, and Guo G. C, Phys. Lett. A 286, 401 (2001).

[18] Guo G. P, Lin Z. R, Tu T, Cao G, Li X. P, and Guo G. C, New J. Phys.11, 123005 (2009).

[19] Güttinger J, Seif J, Stampfer C, Capelli A, Ensslin K, and Ihn T, Phys. Rev. B 83, 165445 (2011).

[20] Güttinger J, Stampfer C, Hellmüller S, Molitor F, Ihn T, and Ensslin K, Appl. Phys. Lett. 93, 212102 (2008).

[21] Hanson R, Kouwenhoven L. P, Petta J. R, Tarucha S, and Vandersypen L. M. K, Rev. Mod. Phys. 79, 1217 (2007).

[22] Hanson R, Witkamp B, Vandersypen L. M. K, van Beveren L. H. W, Elzerman J. M, and Kouwenhoven L. P, Phys. Rev. Lett. 91, 196802 (2003).

[23] Hanson R. and Awschalom D, Nature 453, 1043 (2008)

[24] Hatano T, Stopa M, and Tarucha S, Science 309, 268 (2005).

[25] Holleitner A. W, Blick R. H, Hu° ttel A. K, Eberl K, and Kotthaus J. P, Science 297, 70 (2002).

[26] Holleitner A. W, Decker C. R, Qin H, Eberl K, and Blick R. H, Phys. Rev. Lett. 87, 256802 (2001).

[27] Huard B, Sulpizio J. Stander A, Todd N. K, Yang B, and Goldhaber Gordon D, Phys. Rev. Lett. 98, 236803 (2007).

[28] Ihn T., Guttinger J, Molitor F, Schnez S. Schurtenberger E, Jacobsen A, Hellmu°ller S, Frey T, Droscher S, Stampfer C. et al., Mater. Today 13, 44 (2010).

[29] Knobel G. and Cleland A. N., Nature London 424, 291 (2003).

[30] Kouwenhoven L. P, Marcus C, McEuen P. L, Tarucha S, Westervelt R. M, and N. S. Wingreen, in Mesoscopic Electron Transport, Series E:Applied Sciences Vol. 345, edited by Sohn L. L, Kouwenhoven L. P, and Schon G, Dordrecht Kluwer, 1997, pp. 105–214.

[31] L. Ponomarenko, F. Schedin, Katsnelson M, Yang R, Hill E, Novoselov K, and Geim A, Science 320, 356 (2008).

[32] Ladro′n de Guevara M. L, Claro F, and Orellanal P. A, Phys. Rev. B 67, 195335 (2003).

[33] LaHaye D, Buu O, Camarota B, and Schwab K. C, Science 304, 74 (2004).

[34] Lehnert K. W, Bladh K, Spietz L. F, Gunnarsson D, Schuster D. I, Delsing P, and Schoelkopf R. J, Phys. Rev. Lett. 90, 027002 (2003).

[35] Liu X. L, Hug D, and L. Vandersypen M. K. Nano Lett. 10, 1623 (2010).

[36] Livermore C, Crouch C. H, Westervelt R. M, Campman K. L, and A. Gossard C, Science 274, 1332 (1996).

[37] Lo′pez R, Aguado R, and Platero G, Phys. Rev. Lett. 89, 136802 (2002).

[38] Lu W, Ji Z. Q, Pfeiffer L, K. W. West, and A. J. Rimberg, Nature London 423, 422 (2003).

[39] Mason N, Biercuk M. J, and Marcus C. M, Science 303, 655 (2004).

[40] Molitor F, Droscher S, Guttinger J, Jacobson A, Stampfer C, Ihn T, and Ensslin K, Appl. Phys. Lett. 94, 222107 (2009).

[41] Molitor F, Knowles H, Droscher S, Gasser U, Choi T, Roulleau P, Guttinger J, Jacobsen A, Stampfer C, Ensslin K. et al., Europhys. Lett. 89, 67005 (2010).

[42] Moriyama S, Tsuya D, Watanabe E, Uji S, Shimizu M, T. Mori, T. Yamaguchi, and Ishibashi K, Nano Lett. 9, 2891 (2009).

[43] Novoselov K. S, Geim A. K, Morozov S. V, Jiang D, Zhang Y, Dubonos S. V, Grigorieva I. V, and Firsov A. A, Science 306, 666 (2004).

[44] Orellana P. A, Ladron de Guevara M. L, and Claro F, Phys. Rev. B 70, 233315 (2004).

[45] Petta J. R, Johnson A. C, Taylor J. M, Laird E. A, Yacoby A, Lukin M. D , Marcus C. M, Hanson M. P, and Gossard A. C, Science 309, 2180 (2005).

[46] Ponomarenko L, Schedin F, Katsnelson M, Yang R. Hill E, Novoselov K, and Geim A. K, Science 320, 356 (2008).

[47] Schnez S, Molitor F, Stampfer C, Guttinger J, Shorubalko I, Ihn T, and Ensslin K, Appl. Phys. Lett. 94, 012107 (2009).

[48] Stampfer C, Guttinger J, F. Molitor, D. Graf, T. Ihn, and K. Ensslin, Appl. Phys. Lett. 92, 012102 (2008a).

[49] Stampfer C, Schurtenberger E, Molitor F, Guettinger J, Ihn T, and Ensslin K, Nano Lett. 8, 2378 (2008b).

[50] Stampfer C., Güttinger J., Hellmüller S., Molitor F., Ensslin K., and Ihn T., Phys. Rev. Lett. 102, 056403 (2009).

[51] Todd K, Chou H. Amasha T, S, and Goldhaber-Gordon D, Nano Lett. 9, 416 (2009).

[52] Trauzettel B, Bulaev D. V, Loss D, and Burkard G, Nat. Phys. 3, 192 (2007).

[53] van der Wiel W. de Francheschi G, S, Elzermann J. M, Fujisawa T, Tarucha S, and Kouwenhoven L P, Rev. Mod. Phys. 75, 1 (2003).

[54] Vijay R, M. H. Devoret, and I. Siddiqi, Rev. Sci. Instrum. 80, 111101 (2009).

[55] Vink I. T, Nooitgedagt T, Schouten R. N, and Vandersypen L. M. K, Appl. Phys. Lett. 91, 123512 (2007).

[56] Volk C, Fringes S, Terres B, Dauber J, Engels S, Trellenkamp S, and Stampfer C,.Nano Lett. 11, 3581 (2011).

[57] Wang L. J, Cao G, Li H. O, Tu T, Zhou C, Hao X. J, Guo G. C, and Guo G. P, Chinese .Physics. Letters. 28, 067301 (2011b).

[58] Wang L. J, Cao G, Tu T, Li H. O, Zhou C, Hao X. J, Su Z, Guo G. C, Jiang H. W, and Guo G. P, Appl. Phys. Lett. 97, 262113 (2010).

[59] Wang L. J, Guo G. P, Wei D, Cao G, Tu T, Xiao M, Guo G. C, and Chang A.M, Appl. Phys. Lett. 99, 112117 (2011a).

[60] Wang L. J, Li H. O, Tu T, Cao G, Zhou C, Hao X. J, Su Z, Xiao M, Guo G. C, Chang A.M, and Guo G. P, Appl. Phys. Lett. 100, 022106(2012).

[61] Yang S, Wang X, and Das Sarma S, Phys. Rev. B 83, 161301(R) (2011).

[62] Zhang H, Guo G. P, Tu T, and Guo G. C, Phys. Rev. A 76, 012335 (2007).

[63] Ziegler R, Bruder C, and Schoeller H, Phys. Rev. B 62, 1961 (2000).

Chapter 7

Advances in Resistive Switching Memories Based on Graphene Oxide

Fei Zhuge, Bing Fu and Hongtao Cao

Additional information is available at the end of the chapter

http://dx.doi.org/10.5772/51260

1. Introduction

Memory devices are a prerequisite for today's information technology. In general, two different segments can be distinguished. Random access type memories are based on semiconductor technology. These can be divided into static random access memories (SRAM) and dynamic random access memories (DRAM). In the following, only DRAM will be considered, because it is the main RAM technology for standalone memory products. Mass storage devices are traditionally based on magnetic- and optical storage. But also here semiconductor memories are gaining market share. The importance of semiconductor memories is consequently increasing (Mikolajick et al., 2009). Though SRAM and DRAM are very fast, both of them are volatile, which is a huge disadvantage, costing energy and additional periphery circuitry. Si-based Flash memory devices represent the most prominent nonvolatile data memory (NVM) because of their high density and low fabrication costs. However, Flash suffers from low endurance, low write speed, and high voltages required for the write operations. In addition, further scaling, i.e., a continuation in increasing the density of Flash is expected to run into physical limits in the near future. Ferroelectric random access memory (FeRAM) and magnetoresistive random access memory (MRAM) cover niche markets for special applications. One reason among several others is that FeRAM as well as conventional MRAM exhibit technological and inherent problems in the scalability, i.e., in achieving the same density as Flash today. In this circumstance, a renewed nonvolatile memory concept called resistance-switching random access memory (RRAM), which is based on resistance change modulated by electrical stimulus, has recently inspired scientific and commercial interests due to its high operation speed, high scalability, and multibit storage potential (Beck et al., 2000; Lu & Lieber, 2007; Dong et al., 2008). The reading of resistance states is nondestructive, and the memory devices can be operated without transistors in every cell (Lee et al., 2007; Waser & Aono, 2007), thus making a cross-bar structure feasible. A large variety of solid-state materials have been found to show these resistive switching characteristics, including solid electrolytes such as GeSe and Ag_2S (Waser & Aono, 2007), perovskites such as $SrZrO_3$ (Beck et al., 2000), $Pr_{0.7}Ca_{0.3}MnO_3$ (Liu et al., 2000; Odagawa et al., 2004; Liao et al., 2009), and $BiFeO_3$ (Yang et al., 2009; Yin et al., 2010), binary transition metal oxides such as NiO (Seo et al., 2004; Kim et al., 2006; Son & Shin, 2008), TiO_2 (Kim et al., 2007; Jeong et al., 2009; Kwon et al., 2010), ZrO_2 (Wu et al., 2007; Guan et al., 2008; Liu et al., 2009), and ZnO (Chang et al., 2008; Kim et al., 2009; Yang et al., 2009), organic materials (Stewart et al., 2004), amorphous silicon (a-Si) (Jo and Lu, 2008; Jo et al., 2009), and amorphous carbon (a-C) (Sinitskii & Tour, 2009; Zhuge et al., 2010) (Zhuge et al., 2011).

In last decades, carbon-based materials have been studied intensively as a potential candidate to overcome the scientific and technological limitations of traditional semiconductor devices (Rueckes et al., 2000; Novoselov et al., 2004; Avouris et al., 2007). It is worthy mentioning that most of the work on carbon-based electronic devices has been focused on field-effect transistors (Wang et al., 2008; Burghard et al., 2009). Thus, it would be of great interest if nonvolatile memory can also be realized in carbon so that logic and memory devices can be integrated on a same carbon-based platform. Graphene oxide (GO) with an ultrathin thickness (~1 nm) is attractive due to its unique physical-chemical properties. A GO layer can be considered as a graphene sheet with epoxide, hydroxyl, and/or carboxyl groups attached to both sides. GO can be readily obtained through oxidizing graphite in mixtures of strong oxidants, followed by an exfoliation process. Due to its water solubility, GO can be transferred onto any substrates uniformly using simple methods such as drop-casting, spin coating, Langmuir-Blodgett (LB) deposition and vacuum filtration. The as-deposited GO thin films can be further processed into functional devices using standard lithography processes without degrading the film properties (Eda et al., 2008; Cote et al., 2009). Furthermore, the band structure and electronic properties of GO can be modulated by changing the quantity of chemical functionalities attached to the surface. Therefore, GO is potentially useful for microelectronics production. Considering that although a large variety of solid-state materials have been found to show resistive switching characteristics, none of them can fully meet the requirements of RRAM applications, exploration of new storage media is still a key project for the development of RRAM (Zhuge et al., 2011). This review focuses on GO-based RRAM cells, highlighting their advantages as the next generation memories. Section 2 describes the basic concepts of resistive switching and resistance-switching random access memory and physical storage mechanisms. In section 3, the resistive switching mechanisms of GO thin films and memory properties of GO-based RRAM cells are presented. Detailed current–voltage measurements show that in metal/GO/metal sandwiches, the resistive switching originates from the formation and rupture of conducting filaments. An analysis of the temperature dependence of the ON-state resistance reveals that the filaments are composed of metal atoms due to the diffusion of the top electrodes under a bias voltage. Moreover, the resistive switching is found to occur within confined regions of the metal filaments. The resistive switching effect is also observed in GO/metal structures by conducting atomic force microscopy. It is attributed to the redox reactions between GO and adsorbed water induced by external voltage biases. The GO-based RRAM cells show an ON/OFF ratio >100, a retention time $>10^5$ s, and switching threshold voltages <1 V. In section 4, the resistive switching mechanisms of conjugated-polymer-functionalized GO thin films and memory properties of corresponding RRAM cells are described. In this case, the resistive switching is ascribed to electron/hole transfer between graphene sheets and polymer molecules. The RRAM cells exhibit excellent memory performances, such as large ON/OFF ratio, good endurance, and high switching speed. In the last section, it is proposed that the realization of bidirectional or reversible electron transfer in graphene-based hybrid systems is expected to overcome the "voltage–time dilemma" (i.e., one could not realize high write/erase speed and long retention time simultaneously) in pure electronic mechanism-based RRAM cells (Schroeder et al., 2010). Pure electronic mechanisms in RRAM cells postulate the trapping and detrapping of electron in immobile traps as the reason for the resistance changes, also known as Simmons & Verderber model (Simmons & Verderber, 1967). While in graphene-based hybrid systems, the electron transfer occurs between graphene sheets and functional molecules covalently or non-covalently bonded to graphene, which may avoid the "voltage–time dilemma".

2. Resistive switching and RRAM

A resistive switching memory cell in an RRAM is generally composed of an insulating or resistive material sandwiched between two electron-conductive electrodes to form metal-insulator-metal (MIM) structure. By applying an appropriate voltage, the MIM cell can be switched between a high-resistance state (HRS or OFF-state) and a low-resistance state (LRS or ON-state). Switching from OFF-state to ON-state is called the SET process, while switching from ON-state to OFF-state is called the RESET process. These two states can represent the logic values 1 and 0, respectively. Depending on voltage polarity, the resistive switching behavior of an RRAM device is classified as unipolar and bipolar. For unipolar switching, resistive switching is induced by a voltage of the same polarity but a different magnitude, as shown in Fig. 1(a) (Waser & Aono, 2007). For bipolar switching, one polarity is used to switch from HRS to LRS, and the opposite polarity is used to switch back into HRS, as shown in Fig. 1(b) (Waser & Aono, 2007). A current compliance (CC) is usually needed during the SET process to prevent the device from a permanent breakdown. (Li et al., 2011)

In the simplest approach, the pure memory element can be used as a basic memory cell, resulting in a configuration where parallel bitlines are crossed by perpendicular wordlines with the switching material placed between wordline and bitline at every cross-point. This configuration is called a cross-point cell. Since this architecture will lead to a large parasitic current flowing through nonselected memory cells, the cross-point array has a very slow read access. A selection element can be added to improve the situation. A series connection of a diode in every cross-point allows to reverse bias all nonselected cells. This can be arranged in a similar compact manner as the basic cross-point cell. Finally a transistor device (ideally a MOS Transistor) can be added which makes the selection of a cell very easy and therefore gives the best random access time, but comes at the price of increased area consumption. Figure 2 illustrates the different cell type possibilities. For random access type memories, a transistor type architecture is preferred while the cross-point architecture and the diode architecture open the path toward stacking memory layers on top of each other and therefore are ideally suited for mass storage devices (Mikolajick et al., 2009; Pinnow & Mikolajick, 2004).

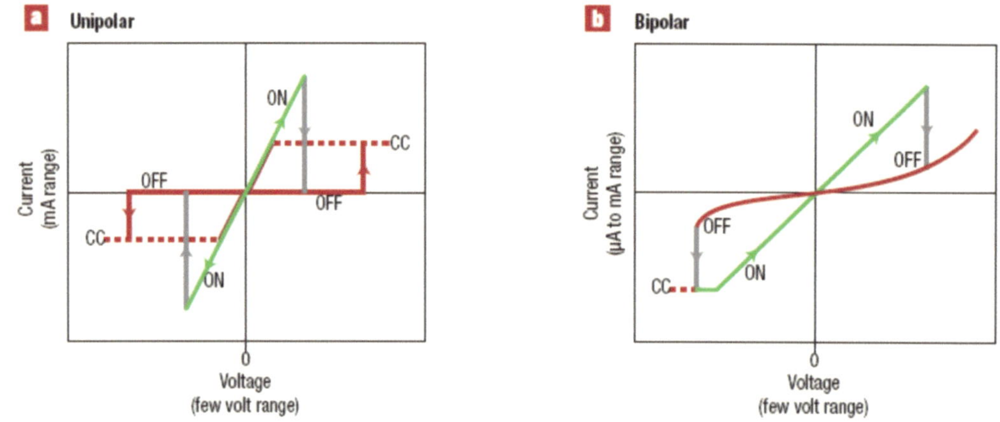

Figure 1. (a) Unipolar switching. The SET voltage is always higher than the voltage at which RESET takes place, and the RESET current is always higher than the CC during SET operation. (b) Bipolar switching. The SET operation takes place on one polarity of the voltage or current, and the RESET operation requires the opposite polarity. (Waser & Aono, 2007)

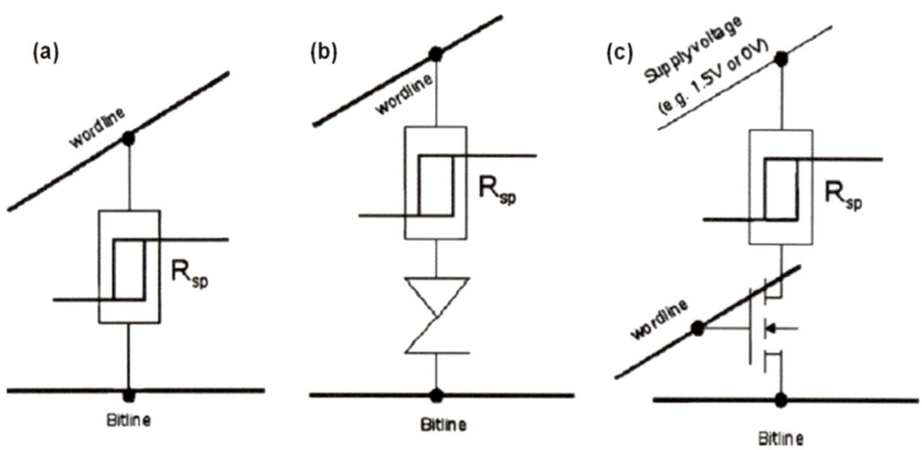

Figure 2. Three different cell architectures for RRAM cells: (a) cross-point cell, (b) diode cell, and (c) transistor cell together with their respective area consumption in F^2. F denotes the minimum feature size of the fabrication technology. (Mikolajick et al., 2009)

Based on the circuit requirements of high-density NVM today such as Flash and taking predictions about technology scaling of the next 15 years into account, one can collect a number of requirements for RRAM cells (Waser et al., 2009):

2.1. Write operation

Write voltages should be in the range of a few hundred mV to be compatible with scaled CMOS to few V (to give an advantage over Flash which suffers from high programming voltages). The length of write voltage pulses is desired to be <100 ns in order to compete with DRAM specifications and to outperform Flash which has a programming speed of some 10 ms, or even <10 ns to approach high-performance SRAM.

2.2. Read operation

Read voltages need to be significantly smaller than write voltages in order to prevent a change of the resistance during the read operation. Because of constraints by circuit design, read voltage cannot be less than approximately one tenth of write voltage. An additional requirement originates from the minimum read current. In the ON-state, read current should not be less than approximately 1mA to allow for a fast detection of the state by reasonably small sense amplifiers. The read time must be in the order of write time or preferably shorter.

2.3. Resistance ratio

Although an ON/OFF (R_{OFF}/R_{ON}) ratio of only 1.2 to 1.3 can be utilized by dedicated circuit design as shown in MRAM, ON/OFF ratios >10 are required to allow for small and highly efficient sense amplifiers and, hence, RRAM devices which are cost competitive with Flash.

2.4. Endurance

Contemporary Flash shows a maximum number of write cycles between 10^3 and 10^7, depending on the type. RRAM should provide at least the same endurance, preferably a better one.

2.5. Retention

A data retention time of >10 years is required for universal NVM. This retention time must be kept at thermal stress up to 85 ºC and small electrical stress such as a constant stream of read voltage pulses.

Despite a bursting body of experimental data that is rapidly becoming available, the precise mechanism behind the physical effect of resistive switching remains elusive. A few qualitative models have been proposed emphasizing different aspects: electric-field-induced defect migration (Baikalov et al., 2003; Nian et al., 2007), phase separation (Tulina et al., 2001), tunneling across interfacial domains (Rozenberg et al., 2004), control of the Schottky barrier's height (Jeong et al., 2009), etc., as shown in Fig. 3. A general consensus has emerged on the empirical relevance of three key features: (i) a highly spatially inhomogeneous conduction in the low resistive state, (ii) the existence of a significant number of defects, and (iii) a pre-eminent role played by the interfaces, namely, the regions of the oxide that are near each of the metallic electrodes which often form Schottky barriers. (Rozenberg et al., 2010)

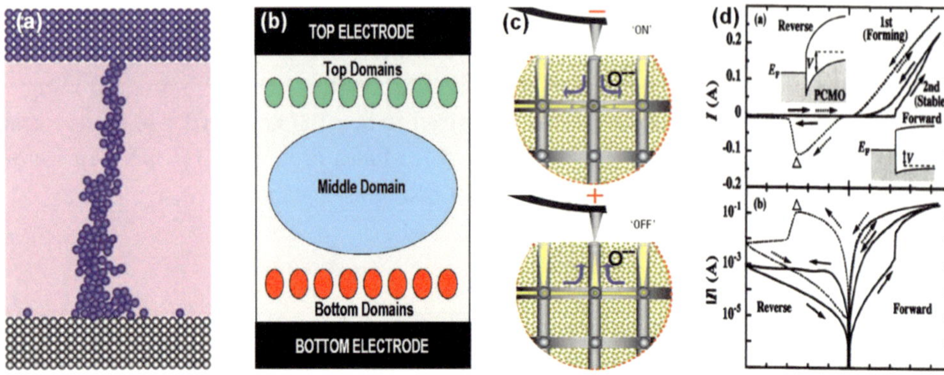

Figure 3. Reported several resistive switching mechanisms. (a) Filamentary model (Yang et al., 2009). (b) Domain model (Rozenberg et al., 2004). (c) Electrical field induced oxygen vacancy migration model (Szot et al., 2006). (d) Schottky barrier modulation model (Sawa et al., 2004).

3. Resistive switching and memory properties in GO-based RRAM cells

He et al. firstly reported reliable and reproducible resistive switching behaviors in GO thin films prepared by the vacuum filtration method on 2009 (He et al., 2009). The Cu/GO/Pt structure showed an ON/OFF ratio of about 20, aretention time of more than 10^4 s, and switching threshold voltages of less than 1 V, as shown in Fig. 4.

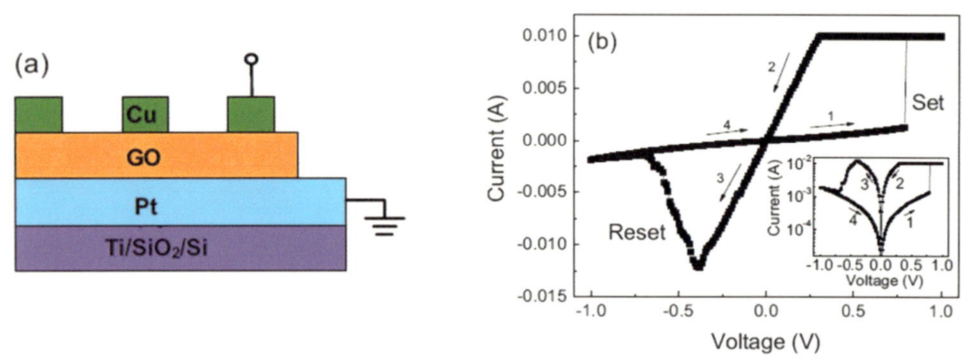

Figure 4. (a) A schematic configuration of the Cu/GO/Pt sandwiched structure. (b) I–V characteristics of the Cu/GO/Pt structure. The arrows indicate the sweep direction. The inset shows the I–V characteristics in semilogarithmic scale. (He et al., 2009)

It indicates that GO is potentially useful for future nonvolatile memory applications. At a later time, Zhuge et al. achieved larger ON/OFF ratios of more than 100 in metal (Cu, Ag, Ti, and Au)/GO/Pt devices (Zhuge et al., 2011). They considered that the moisture in air affects the ON/OFF ratio of metal/GO/Pt memory cells severely. Furthermore, Jeong et al. presented a GO based memory that can be easily fabricated using a room temperature spin-casting method on flexible substrates and has reliable memory performance in terms of retention and endurance, as shown in Fig. 5 (Jeong et al., 2010). Therefore, the GO memory is an excellent candidate to be a memory device for future flexible electronics (Hong et al., 2010).

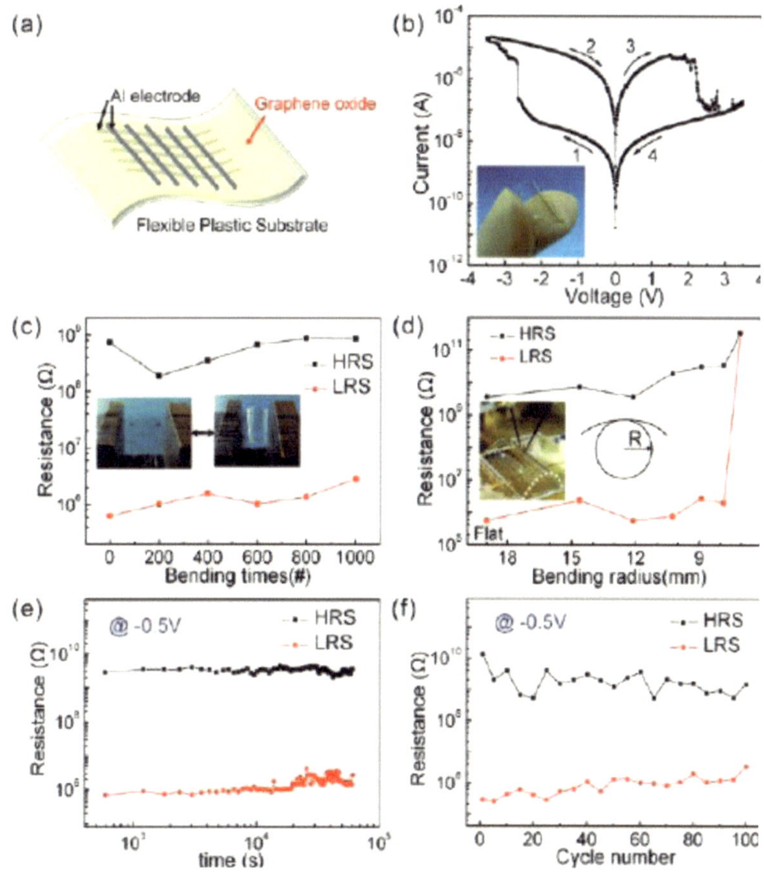

Figure 5. (a) A schematic illustration of a GO based flexible crossbar memory device. (b) Typical I–V curve of a Al/GO/Al/PES device plotted on a semilogarithmic scale. The arrows indicate the voltage sweep direction. The left inset is a real photo image of a device. (c) Continuous bending effect of a Al/GO/Al/PES device. The insets show photographs of repeated two bending states. (d) The resistance ratio between the HRS and LRS as a function of the bending radius (*R*). The inset is a photograph of an I-V measurement being performed under a flexed condition. (e) Retention test of Al/GO/Al/PES device read at −0.5 V. (f) Endurance performance of an Al/GO/Al/PES device measured during 100 sweep cycles. (Jeong et al., 2010)

As to the mechanism of the resistive switching effect in GO thin films, Zhuge et al. pointed out that in metal/GO/Pt sandwiches, the resistive switching originates from the formation and rupture of conducting filaments, as schematically shown in Fig. 6 (Zhuge et al., 2011).

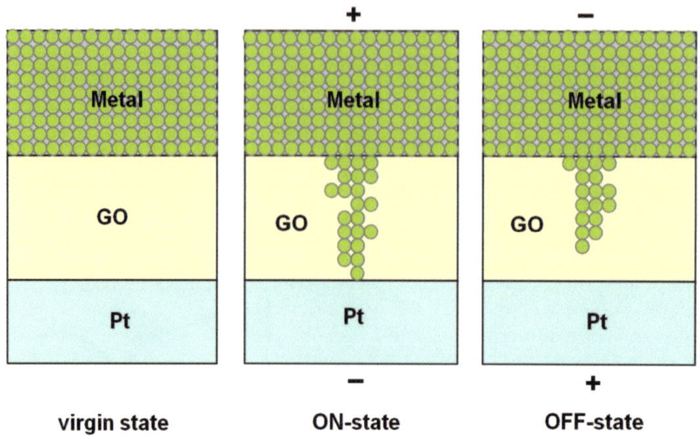

Figure 6. A schematic diagram for the mechanism of the resistive switching in metal/GO/Pt memory cells. (Zhuge et al., 2011)

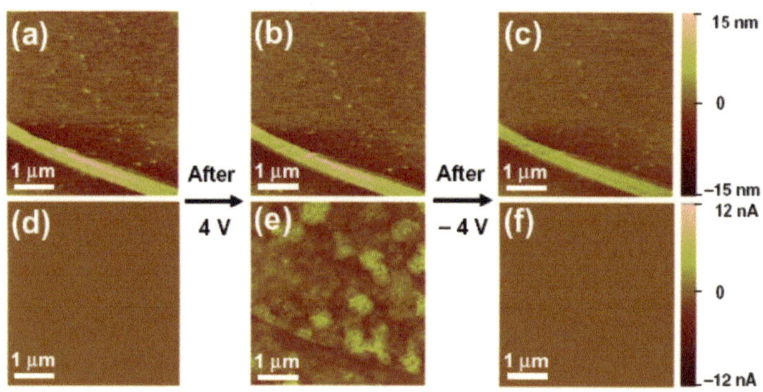

Figure 7. (a), (b) and (c) AFM images of virgin GO films, GO films in LRS, and GO films in HRS. The light-colored ribbons represent folded regions. (d), (e) and (f) the corresponding CAFM images under a read voltage of 1 V. (Zhuge et al., 2011)

An analysis of the temperature dependence of the ON-state resistance reveals that the filaments are composed of metal atoms due to the diffusion of the top electrodes under a bias voltage. Tsuruoka et al. pointed out that the formation of a metal filament is due to inhomogeneous nucleation and subsequent growth of metal, based on the migration of metal ions in the oxide matrix (Tsuruoka et al., 2010). Recently, they reported that the ionization of metal at the anode interfaces is likely to be attributed to chemical oxidation via residual water in the oxide layer, and metal ions migrate along grain boundaries in the oxide layer, where a hydrogen-bond network might be formed by moisture absorption (Tsuruoka et al., 2012). Moreover, the switching occurs within confined regions of the metal filaments. The RESET process is considered to consist of the Joule-heating-assisted oxidation of metal atoms at the thinnest part of the metal filament followed by diffusion and drift of the metal ions under their own concentration gradient and the applied electric field, disconnecting the metal filament (Tsuruoka et al., 2010). Zhuge et al. also observed the resistive switching effect in GO/Pt structures by conducting atomic force microscopy (CAFM), as shown in Fig. 7 (Zhuge et al., 2011). It is attributed to the redox reactions between GO and adsorbed water induced by external voltage biases. While for Al/GO/Al memory cells, Jeong et al. attributed the bipolar resistive switching behavior to rupture and formation of conducting filaments at the top amorphous interface layer formed between the GO film and the top Al metal electrode, as shown in Fig. 8 (Jeong et al, 2010).

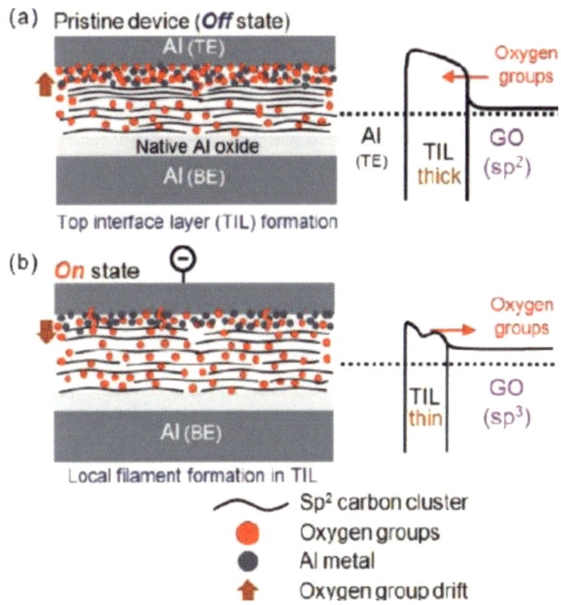

Figure 8. Schematic of the proposed bipolar resistive switching model for Al electrode/GO/Al electrode crossbar memory device. (a) The pristine device is in the OFF-state due to the (relatively) thick insulating top interface layer formed by a redox reaction between vapor deposited Al and the GO thin film. (b) The ON-state is induced by the formation of local filaments in the top interface layer due to oxygen ion diffusion back into the GO thin film by an external negative bias on the top electrode. (Jeong et al., 2010)

Furthermore, Hong et al. pointed out that for Al/GO/metal memory devices, the resistive switching operation is governed by dual mechanism of oxygen migration and Al diffusion (Hong et al., 2011). The Al diffusion into the graphene oxide is the main factor to determine the switching endurance property which limits the long term lifetime of the device. The electrode dependence on graphene oxide RRAM operation has been analyzed as well and is attributed to the difference in surface roughness of graphene oxide for the different bottom electrodes, as shown in Fig. 9 (Hong et al., 2011). Interestingly, Panin et al. observed both diode-like (rectifying) and resistor-like (nonrectifying) resistive switching behaviors in an Al/GO/Al planar structure, as shown in Fig. 10 (Panin et al., 2011). Electrical characterization of the Al/GO interface using the induced current identifies a potential barrier near the interface and its spatial modulation, caused by local changes of resistance at a bias voltage, which correlated well with the resistive switching of the whole structure. Recently, Wang et al. found that the speed of the SET and RESET operations of the Al/GO/ITO resistive memories is significant asymmetric (Wang et al., 2012). The RESET speed is in the order of 100 ns under a –5 V voltage while the SET speed is three orders of magnitude slower (100 μs) under a 5 V bias. The behavior of resistive switching speed difference is elucidated by voltage modulated oxygen diffusion barrier change, as shown in Fig. 11 (Wang et al., 2012).

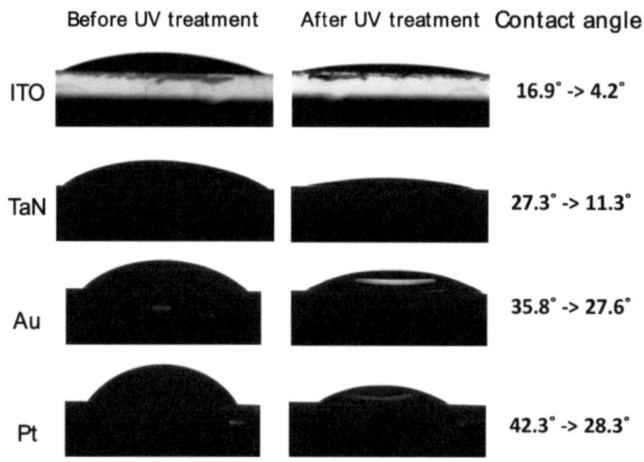

Figure 9. The contact angles of graphene oxide solution on four different surfaces of ITO, TaN, Su, and Pt. UV treatment is done to promote adhesion. (Hong et al., 2011)

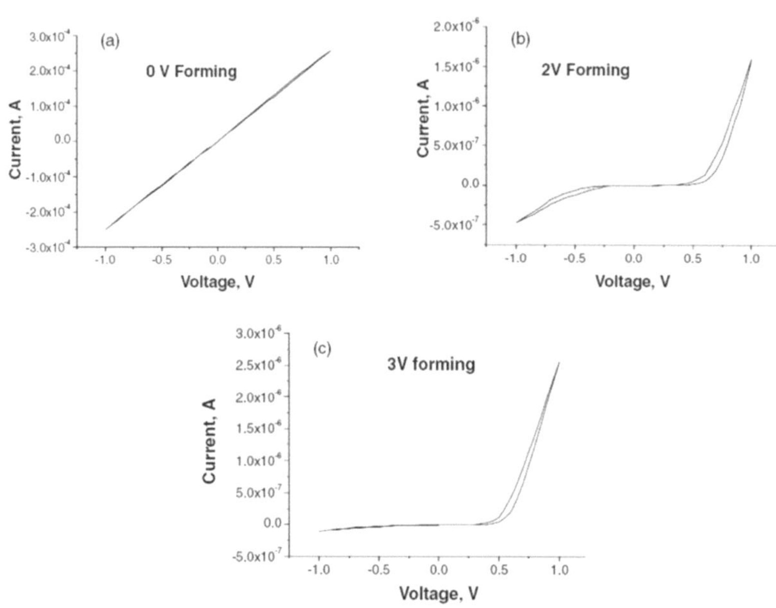

Figure 10. I–V curves of Al/GO/Al structures pre-formed at different forming voltages. (Panin et al., 2011)

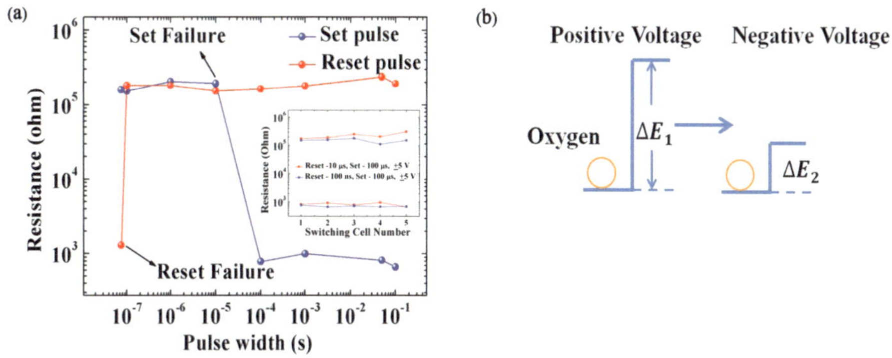

Figure 11. Pulse behavior of the Al/GO/ITO/PET memory cell, HRS and LRS is read at 0.3 V. (a) the SET and RESET operations of the devices with different pulsing width at ±5. The HRS and LRS of the devices are measured at 0.3V and the SET operation is found to be three orders of magnitude slower than the RESET operation; (b) a schematic of oxygen hopping barrier change model. (Wang et al., 2012)

4. Resistive switching and memory properties in GO-polymer hybrid RRAM cells

Liu et al. prepared a solution-processable and electroactive complex of poly(*N*-vinylcarbazole)-derivatized graphene oxide (GO-PVK) via amidation of end-functionalized PVK, from reversible addition fragmentation chain transfer polymerization, with tolylene-2,5-diisocyanate-functionalized graphene oxide (Liu et al., 2009). The Al/GO-PVK/ITO device exhibits bistable electrical conductivity switching and nonvolatile rewritable memory effects. The resistive switching is attributed to electron transfer between GO and PVK, as shown in Fig. 12 (Liu et al., 2009).

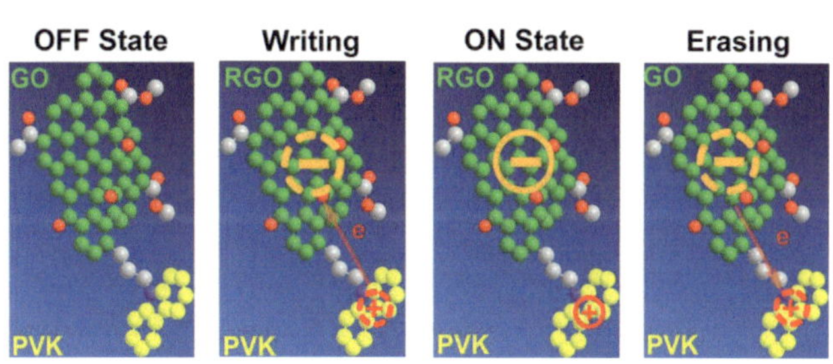

Figure 12. Plausible switching mechanism of GO-PVK. RGO stands for reduced graphene oxide. (Liu et al., 2009)

Zhuang et al. synthesized a novel conjugated-polymer-modified graphene oxide (TPAPAM-GO), which was successfully used to fabricate a TPAPAM-GO-based RRAM device (Zhuang et al., 2010). The device exhibits a typical bistable electrical switching and nonvolatile rewritable memory effect, with a SET voltage of about–1V and an ON/OFF ratio of more than 10^3, as shown in Fig. 13 (Zhuang et al., 2010).

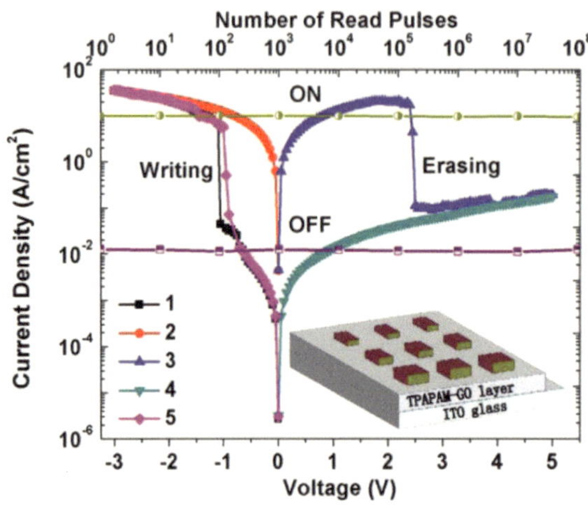

Figure 13. J–V characteristics and stability tested in either ON- or OFF-state under stimulus by read pulses of a 0.16-mm^2 ITO/TPAPAM-GO/Al device. Inset: schematic diagram of the single-layer memory devices. (Zhuang et al., 2010)

Both the ON- and OFF-state are stable under a constant voltage stress and survive up to 10^8 read cycles at a read voltage of –1 V. As to the switching mechanism, they deduced that at the switching threshold voltage, electrons transit from the hole transporting (electron donating) polymer TPAPAM (highest occupied molecular orbital, HOMO) into the graphene monoatom layer (lowest unoccupied molecular orbital, LUMO) via intramolecular charge-transfer (CT) interaction (Ling et al., 2008). The transferred electrons can delocalize effectively in the giant p-conjugation system, and reduce graphene oxide to graphene (Elias et al., 2009; Robinson et al., 2008). Upon electrochemical reduction of the functionalized graphene oxide, electrons can propagate with less scattering, giving rise to a substantially enhanced room temperature conductivity (~10^2 S m^{-1}) of the composite material (Robinson et al., 2008; Stankovich et al., 2006). Along with the increase of CT interaction, dual-channel charge-transport pathways will form via interplane hopping in graphene films and switch the ITO/TPAPAM-GO/Al device from the OFF-state to the ON-state (Ling et al., 2008). The application of a reverse positive bias to the device can, however, extract electrons from the reduced graphene nanosheet, returning it to the initial less-conductive form and programming the device back to the OFF-state (Zhuang et al., 2010). Wu et al. fabricated GO-polyimide (PI) hybrid RRAM cells, as shown in Fig. 14 (Wu et al., 2011). The functionalization of GO sheets with PI enables the layer-by-layer fabrication of a GO-PI hybrid resistive-switch device and leads to high reproducibility of the memory effect. The current-voltage curves for the as-fabricated device exhibit multilevel resistive-switch properties under various reset voltages. The capacitance-voltage characteristics for a capacitor based on GO-PI nanocomposite indicate that the electrical switching may originate from the charge trapping in GO sheets. The high device-to-device uniformity and unique memory properties of the device make it an attractive candidate for applications in next-generation high-density nonvolatile flash memories (Wu et al., 2011). Yu et al. reported bistable resistive switching characteristics for write-once-read-many-times (WORM) memory devices using a supramolecular hybrid route to hydrogen-bonded block copolymers (BCP) and GO as charge storage materials (Yu et al., 2012). The ITO/7 wt% GO composite/Al device exhibits a one-time programmable effect with an ON/OFF ratio of 10^5 at –1.0 V, a retention of 10^4 s and a 10^8 read pulse of –1.0 V, as shown in Fig. 15 (Yu et al., 2012). The switching phenomena were attributed to the charge trapping environment operating across the BCP/GO interface and from the GO intrinsic defect. Controlling the physical interaction of BCP and functional GO sheets can generate a well-dispersed charge storage composite device for future flexible information technology.

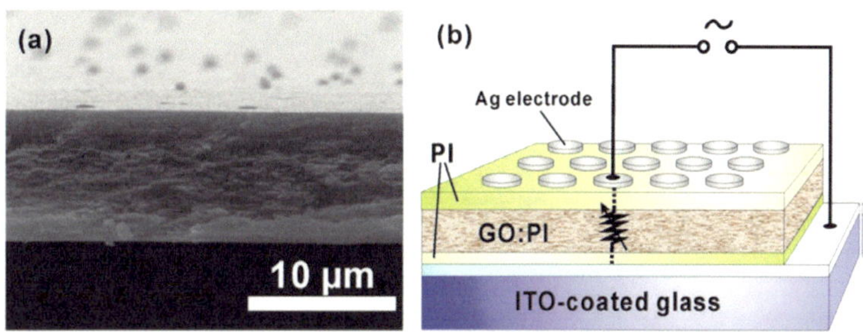

Figure 14. (a) FE-SEM image of the cross sectional view for the GO-PI film. (b) Schematic of Ag/PI/GO:PI/PI/ITO memory device. (Wu et al., 2011)

Recently, Hu et al. prepared a novel RRAM device based on reduced GO noncovalently functionalized by thionine (Hu et al., 2012). The device shows nonvolatile resistive switching behaviors with an ON/OFF ratio of more than 10^4, fast switching speed of <5 ns, long retention time of >10^5 s, and good endurance. The resistive switching in such memory device is attributed to electron transfer reaction between reduced graphene oxide sheets and thionine molecules.

Noting that besides GO, graphene can also be used for resistive switching memory devices. Standley et al. developed a nonvolatile resistive memory element based on graphene break junctions which demonstrates thousands of writing cycles and long retention times (Standley et al., 2008). They proposed a model for device operation based on the formation and breaking of carbon atomic chains that bridges the junctions, as shown in Fig. 16 (Standley et al., 2008). Recently, He et al. reported a planar graphene/SiO2 nanogap structure for multilevel resistive switching (He et al., 2012). Such two-terminal devices exhibited excellent memory characteristics with good endurance up to 10^4 cycles, long retention time more than 10^5 s, and fast switching speed down to 500 ns. At least five conduction states with reliability and reproducibility were demonstrated in these memory devices, as shown in Fig. 17 (He et al., 2012). The mechanism of the resistive switching was attributed to a reversible thermal-assisted reduction and oxidation process that occurred at the breakdown region of the SiO_2 substrate.

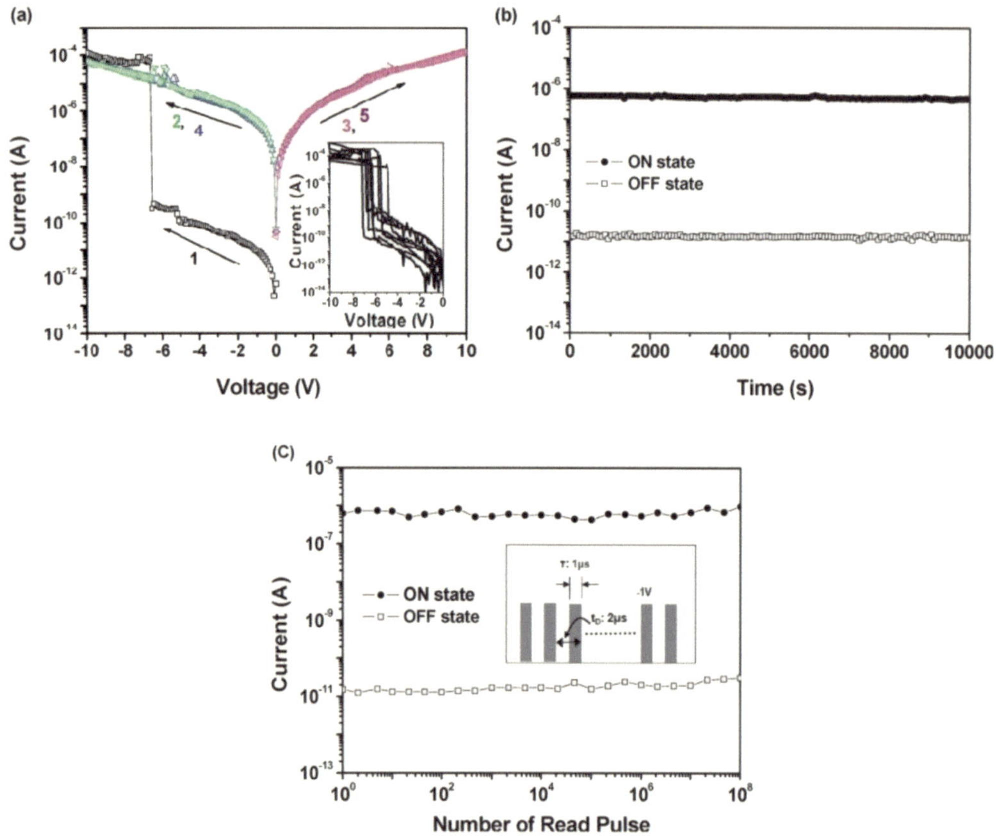

Figure 15. (a) I–V characteristics of 7 wt% GO composite device. The inset shows the switching behavior in different memory cells. (b) Retention time test. (c) Stimulus effect of read pulses. (Yu et al., 2012)

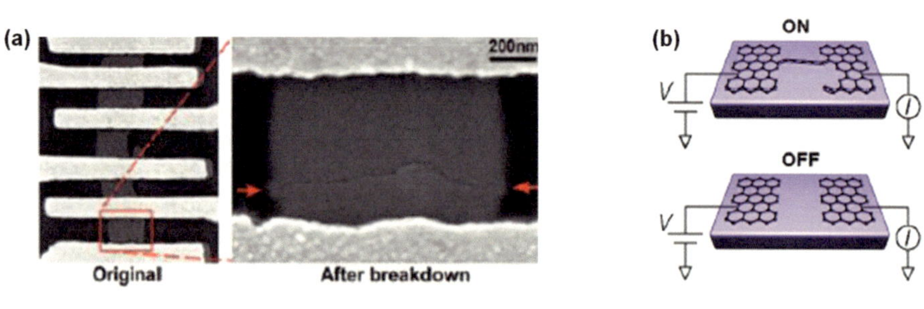

Figure 16. (a) SEM image of the device before (left panel) and after breakdown (right panel). The arrows indicate the edges of the nanoscale gap. (b) Proposed schematic atomic configurations in the ON and OFF states. (Standley et al., 2008)

http://dx.doi.org/10.5772/51260

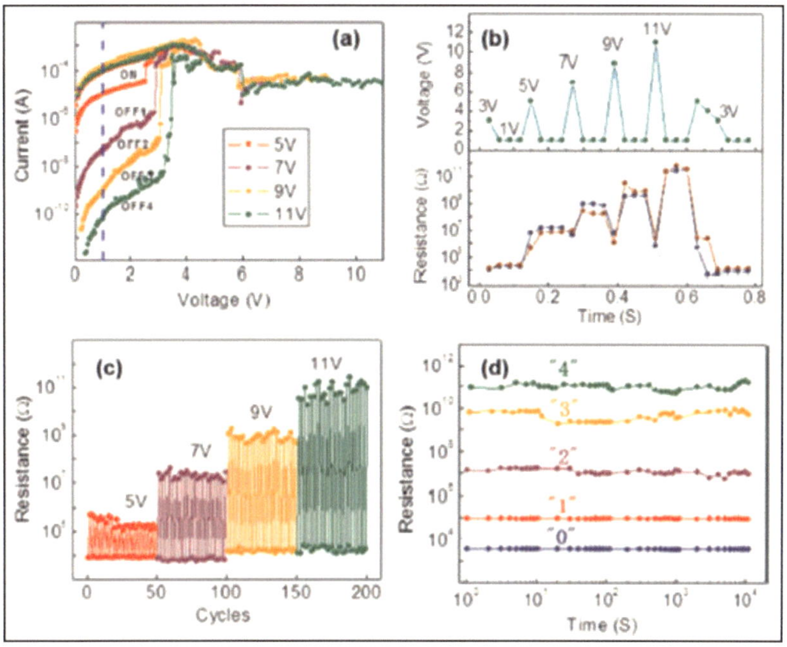

Figure 17. Multilevel resistive switching properties of graphene/SiO_2 nanogap structures. (a) Typical I–V characteristics of a device with a width of 1 μm, length of 0.4 μm, and thickness of 2.3 nm. The vertical line cut at 1 V indicates five resistance states. By sweeping the reset voltage from 0 to 5 V, the OFF1 state (red) was established. The subsequent reset voltages sweep up to higher voltage of 7 V (purple), 9 V (orange), and even higher to 11 V (olive) from 0 V, and lower conduction states of OFF2, OFF3, and OFF4 were achieved subsequently. (b) Top: series of bias pulses with different magnitudes of 3, 5, 7, 9, and 11 V, corresponding to the sweep voltages in (a) with three reading pulses of 1 V after each programming pulse was applied. Bottom: resistance changes corresponding to the each voltage pulse in the top panel. (c) Cycled switching of the device under various reset voltages. (d) Retention time of more than 10^4 s for each conduction state tested by a continuous 1 V pulse. (He et al., 2012)

5. Summary and prospect

Resistive random access memory based on the resistive switching effect induced by electrical stimulus has inspired scientific and commercial interests due to its high operation speed, high scalability, and multibit storage potential. The reading of resistance states is nondestructive, and the memory devices can be operated without transistors in every cell, thus making a cross-bar structure feasible. Although a large variety of solid-state materials have been found to exhibit the resistive switching effect, GO is a very promising material for RRAM applications since due to an ultrathin thickness (~1 nm) and its unique physical–chemical properties. Both GO and GO-polymer hybrid exhibit good memory performances, such as high ON/OFF ratio and long retention time. The resistive switching of GO is always related to defect migration, such as metal ions and oxygen vacancies, whereas the switching of GO-polymer hybrid is considered to be attributed to charge transfer reaction between GO sheets and polymer molecules. Since both high switching speed and good retention could be simultaneously achieved in GO-polymer hybrid RRAM device, such memory device is expected to overcome the "voltage–time dilemma" (i.e., one could not realize high write/erase speed and long retention time simultaneously in pure electronic mechanism-based RRAM cells). Pure electronic mechanisms in RRAM cells postulate the trapping and detrapping of electron in immobile traps as the reason for the resistance changes, also known as Simmons & Verderber model. While in GO-polymer hybrid systems, the electron transfer occurs between graphene sheets and functional molecules covalently or non-covalently bonded to graphene, which may avoid the "voltage–time dilemma".

However, to meet the requirements of future memory applications, GO-based resistance memories should overcome several hurdles. Firstly, the size and chemical composition of GO sheets must be controllable, for example, the type, number and distribution of oxygen functional groups attached to both sides of graphene sheets. Secondly, the resistive switching mechanism of GO is still not clear. For metal/GO/metal sandwiches, although the formation/rupture of metal filaments is considered to be responsible for the resistive switching, the filament growth and inhibition kinetics remains ambiguous. As to the switching of GO-polymer hybrid, no direct evidences have been provided to support the charge transfer hypothesis so far. Thirdly, it is a real challenge to improve the thermal stability of GO and GO-polymer hybrid since memory devices may work at elevated temperature.

Author details

Fei Zhuge[1,2*], Bing Fu[1] and Hongtao Cao[1*]

*Address all correspondence to: zhugefei@nimte.ac.cn and h_cao@nimte.ac.cn

1 Ningbo Institute of Materials Technology and Engineering, Chinese Academy of Sciences, People's Republic of China

2 State Key Laboratory of Silicon Materials, Zhejiang University, People's Republic of China

References

[1] Avouris, P., Chen, Z. H., & Perebeinos, V. (2007). Carbon-based electronics. *Nature Nanotechnology*, 2(10), 1748-3387.

[2] Baikalov, A., Wang, Y. Q., Shen, B., Lorenz, B., Tsui, S., Sun, Y. Y., Xue, Y. Y., & Chu, C. W. (2003). Field-driven hysteretic and reversible resistive switch at the Ag-$Pr_{0.7}Ca_{0.3}MnO_3$ interface. *Applied Physics Letters*, 83(5), 957-959, 0003-6951.

[3] Beck, A., Bednorz, J. G., Gerber, Ch., Rosse, C. L., & Widmer, D. (2000). Reproducible switching effect in thin oxide films for memory applications. *Applied Physics Letters*, 77(1), 139-141, 0003-6951.

[4] Burghard, M., Klauk, H., & Kern, K. (2009). Carbon-Based Field-Effect Transistors for Nanoelectronics. *Advanced Materials*, 21(25-26), 2586-2600, 0935-9648.

[5] Chang, W. Y., Lai, Y. C., Wu, T. B., Wang, S. F., Chen, F., & Tsai, M. J. (2008). Unipolar resistive switching characteristics of ZnO thin films for nonvolatile memory applications. *Applied Physics Letters*, 92(2), 22110, 0003-6951.

[6] Cote, L. J., Kim, F., & Huang, J. X. (2009). Langmuir–Blodgett Assembly of Graphite Oxide Single Layers. *Journal of the American Chemical Society*, 131(3), 1043-1049, 0002-7863.

[7] Dong, Y., Yu, M. G., Mc Alpine, C., Lu, W., & Lieber, C. M. (2008). Si/α-Si core/shell nanowires as nonvolatile crossbar switches. *Nano Letters*, 8(2), 386-391, 1530-6984.

[8] Eda, G., Fanchini, G., & Chhowalla, M. (2008). Large-area ultrathin films of reduced graphene oxide as a transparent and flexible electronic material. *Nature Nanotechnology*, 3(5), 270-274, 1748-3387.

[9] Elias, D. C., Nair, R. R., Mohiuddin, T. M. G., Morozov, S. R., Blake, P., Halsall, M. P., & Ferrari, A. C. (2009). Control of graphene's properties by reversible hydrogenation: Evidence for graphane. *Science*, 323(5914), 610-613, 0036-8075.

[10] Guan, W. H., Long, S. B., Jia, R., & Liu, M. (2007). Nonvolatile resistive switching memory utilizing gold nanocrystals embedded in zirconium oxide. *Applied Physics Letters*, 91(6), 062111, 0003-6951.

[11] Guan, W. H., Liu, M., Long, S. B., Liu, Q., & Wang, W. (2008). On the resistive switching mechanisms of Cu/ZrO_2:Cu/Pt. *Applied Physics Letters*, 93(22), 223506, 0003-6951.

[12] He, C. L., Zhuge, F., Zhou, X. F., Li, M., Zhou, G. C., Liu, Y. W., Wang, J. Z., Chen, B., Su, W. J., Liu, Z. P., Wu, Y. H., Cui, P., & Li, R. W. (2009). Nonvolatile resistive switching in graphene oxide thin films. *Applied Physics Letters*, 95(23), 232101, 0003-6951.

[13] He, C. L., Shi, Z. W., Zhang, L. C., Yang, W., Yang, R., Shi, D. X., & Zhang, G. Y. (2012). Multilevel resistive switching in planar graphene/SiO2 nanogap structures. *ACS Nano*, 6(5), 4214-4221, 1936-0851.

[14] Hong, S. K., Kim, J. E., Kim, S. O., Choi, S. Y., & Cho, B. J. (2010). Flexible resistive switching memory device based on graphene oxide. *IEEE Electron Device Letters*, 31(9), 1005-1007, 0741-3106.

[15] Hong, S. K., Kim, J. E., Kim, S. O., & Cho, B. J. (2011). Analysis on switching mechanism of graphene oxide resistive memory device. *Journal of Applied Physics*, 110(4), 044506, 0021-4922.

[16] Hu, B. L., Quhe, R. G., Chen, C., Zhuge, F., Zhu, X. J., Peng, S. S., Chen, X. X., Pan, L., Wu, Y. Z., Zheng, W. G., Yan, Q., Lu, J., & Li, R. W. (). *Electrically controlled electron transfer and resistance switching in graphene oxide noncovolently functionalized with dye*, unpublished.

[17] Jeong, D. S., Schroeder, H., & Waser, R. (2009). Abnormal bipolar-like resistance change behavior induced by symmetric electroforming in Pt/TiO_2/Pt resistive switching cells. *Nanotechnology*, 20(37), 375201, 0957-4484.

[18] Jeong, D. S., Schroeder, H., & Waser, R. (2009). Mechanism for bipolar switching in a Pt/TiO_2/Pt resistive switching cell. *Physical Review B*, 79(19), 195317, 1098-0121.

[19] Jeong, H. Y., Kim, J. Y., Kim, J. W., Hwang, J. O., Kim, J. E., Lee, J. Y., Yoon, T. H., Cho, B. J., Kim, S. O., Ruoff, R. S., & Choi, S. Y. (2010). Graphene oxide thin films for flexible nonvolatile memory applications. *Nano Letters*, 10(11), 4381-4386, 1476-1122.

[20] Jo, S. H., & Lu, W. (2008). CMOS Compatible Nanoscale Nonvolatile Resistance Switching Memory. *Nano Letters*, 8(2), 392-397, 1476-1122.

[21] Jo, S. H., Kim, K. H., & Lu, W. (2009). Programmable Resistance Switching in Nanoscale Two-Terminal Devices. *Nano Letters*, 9(1), 496-500, 1476-1122.

[22] Jo, S. H., Kim, K. H., & Lu, W. (2009). High-Density Crossbar Arrays Based on a Si Memristive System. *Nano Letters*, 9(2), 870-874, 1476-1122.

[23] Kim, D. C., Lee, M. J., Ahn, S. E., Seo, S., Park, J. C., Yoo, I. K., Baek, I. G., Kim, H. J., Yim, E. K., Lee, J. E., Park, S. O., Kim, H. S., In, Chung. U., Moon, J. T., & Ryu, B. I. (2006). Improvement of resistive memory switching in NiO using IrO_2 . *Applied Physics Letters*, 88(23), 232106, 0003-6951.

[24] Kim, K. M., Choi, B. J., Shin, Y. C., Choi, S., & Hwang, C. S. (2007). Anode-interface localized filamentary mechanism in resistive switching of TiO_2 thin films. *Applied Physics Letters*, 91(1), 012907, 0003-6951.

[25] Kim, S., Moon, H., Gupta, D., Yoo, S., & Choi, Y. K. (2009). Resistive Switching Characteristics of Sol-Gel Zinc Oxide Films for Flexible Memory Applications. *IEEE Transactions on Electron Devices*, 56(4), 696-699, 0018-9383.

[26] Kwon, D. H., Kim, K. M., Jang, J. H., Jeon, J. M., Lee, M. H., Kim, G. H., Li, X. S., Park, G. S., Lee, B., Han, S., Kim, M., & Hwang, C. S. (2010). Atomic structure of conducting nanofilaments in TiO_2 resistive switching memory. *Nature Nanotechnology*, 5(2), 148-153, 1748-3387.

[27] Lee, M. J., Park, Y., Suh, D. S., Lee, E. H., Seo, S., Kim, D. C., Jung, R., Kang, B. S., Ahn, S. E., Lee, C. B., Seo, D. H., Cha, Y. K., Yoo, I. K., Kim, J. S., & Park, B. H. (2007). Two Series Oxide Resistors Applicable to High Speed and High Density Nonvolatile Memory. *Advanced Materials*, 19(22), 3919-3923, 0935-9648.

[28] Li, Y. T., Long, S. B., Liu, Q., Lv, H. B., Liu, S., & Liu, M. (2011). An overview of resistive random access memory devices. *Chinese Science Bulletin*, 56(28-29), 3072-3078, 1001-6538.

[29] Liao, Z. L., Wang, Z. Z., Meng, Y., Liu, Z. Y., Gao, P., Gang, J. L., Zhao, H. W., Liang, X. J., Bai, X. D., & Chen, D. M. (2009). Categorization of resistive switching of metal-$Pr_{0.7}Ca_{0.3}MnO_3$-metal devices. *Applied Physics Letters*, 94(25), 253503, 0003-6951.

[30] Ling, Q. D., Liaw, D. J., Zhu, C. X., Chan, D. S. H., Kang, E. T., & Neoh, K. G. (2008). Polymer electronic memories: Materials, devices and mechanisms. *Progress in Polymer Sciecne*, 33(10), 917-978, 0079-6700.

[31] Liu, S. Q., Wu, N. J., & Ignatiev, A. (2000). Electric-pulse-induced reversible resistance change effect in magnetoresistive films. *Applied Physics Letters*, 76(19), 2749-2751, 0003-6951.

[32] Liu, Q., Long, S. B., Wang, W., Zuo, Q. Y., Zhang, S., Chen, J. N., & Liu, M. (2009). Improvement of resistive switching properties in ZrO_2-based ReRAM with implanted Ti ions. *IEEE Electron Device Letters*, 30(12), 1335-1337, 0741-3106.

[33] Liu, G., Zhuang, X. D., Chen, Y., Zhang, B., Zhu, J. H., Zhu, C. X., Neoh, K. G., & Kang, E. T. (2009). Bistable electrical switching and electronic memory effect in a solution-processable graphene oxide-donor polymer complex. *Applied Physics Letters*, 95(25), 253301, 0003-6951.

[34] Lu, W., & Lieber, C. M. (2007). Nanoelectronics from the bottom up. *Nature Materials*, 6(11), 841-850, 1476-1122.

[35] Mikolajick, T., Salinga, M., Kund, M., & Kever, T. (2009). Nonvolatile memory concepts based on resistive switching in inorganic materials. *Advanced Engineering Materials*, 11(4), 235-240, 1438-1656.

[36] Nian, Y. B., Strozier, J., Wu, N. J., Chen, X., & Ignatiev, A. (2007). Evidence for an oxygen diffusion model for the electric pulse induced resistance change effect in transition-metal oxides. *Physical Review Letters*, 98(14), 146403, 0031-9007.

[37] Novoselov, K. S., Geim, A. K., Morozov, S. V., Jiang, D., Zhang, Y., Dubonos, S. V., Grigorieva, I. V., & Firsov, A. A. (2004). Electric field effect in atomically thin carbon films. *Science*, 306(5696), 666-669, 0036-8075.

[38] Odagawa, A., Sato, H., Inoue, I. H., Akoh, H., Kawasaki, M., & Tokura, Y. (2004). Colossal electroresistance of a $Pr_{0.7}Ca_{0.3}MnO_3$ thin film at room temperature. *Physical Review B*, 70(22), 224403, 1098-0121.

[39] Panin, G. N., Kapitanova, O. O., Lee, S. W., Baranov, A. N., & Kang, T. W. (2011). Resistive switching in Al/graphene oxide/Al structure. *Japanese Journal of Applied Physics*, 50(7), 070110, 0021-4922.

[40] Pinnow, C. U., & Mikolajick, T. (2004). Material aspects in emerging nonvolatile memories. *Journal of the Electrochemical Society*, 151(6), K13-K19, 0013-4651.

[41] Robinson, J. T., Perkins, F. K., Snow, E. S., Wei, Z., & Sheehan, P. E. (2008). Reduced graphene oxide molecular sensors. *Nano Letters*, 8(10), 3137-3140, 1476-1122.

[42] Rozenberg, M. J., Inoue, I. H., & Sanchez, M. J. (2004). Nonvolatile memory with multilevel switching: A basic model. *Physical Review Letters*, 92(17), 178302, 0031-9007.

[43] Rozenberg, M. J., Sanchez, M. J., Weht, R., Acha, C., Gomez-Marlasca, F., & Levy, P. (2010). Mechanism for bipolar resistive switching in transition-metal oxides. *Physical Review B*, 81(11), 115101, 1098-0121.

[44] Rueckes, T., Kim, K., Joselevich, E., Tseng, G. Y., Cheung, C. L., & Lieber, C. M. (2000). Carbon nanotube-based nonvolatile random access memory for molecular computing. *Science*, 289(5476), 94-97, 0036-8075.

[45] Sawa, A., Fujii, T., Kawasaki, M., & Tokura, Y. (2004). Hysteretic current-voltage characteristics and resistance switching at a rectifying $Ti/Pr_{0.7}Ca_{0.3}MnO_3$ interface. *Applied Physics Letters*, 85(18), 4073-4075, 0003-6951.

[46] Schroeder, H., Zhirnov, V. V., Cavin, R. K., & Waser, R. (2010). Voltage-time dilemma of pure electronic mechanisms in resistive switching memory cells. *Journal of Applied Physics*, 107(5), 054517, 0021-4922.

[47] Seo, S., Lee, M. J., Seo, D. H., Jeoung, E. J., Suh, D. S., Joung, Y. S., Yoo, I. K., Hwang, I. R., Kim, S. H., Byun, I. S., Kim, J. S., Choi, J. S., & Park, B. H. (2004). Reproducible resistance switching in polycrystalline NiO films. *Applied Physics Letters*, 85(23), 5655-5657, 0003-6951.

[48] Simmons, J. G., & Verderber, R. R. (1967). New conduction and reversible memory phenomena in thin insulating films. *Proceedings of the Royal Society of London Series A-Mathematical and Physical Sciences*, 301(1464), 77, 1364-5021.

[49] Sinitskii, A., & Tour, J. M. (2009). Lithographic graphitic memories. *ACS Nano*, 3(9), 2760-2766, 1936-0851.

[50] Son, J. Y., & Shin, Y. H. (2008). Direct observation of conducting filaments on resistive switching of NiO thin films. *Applied Physics Letters*, 92(22), 222106, 0003-6951.

[51] Standlety, B., Bao, W. Z., Zhang, H., Bruck, J., Lau, C. N., & Bockrath, M. (2008). Graphene-based atomic-scale switches. *Nano Letters*, 8(10), 3345-3349, 1476-1122.

[52] Stankovich, S., Dikin, D. A., Dommett, G. H. B., Kohlhaas, K. M., Zimney, E. J., Stach, E. A., Piner, R. D., Nguyen, S. T., & Ruoff, R. S. (2006). Graphene-based composite materials. *Nature*, 442(7100), 282-286, 0028-0836.

[53] Stewart, D. R., Ohlberg, D. A. A., Beck, P. A., Chen, Y., Williams, R. S., Jeppesen, J. O., Nielsen, K. A., & Stoddart, J. F. (2004). Molecule-Independent Electrical Switching in Pt/Organic Monolayer/Ti Devices. *Nano Letters*, 4(1), 133-136, 1476-1122.

[54] Szot, K., Speier, W., Bihlmayer, G., & Waser, R. (2006). Switching the electrical resistance of individual dislocations in single-crystalline $SrTiO_3$. *Nature Materials*, 5(4), 312-320, 1476-1122.

[55] Tsuruoka, T., Terabe, K., Hasegawa, T., & Aono, M. (2010). Forming and switching mechanisms of a cation-migration-based oxide resistive memory. *Nanotechnology*, 21(42), 425205, 0957-4484.

[56] Tsuruoka, T., Terabe, K., Hasegawa, T., Valov, I., Waser, R., & Aono, M. (2012). Effects of moisture on the switching characteristics of oxide-based, gapless-type atomic switches. *Advanced Functional Materials*, 22(1), 70-77, 1616-301X.

[57] Tulina, N. A., Zver'kov, S. A., Mukovskii, Y. M., & Shulyatev, D. A. (2001). Current switching of resistive states in normal-metal manganite single-crystal point contacts. *Europhysics Letters*, 56(6), 836-841, 1286-4854.

[58] Wang, L. H., Yang, W., Sun, Q. Q., Zhou, P., Lu, H. L., Ding, S. J., & Zhang, D. W. (2012). The mechanism of the asymmetric SET and RESET speed of graphene oxide based flexible resistive switching memories. *Applied Physics Letters*, 100(6), 063509, 0003-6951.

[59] Wang, X. R., Ouyang, Y. J., Li, X. L., Wang, H. L., Guo, J., & Dai, H. J. (2008). Room-Temperature All-Semiconducting Sub-10-nm Graphene Nanoribbon Field-Effect Transistors. *Physical Review Letters*, 100(20), 206803, 0031-9007.

[60] Waser, R., & Aono, M. (2007). Nanoionics-based resistive switching memories. *Nature Materials*, 6(11), 833-840, 1476-1122.

[61] Waser, R., Dittmann, R., Staikov, G., & Szot, K. (2009). Redox-based resistive switching memories-Nanoionic mechanisms, prospects, and challenges. *Advanced Materials*, 21(25-26), 2632-2663, 1476-1122.

[62] Wu, X., Zhou, P., Li, J., Chen, L. Y., Lv, H. B., Lin, Y. Y., & Tang, T. A. (2007). Reproducible unipolar resistance switching in stoichiometric ZrO_2 films. *Applied Physics Letters*, 90(18), 183507, 0003-6951.

[63] Wu, C. X., Li, F. S., Zhang, Y. A., Guo, T. L., & Chen, T. (2011). Highly reproducible memory effect of organic multilevel resistive-switch device utilizing graphene oxide sheets/polyimide hybrid nanocomposite. *Applied Physics Letters*, 99(4), 042108, 0003-6951.

[64] Yang, C. H., Seidel, J., Kim, S. Y., Rossen, P. B., Yu, P., Gajek, M., Chu, Y. H., Martin, L. W., Holcomb, M. B., He, Q., Maksymovych, P., Balke, N., Kalinin, S. V., Baddorf, A. P., Basu, S. R., et al. (2009). Electric modulation of conduction in multiferroic Ca-doped $BiFeO_3$ films. *Nature Materials*, 8(6), 485-493, 1476-1122.

[65] Yang, Y. C., Pan, F., Liu, Q., Liu, M., & Zeng, F. (2009). Fully Room-Temperature-Fabricated Nonvolatile Resistive Memory for Ultrafast and High-Density Memory Application. *Nano Letters*, 9(4), 1636-1643, 1476-1122.

[66] Yin, K. B., Li, M., Liu, Y. W., He, C. L., Zhuge, F., Chen, B., Lu, W., Pan, X. Q., & Li, R. W. (2010). Resistance switching in polycrystalline $BiFeO_3$ thin films. *Applied Physics Letters*, 97(4), 042101, 0003-6951.

[67] Yu, A. D., Liu, C. L., & Chen, W. C. (2012). Supramolecular block copolymers: graphene oxide composites for memory device applications. *Chemical Communications*, 48(3), 383-385, 1359-7345.

[68] Zhuang, X. D., Chen, Y., Liu, G., Li, P. P., Zhu, C. X., Kang, E. T., Neoh, K. G., Zhang, B., Zhu, J. H., & Li, Y. X. (2010). Conjugated-polymer-functionalized graphene oxide: Synthesis and nonvolatile rewritable memory effect. *Advanced Materials*, 22(15), 1731-1735, 1476-1122.

[69] Zhuge, F., Dai, W., He, C. L., Wang, A. Y., Liu, Y. W., Li, M., Wu, Y. H., Cui, P., & Li, R. W. (2010). Nonvolatile resistive switching memory based on amorphous carbon. *Applied Physics Letters*, 96(16), 163505, 0003-6951.

[70] Zhuge, F., Hu, B. L., He, C. L., Zhou, X. F., Liu, Z. P., & Li, R. W. (2011). Mechanism of nonvolatile resistive switching in graphene oxide thin films. *Carbon*, 49(12), 3796-3802, 0008-6223.

[71] Zhuge, F., Li, R. W., He, C. L., Liu, Z. P., & Zhou, X. F. (2011, Mar.) Non-volatile resistive switching in graphene oxide thin films. *Physics and Applications of Graphene-Experiments*, 421-438, Print, 978-953-307-217-3, Croatia, InTech.

Chapter 8

Surface Functionalization of Graphene with Polymers for Enhanced Properties

Wenge Zheng, Bin Shen and Wentao Zhai

Additional information is available at the end of the chapter

http://dx.doi.org/10.5772/50490

1. Introduction

Graphene, a single-atom-thick sheet of hexagonally arrayed sp^2 bonded carbon atoms, has been under the spotlight owning to its intriguing and unparalleled physical properties [1]. Because of its novel properties, such as exceptional thermal conductivity, [2] high Young's modulus, [3] and high electrical conductivity,[4] graphene has been highlighted in fabricating various micro-electrical devices, batteries, supercapacitors, and composites [5, 7]. Especially, integration of graphene and its derivations into polymer has been highlighted, from the point views of both the spectacular improvement in mechanical, electrical properties, and the low cost of graphite [8, 9]. Control of the size, shape and surface chemistry of the reinforcement materials is essential in the development of materials that can be used to produce devices, sensors and actuators based on the modulation of functional properties. The maximum improvements in final properties can be achieved when graphene is homogeneously dispersed in the matrix and the external load is efficiently transferred through strong filler/polymer interfacial interactions, extensively reported in the case of other nanofillers. However, the large surface area of graphene and strong van der Waals force among them result in severe aggregation in the composites matrix. Furthermore, the carbon atoms on the graphene are chemically stable because of the aromatic nature of the bond. As a result, the reinforcing graphene are inert and can interact with the surrounding matrix mainly through van der Waals interactions, unable to provide an efficient load transfer across the graphene/matrix interface. To obtain satisfied performance of the final graphene/polymer composites, the issues of the strong interfacial adhesion between graphene–matrix and well dispersion of graphene should be addressed.

To date, the mixing of graphene and functionalized graphene with polymers covers the most of the published studies, and the direct modification of graphene with polymers is a somewhat less explored approach. However, in many cases, to achieve stable dispersions of graphene and adequate control of the microstructure of the nanocomposites, non-covalent or covalent functionalization of graphene with polymers may be necessary. The non-covalent functionalization, which relies on the van der Waals force, electrostatic interaction or π-π stacking [10, 12], is easier to carry out without altering the chemical structure of the graphene sheets, and provides effective means to tailor the electronic/optical property and solubility of the nanosheets [13]. The covalent functionalization of graphene derivatives is mainly based on the reaction between the functional groups of the molecules and the oxygenated groups on graphene oxide (GO) or reduced GO (r-GO) surfaces [14, 15], such as epoxides and hydroxyls on their basal planes and carboxyls on the edges [16]. Compared with non-covalent functionalization, the covalent functionalization of graphene-based sheets holds versatile possibility due to the rich surface chemistry of GO/r-GO. However, it should be pointed out that the non-covalent or covalent attachment of graphene to polymer chains can improve some properties, but may be negative for others, especially those related to the movement of electrons or phonons. Although the functionalization of graphene with polymers is generally attempted with a view to conferring to the polymer new or improved properties, the polymer may also prevent the aggregation of the graphene sheets, where the graphene-polymer size ratio and molecular weight play important roles. For general bibliography on typical graphene-based nanocomposites, the reader can consult several monographs, reviews, and feature articles that summarize the state of the art of the field.

The objective of the present work is to provide a broad overview on the methods developed to non-covalently or covalently bind graphene to polymers. The covalent linking of polymeric chains to graphene is at its initial stages and there is significant room for the development of new and improved strategies.

2. The precursor of functionalized graphene

As we know, GO is the main precursor for the functionalization of graphene with polymers. It is because that there are multiple oxygen-containing functionalities, such as hydroxyl, epoxy and carboxyl groups on GO sheets [16]. GO is usually produced using different variations of the Staudenmaier [17] or Hummers [18] method in which graphite is oxidized using strong oxidants such as $KMnO_4$, $KClO_3$, and $NaNO_2$ in the presence of nitric acid or its mixture with sulfuric [19, 20]. For more details about GO, we refer the reader to the extensive review of GO preparation, structure, and reactivity by Dreyer et al and Zhu et al. [19, 20].

Furthermore, the reduction of GO will remove most, but not all, of the oxygen-containing functionalities such as hydroxyl, carboxylic acid and epoxy groups. Therefore, some functionalization reactions are based on the reduced GO. Generally, GO can be exfoliated using a variety of methods, most commonly by solvent-based exfoliation and reduction in appropriate media or thermal exfoliation and reduction [16, 21] In the former route, the hydrophilic nature and increased interlayer spacing of GO facilitates direct exfoliation into solvents

http://dx.doi.org/10.5772/50490

(water, alcohol, and other protic solvents) assisted by mechanical exfoliation, such as ultrasonication and/or stirring, forming colloidal suspensions of "graphene oxide". The chemically reduced graphene oxide is produced by chemical reduction of the exfoliated graphene oxide sheets using hydrazine, [22, 24] dimethylhydrazine, [25] sodium borohydride followed by hydrazine, [26] hydroquinone, [27] vitamin C, [28] etc. However, the hazardous nature and cost of the chemicals used in reduction may limit its application. The most promising methods for large scale production of graphene is the thermal exfoliation and reduction of GO. Thermally reduced graphene oxide can be produced by rapid heating of dry GO under inert gas and high temperature [29, 31]. Heating GO in an inert environment at 1050°C for 30 s leads to reduction and exfoliation of GO, producing low-bulk-density TRG sheets, which are highly wrinkled [32]. In the thermal process, the epoxy and hydroxyl sites of GO decompose to produce gases like H_2O and CO_2, yielding pressures that exceed van der Waals forces holding the graphene sheets together, causing the occurrence of exfoliation.

3. Covalent functionalization of graphene with polymers

It is desirable that stronger bonds are usually formed between the graphene and the polymers by covalent functionalization of graphene with polymers. However, it is usually difficult to realize because ideal graphene lacks functional groups that can be conjugated with. In some cases, when the graphene sheets were exfiolated from GO, incomplete reduction process leaves oxygen-containing functionalities that are then available for further functionalizations. Other covalent functionalization strategies typically involve further disruption of the conjugation of the graphene sheets. Although covalent functionalization of graphene will compromise some of its natural conductivity, this method is still valuable in some cases when graphene's other properties are desirable. More details of graphene functionalization via covalent bonds will be discussed below.

3.1. Functionalizations via "grafting from" method

Until now, "grafting to" and "grafting from" techniques have been developed to graft the polymer chains onto the graphene surface. The "grafting from" method relies on the immobilization of initiators at the surface of graphene, followed by in situ surface-initiated polymerization to generate tethered polymer chains. A number of studies of polymer-functionalized graphene by the "grafting from" method have been reported.

3.1.1. Atom transfer radical polymerization (ATRP)

Among the types of "grafting from" polymerization [33, 46], ATRP is the most widely used, and represents the majority of the studies reported. ATRP is almost certainly chosen because it offers the advantages of radical polymerization, that is, a fast initiation process and the development of a dynamic equilibrium between dormant and growing radicals [47]. In addition, a wide range of monomers can be polymerized by ATRP with controlled chain length. Moreover, block copolymers can be prepared by ATRP because of the living radical process. Furthermore, ATRP is probably the most practical technique for preparation of functional polymers because the terminal alkyl halide can be converted to a wide variety of functionalities by using conventional organic synthetic procedures.

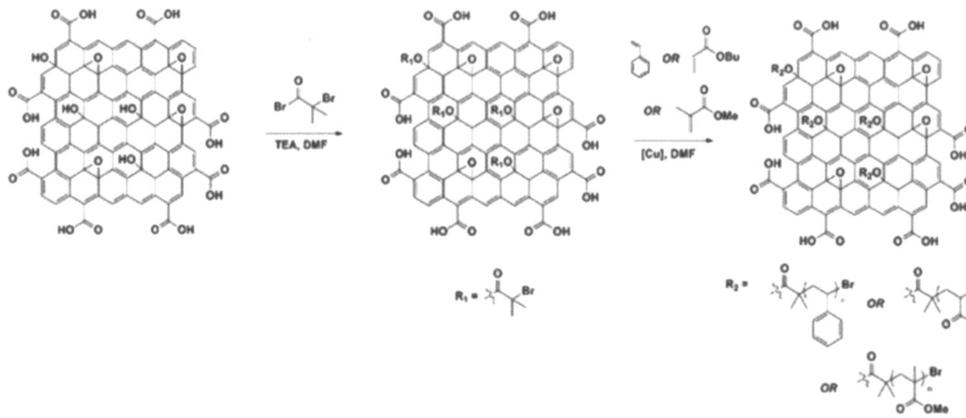

Figure 1. Synthesis of surface-functionalized GO via attachment of an ATRP initiator (a-bromoisobutyryl bromide) followed by polymerization of styrene, butyl acrylate, or methyl methacrylate [38].

Lee et al. [38] have reported a new method for attaching polymer brushes to GO sheet using surface-initiated ATRP. The hydroxyl groups present on the surface of GO were first functionalized with a wellknown ATRP initiator (a-bromoisobutyryl bromide), and then polymers of styrene, butyl acrylate, or methyl methacrylate were grown directly via a surface-initiated polymerization (SIP) (Figure 1). The authors studied the case of polystyrene (PS) in detail and presented two main conclusions. First, they suggested that the polymer chain length can be tunable by changing the ratio of monomer and initiator modified GO. Second, they reported that the monomer loading can vary the molecular weight of the grafted PS, which was obtained by gel permeation chromatography (GPC) after detaching by saponification, and the polydispersity was low, which suggested the polymerization proceeds in a controlled manner. Furthermore, the PS-functionalized GO was shown to significantly increase the solubility in N,N-dimethylformamide (DMF), toluene, chloroform, and dichloromethane, improving the processing potential of these materials for applications in polymer composites.

Fang et al. [43, 44] demonstrated the ability to systematically tune the grafting density and chain length of PS covalently bonded to graphene sheets by combining diazonium addition and ATRP. After reduction, r-GO was functionalized with 2-(4-aminophenyl) ethanol, reacted with 2-bromo-2-methylpropi-onyl bromide (BMPB), and subsequently the polymerization of styrene was carried out (Figure 2). Their results showed that the polydispersity of the high grafting density sample was more uniform than that of the low grafting density, which was attributed to the the degree of functionalization of r-GO sheets with the initiator because the diazonium coupling to graphene follows an identical radi-

http://dx.doi.org/10.5772/50490

cal mechanism as that for carbon nanotubes (CNTs). The relaxation of the polymer chains covalently bonded to the r-GO surface was strongly confined, particularly for segments in close proximity to the r-GO surface. This confinement effect could enhance the thermal conductivity of the polymer nanocomposites. The significant increases in thermal conductivity were observed for only 2.0 wt% functionalized r-GO in PS composites. Also, the resulting PS nanocomposites with 0.9 wt% functionalized r-GO revealed around 70% and 57% increases in tensile strength and Young's modulus.

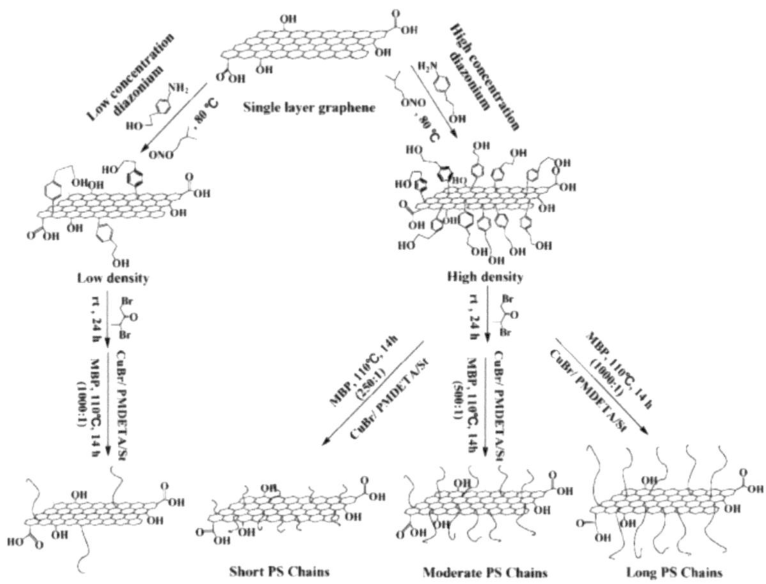

Figure 2. Synthetic routes for achieving controllable functionalization of graphene. [28]

Furthermore, Gonçalves et al. [40] developed the use of poly(methyl methacrylate) (PMMA) grafted from carboxylic groups in GO as a reinforcement filler. Here the BMPB initiators were immobilized by two esterification reactions: the carboxylic groups of GO were esterified with ethylene glycol, followed by reacting with BMPB using the same type of reaction as Lee et al. [38]. In this case, the polydispersity of the grafted PMMA, removed from the GO by hydrolysis, was found to be close to unity, once more suggesting a well-controlled process irrespective of the under estimated molecular weights. This PMMA-functionalized GO showed a good solubility in organic solvents such as chloroform and could be used as reinforcement filler in the preparation of PMMA composite films. Due to the strong interfacial interactions between the PMMA-functionalized GO and PMMA matrix caused by the presence of short PMMA chains covalently bonded to GO, an efficient load transfer from the GO to the matrix was formed, thus improving the mechanical properties of their nanocomposites, which were more stable and tougher than pure PMMA and its nanocomposites with unmodified GO (Figure 3). For example, addition of 1 wt% PMMA-functionalized GO clearly led to a significant improvement of the elongation at break, yielding a much more ductile and tougher material. In addition, the presence of PMMA-functionalized GO also stabilized the nanocomposites increasing the onset of thermal decomposition by around 50 °C.

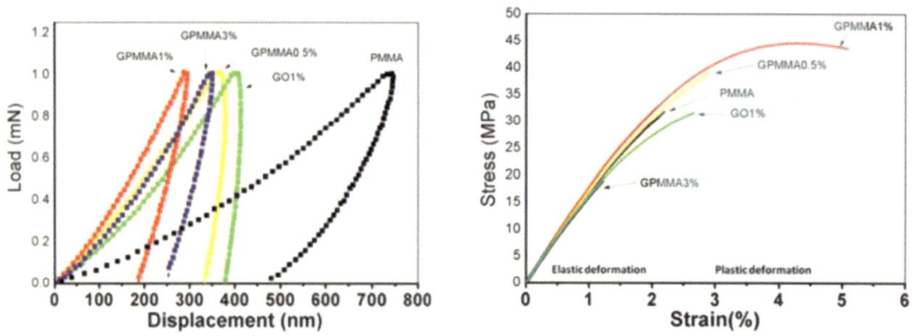

Figure 3. Load-displacement nanoidentation (left) and stress-strain curves (right) of films of PMMA and its nanocomposites with PMMA-functionalized graphene. [40]

Yang et al. [42] also took advantage of the carboxylic groups to graft poly-(2-dimethylaminoethyl methacrylate) (PDMAEMA) onto GO sheets. Here, the BMPB initiators were attached onto GO sheets by two steps involving amidation reactions. This functionalization of GO with PDMAEMA not only enhanced the solubility in acidic aqueous solutions (pH = 1), but also in short chain alcohols. Moreover, this solubility allowed this functionalized-GO to be mixed with spherical particles of poly(ethylene glycol dimethacrylate-co-methacrylic acid) to generate decorated GO sheets.

3.1.2. Other polymerization methods apart from ATRP

Besides the ATRP method, polycondensation, [33] ring opening polymerization [34], reversible addition-fragmentation chain transfer (RAFT) mediated mini-emulsion polymerization [35], direct electrophilic substitution [36], and Ziegler–Natta polymerization [37] have also been used to functionalize the graphene sheets with various polymer chains, which will be discussed below.

Wang et al. [33] functionalized GO sheets with polyurethane (PU) by using the polycondensation method. Here GO sheets were reacted with 4,4′-diphenylmethane diisocyanate followed by polycondensation of poly(tetramethylene glycol) and ethylene glycol. The presence of PU chains linked to GO remarkably improved the dispersion of GO in PU matrix, which was confirmed by the morphological study, and make it compatible with pure PU forming strong interfacial interactions that provide an enhanced load transfer between the matrix and the GO sheets thus improving their mechanical properties, as well as their thermal properties. With the incorporation of 2.0 wt% PU-functionalized GO, the tensile strength and storage modulus of the PU nanocomposites increased by 239% and 202%, respectively (Figure 4). Furthermore, the nanocomposites displayed high electrical conductivity, and improved thermal stability of PU was also achieved (Figure 4).

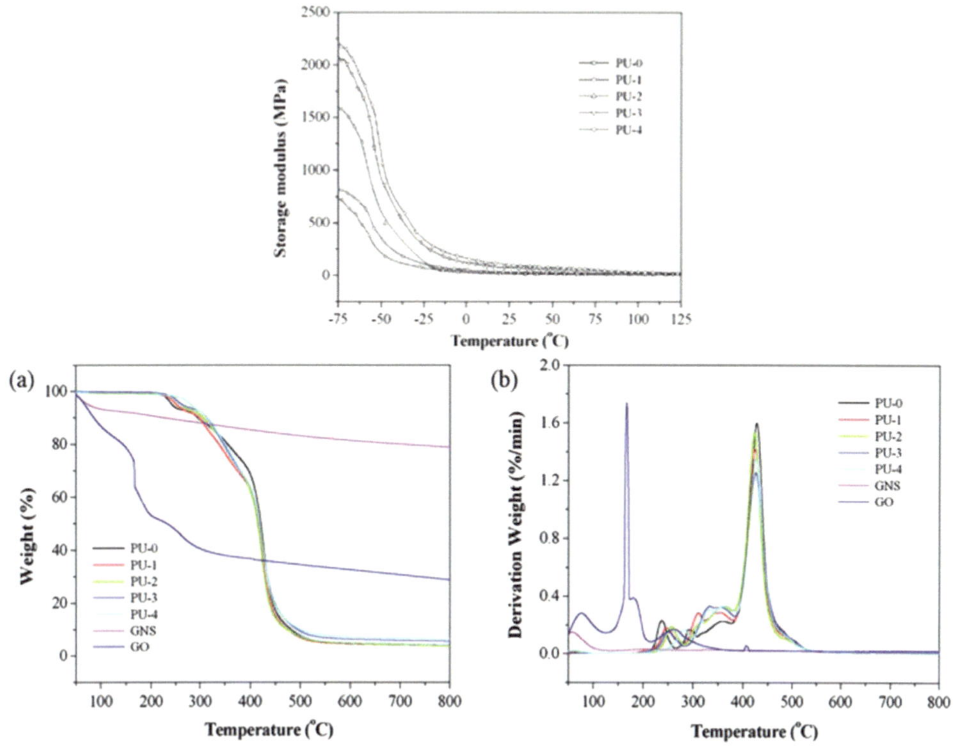

Figure 4. The improved thermal and mechanical propertied of PU/GO nanocomposites [33].

Etmimi et al. [35] investigated the preparation of PS/GO nanocomposites via RAFT mediated mini-emulsion polymerization. In this process, dodecyl isobutyric acid trithiocarbonate (DIBTC) RAFT agent was attached to the hydroxyl groups of GO through an esterification reaction (Figure 5). The resultant RAFT-grafted GO was used for the preparation of PS/GO nanocomposites in miniemulsion polymerization. The stable miniemulsions were obtained by sonicating RAFT-grafted GO in styrene monomer in the presence of a surfactant, followed by polymerizing using AIBN as the initiator to yield encapsulated PS-GO nanocomposites. The molecule weight and polydispersity of PS in the nanocomposites depended on the amount of RAFT-grafted GO in the system, in accordance with the features of the RAFT polymerization method. The thermal stability of the obtained PS/GO nanocomposites was improved, which may be attributed to the intercalation of PS into the lamellae of graphite. Furthermore, the increased RAFT-grafted GO significantly resulted in the improved mechanical properties of the nanocomposites. The storage and loss modulus of the nanocomposites were higher than those of the standard PS when the GO loadings reached 3 and 6%, respectively. Oppositely, as RAFT-grafted GO content increased, the T_g values of the sample decreased. This was attributed to the change in the molecule weight of the PS chains in the nanocomposites.

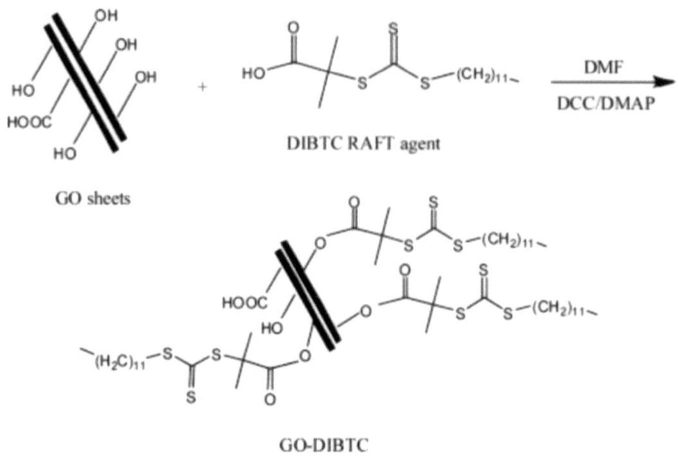

Figure 5. The overall synthesis route for the preparation of RAFT immobilized GO nanosheets [35].

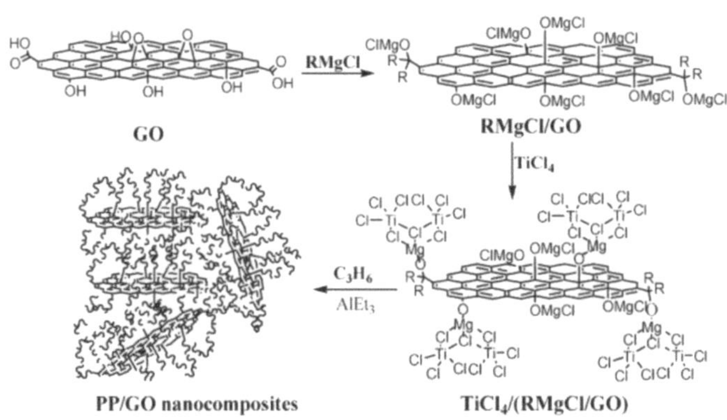

Figure 6. Fabrication of PP/GO nanocomposites by in situ Ziegler-Natta polymerization [37].

Huang et al. [37] reports the first example of preparation of polypropylene/GO (PP/GO) nanocomposites via in situ Ziegler–Natta polymerization. As illustrated in Figure 6, a Mg/Ti catalyst species was immobilized onto GO sheets by reacting with the surface functional groups including –OH and –COOH. Subsequent propylene polymerization led to the in situ formation of PP matrix, which was accompanied by the nanoscale exfoliation of GO. Independent of the opposing nature of the polymer and GO, a good dispersion of GO sheets in PP matrix was observed, which was verified by morphological examination through TEM and SEM observation. Furthermore, high electrical conductivity was discovered with thus prepared PP/GO nanocomposites, this being the only paper reporting conductive materials prepared by grafting a polymer from graphene sheets. For example, at a GO loading of 4.9

http://dx.doi.org/10.5772/50490

wt %, σ_c was measured at 0.3 S m^{-1}. We believe that this must originate from a side reaction involving the reduction of GO sheets that occurs in one of the synthetic steps.

3.1.3. Irradiation-induced polymerization

Ultrasound has found important applications in a diverse range of materials and chemical syntheses [48, 51]. Both the physical and chemical effects of ultrasound arise from acoustic cavitation: the formation, growth, and collapse of bubbles in liquids irradiated with high intensity ultrasound [48 - 50, 52]. Localized hot spots with temperature of ~5000K and pressures of hundreds of bars are generated during the bubble collapse within liquid, which can induce some chemical reactions that can't take place under normal conditions. Xu et al. [53] reported a convenient single-step sonication-induced approach for the preparation of polymer functionalized graphenes from graphite flakes and a reactive monomer, styrene. In this work, they showed that by choosing a reactive medium as the solvent, the combined mechanochemical effects of high intensity ultrasound can, in a single step, readily induce exfoliation of graphite to produce functionalized graphenes. Ultrasonic irradiation of graphite in styrene results in the mechanochemical exfoliation of graphite flakes to single-layer and few-layer graphene sheets combined with functionalization of the graphene with PS chains (Figure 7). The PS chains are formed from sonochemically initiated radical polymerization of styrene. They also tested a variety of other solvents, including toluene, ethylbenzene, 1-dodecene, and 4-vinylpyridine to prepare functionalized graphenes. Only the easily polymerizable reactants containing vinyl groups, styrene and 4-vinylpyridine, lead to stable functionalized graphene. Such functionalized graphene have good stability and solubility in common organic solvents and have great potential for graphene-based composite materials.

Moreover, direct photografting reactions of vinyl monomers came into focus for the preparation of stable polymer brushes. Two approaches in particular, the sequential "living" photopolymerization and the self-initiated photografting and photopolymerization (SIPGP), attracted the attention of numerous research groups because of the facile preparation and broad applicability. Steenackers et al. [54] showed that PS chains could covalently bound to graphene by the UV-induced polymerization of styrene (Figure 8). Photopolymerization occurs at existing defect sites and that there is no detectable disruption of the basal plane conjugation of graphene. This method thus offers a route to define graphene functionality without degrading its electronic properties. Furthermore, photopolymerization with styrene results in self-organized intercalative growth and exfoliation of few layer graphene sheets. Under these reaction conditions, a range of other vinyl monomers exhibits no reactivity with graphene. However, the authors demonstrate an alternative route by which the surface reactivity can be precisely tuned, and these monomers can be locally grafted via electron-beam-induced carbon deposition on the graphene surface.

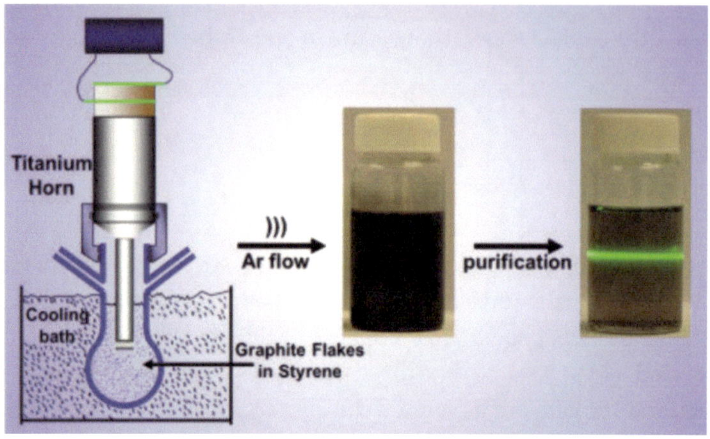

Figure 7. Experimental setup of the one-step mechanochemical process for exfoliation of graphite and sonochemical functionalization of graphene [53].

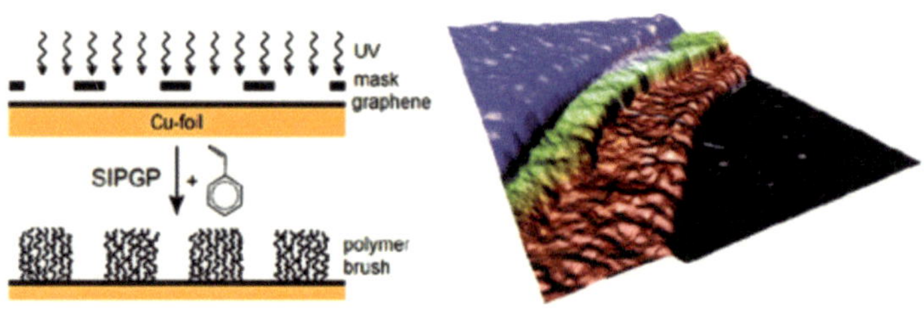

Figure 8. Patterned polymer brush layers on CVD-grown single layer graphene are prepared by UV illumination through a mask in bulk styrene. Surface photopolymerization occurs selectively in illuminated regions of the material [54].

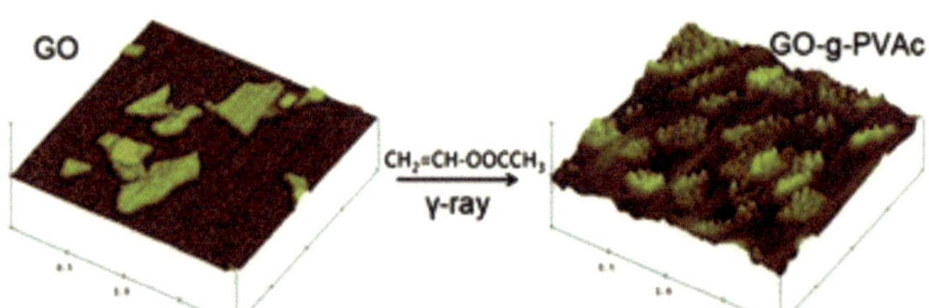

Figure 9. The preparation of PVAc grafted GO by γ-ray irradiation-induced graft polymerization [55].

γ-ray radiation-induced graft polymerization has many advantages, including being a single-step chemical reaction, needing no additives or catalysts, being conducted at room temperature, cost-effective, and so on. Above all, it is versatile for vinyl monomers that undergo free radiation polymerization, and production can be easily scaled-up. Zhang et al. [55] reported a facile approach to functionalize GO sheets with poly(vinyl acetate) (PVAc) by

http://dx.doi.org/10.5772/50490

γ-ray irradiation-induced graft polymerization (Figure 9). Due to the full coverage of PVAc chains and solvated layer formation on GO sheets surface, which weakens the interlaminar attraction of GO sheets, PVAc-functionalized GO was well dispersed in common organic solvents, and the dispersions obtained were extremely stable at room temperature without any aggregation.

Furthermore, Lee at al. [56] have developed a method to selectively fluorinate graphene by irradiating fluoropolymer-covered graphene with a laser (Figure 10). Here the sp^2-hybrized graphene would react with the active fluorine radicals, which was produced by photon-induced decomposition of the fluoropolymer under laser-irradiation, and form C-F bonds. However, this reaction only occurred in the laser-irradiated region. The kinetics of C–F bond formation is dependent on both the laser power and fluoropolymer thickness. Furthermore, the resistance of the graphene dramatically increased due to the fluorination, while the basic skeletal structure of the carbon bonding network is maintained. This is an efficient method for isolating graphene devices because the laser irradiation on fluoropolymer-covered graphene process produces fluorinated graphene with highly insulating properties in a single step.

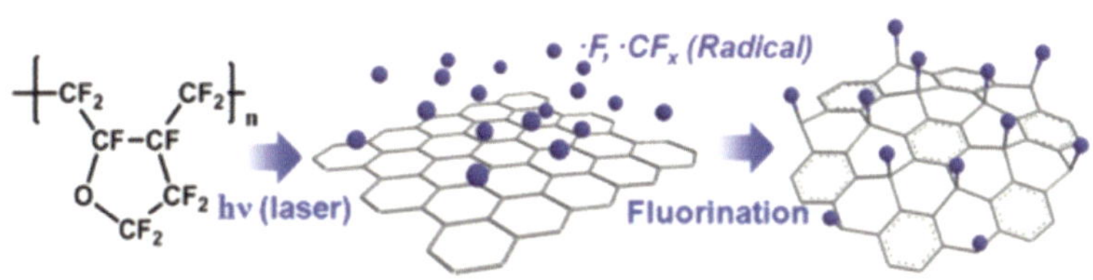

Figure 10. The scheme showing a mechanism for fluorination under laser-irradiation.[56]

3.2. Functionalizations via "grafting to" method

As commented previously, the "grafting from" method relies on the immobilization of initiators at the surface of graphene, followed by in situ surface-initiated polymerization to generate tethered polymer chains. However, this may not be possible in certain cases, where the covalent linkage between the presynthesized polymer and graphene emerges as the only alternative. In order to expand the type of polymers that can be bound to graphene, the category of "grafting-to" method can be employed to achieve this purpose. The "grafting to" technique involves the bonding of preformed end functionalized polymer chains to the surface of graphene. Therefore, the prepared graphene require adequate functional groups, which could react with specific polymers. Or the polymer has functionalities capable of reacting with either graphene or its chemically broader cousin, GO. In the following part, we will summarize the type of reactions and the families of polymers that have been grafted to the graphene.

3.2.1. Esterification/amidation reactions

Esterification/amidation reactions between carboxylic groups in GO and hydroxyl or amine groups in the polymer have been widely investigated [14, 15, 44, 57, 60]. In this respect, poly(vinyl alcohol) (PVA) was covalently bonded to GO [14, 57] and r-GO [14] by using a typical catalytic system for esterification (Figure 11). After functionalization with PVA chains, the solution processability of graphene was significantly improved. And the degree of functionalization was shown to be low, probably due to steric hindrance caused by the huge volume of GO. However, due to the presence of the huge graphene sheets, significant changes in the crystalline properties as well as in the tacticity of the polymer were observed. The originally semicrystalline PVA became completely amorphous, and the T_g increased by 35 °C after bonded to GO sheets. The decrease in crystallinity was attributed to the intercalation of PVA chains between the graphene sheets as well as the formation of "secondary" bonds, for example, hydrogen bonding that breaks intra- and interchain bonds. Finally, it has been demonstrated that the reaction is favoured at specific conformations at the isotactic sequences where the hydroxyl groups are more exposed (lower internal steric hindrance) than in the syndiotactic counterpart.

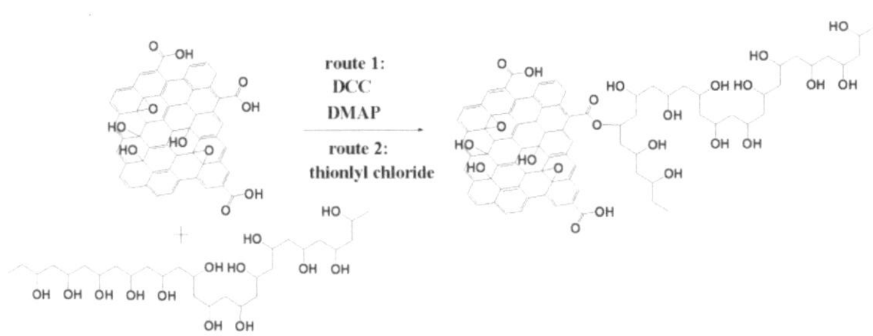

Figure 11. Schematic Illustration of the Esterification of GO with PVA [14].

A similar strategy has been approached to functionalize r-GO with poly(vinyl chloride) (PVC) [59]. In this step, the susceptible groups in PVC chains could react with the functional groups on r-GO sheets by esterification, which was provided by a nucleophilic substitution reaction [61, 62]. Furthermore, several methodologies to prepare r-GO/PVC nanocomposites and the optimum conditions have been established. The covalent attachment of r-GO to appropriately functionalized PVC is the only effective method to produce nanocomposites with improved thermal and mechanical properties (Figure 12). The absolute values of the mechanical and thermal properties of PVC-functionalized GO nanocomposites are higher than those for a similar system using MWNTs as reinforcement because of the higher aspect ratio of the r-GO sheets with respect to the MWNTs. The introduction of r-GO also increases the T_g of the composites, reflecting the changes in the mobility of the PVC chains. However, due to the lower strength of the "secondary bonds", such as halogen bonding and hydrogen bonding for PVC and PVA respectively, the changes in T_g for PVC were much lower than those reported for PVA, which had a similar degree of functionalization with PVC-functionalized GO. The existence of these secondary bonds can lead to some additional ordering that alters the segmental mobility and consequently the final properties.

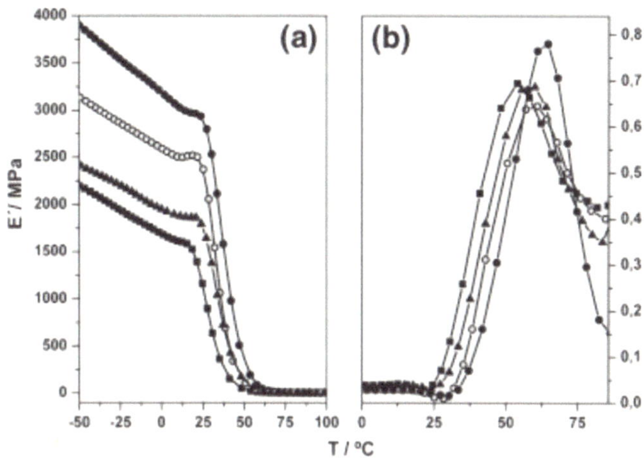

Figure 12. Comparison of (a) storage modulus and (b) tan δ curves for neat PVC (square), PVC functionalized CNTs (triangle), PVC functionalized GO (solid circle), and PVC functionalized r-GO (open circle) [59].

Furthermore, conjugated polymer-functionalized graphene materials have also been prepared by esterification/amidation reactions [60, 63]. In these cases, the ends of the conjugated polymers were bonded to the functional groups on the graphene sheets. As a result, the solubility of the obtained functionalized graphene was significantly improved in common solvents, enabling device preparation by solution processing. Thus, GO functionalized with both triphenylamine-based polyazomethine-modified GO (TPAPAM-GO) and poly(3-hexylthiophene) modified GO (P3HT-GO) (Figure 13) can be incorporated into specific devices by simple spin coating to obtain composites that exhibit non-volatile memory effect as well as higher power conversion efficiency for solar cells, demonstrated in the cases of TPAPAM GO and P3HT-GO, respectively.

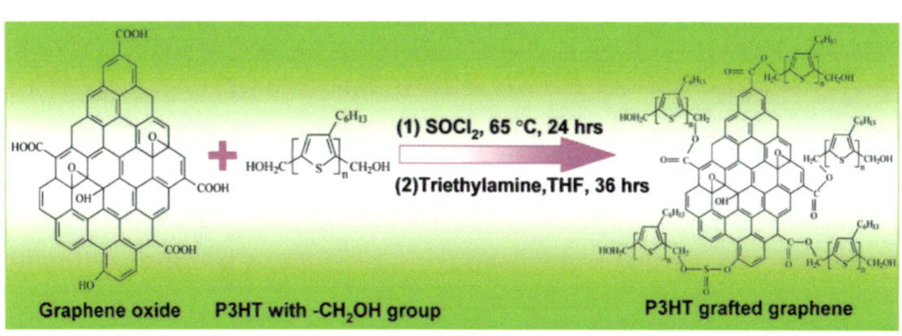

Figure 13. Synthesis procedure for chemical grafting of CH_2OH-terminated P3HT chains onto graphene, which involves the $SOCl_2$ treatment of GO (step 1) and the esterification reaction between acyl-chloride functionalized GO and MeOH-terminated P3HT (step 2) [47].

3.2.2. Nitrene cycloaddition

Nitrene chemistry, an approach used to functionalize graphene with single molecules [64, 65], has also been extended to polymers [66, 67].Nitrene chemistry is a versatile tool that allows the functionalization of graphene with a pool of functionalities, potentiating graphene solubility, and dispersion in a wide variety polymeric matrices. By using of this technology, He and Gao [67] have reported a general and versatile approach to graft of polymers onto graphene sheets. In their experiment, a wide range of immobilized functional groups were used to graft specific polymers from it (Figure 14). Though the cycloadditions of nitrene radicals and thermal reduction of GO occurs simultaneously, the conductivity of graphene diminished after functionalization, values of around 300-700 S/m for PS and poly(ethylene glycol) (PEG)-functioanlzied graphene were obtained. This was mainly due to the high amounts of graphene in the final product, which is reasonable because although the polymer is linked to graphene by the ends, the molecular weight of the polymers is low, making the mass percentage of graphene high.

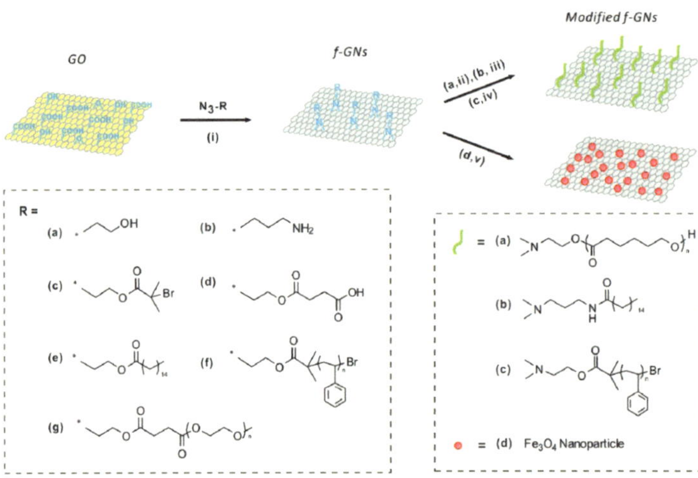

Figure 14. General strategy for the preparation of functionalized graphene sheets by nitrene chemistry and the further chemical modifications [67].

Besides the methods mentioned above, other "grafting to" approaches have also been investigated, such as the opening of maleic rings in maleic acid (MA) grafted polyethylene by amine functionalized graphene [68], nucleophilic epoxy-ring opening in GO by amine groups in biocompatible poly-l-lysine [69], atom transfer nitroxide radical coupling (ATNRP) of PNIPAM and 2,2,6,6-tetramethylpiperidine-1-oxyl-modified graphene [70], and simultaneous reduction of GO and radical grafting of PMMA by phase transfer [71].

3.2.3. Irradiation-induced radical grafting

Shen et al. [72] have reported PVA-functionalized graphene (f-G) could be prepared by ultrasonication of pristine graphene (p-G) in a PVA aqueous solution (Figure 15). Ultrasonic irradiation of graphene can cause a considerable amount of defects on the graphene surface, which might produce reactive sites in situ as a result of the high temperature and pressure during bubble collapse. Moreover, the original defects on pristine graphene can also be easily destroyed and produce reactive sites during ultrasonic irradiation. The PVA chain radicals produced by sonochemical degradation of the PVA solution can react easily with graphene, because of the reactive sites formed on the graphene surface, and readily functionalized them via "grafting to" method. The content of PVA on graphene was estimated to be ~35%. The f-G could be well dispersed in the PVA matrix by a simple solution mixing and casting procedure. Due to the effective load transfer between f-G and PVA matrix, the mechanical properties of the f-G/PVA films were significantly improved. Compared with the p-G/PVA films, a 12.6% increase in tensile strength and a 15.6% improvement of Young's modulus were achieved by addition of only 0.3 wt% f-G. Moreover, this simple ultrasonication technique could enable us to functionalize graphene with other polymers.

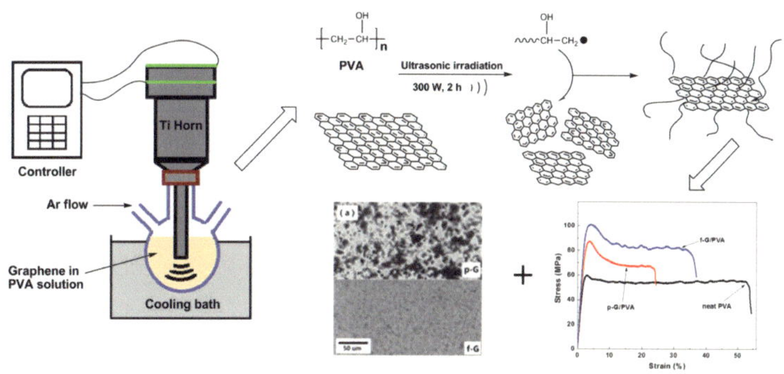

Figure 15. Schematic illustration of the sonochemical preparation process of PVA-functionalized graphene [72].

As a summary, "grafting-to" methods are highly versatile since they take advantage of the chemistry of GO that can be appropriately modified with a wide variety of functional groups providing a capacity for reaction with almost any type of polymers. In addition, "grafting-to" method allows the selection of the location of graphene, that is, at the end or as part of the main chain that can be directly related to changes in the final properties.

4. Non-covalent functionalization of graphene with polymers

As shown in the aforementioned examples, the covalent functionalization of polymers on graphene-based sheets holds versatile possibility due to the rich surface chemistry of GO/r-GO. Nevertheless, the non-covalent functionalization, which almost relies on hydrogen bonding or π–π stacking, is easier to carry out without altering the chemical structure of the capped r-GO sheets, and provides effective means to tailor the electronic/optical property and solubility of the nanosheets. The first example of non-covalent functionalization of r-GO sheets was demonstrated by the in situ reduction of GO with hydrazine in the presence of poly(sodium 4-styrenesulfonate) (PSS) [12], in which the hydrophobic backbone of PSS stabilizes the r-GO, and the hydrophilic sulfonate side groups maintains a good dispersion of the hybrid nanosheets in water.

4.1. Functionalizations via π–π stacking interactions

π–π stacking interactions usually occur between two relatively large non-polar aromatic rings having overlapping π orbitals. They can be comparable to covalent attachment in strength and hence provide more stable alternatives to the weaker hydrogen bonding, electrostatic bonding and coordination bonding strategies. Furthermore, π–π stacking functionalization does not disrupt the conjugation of the graphene sheets, and hence preserves the electronic properties of graphene.

4.1.1. Polymers with pyrene end-groups

In order to functionalize graphene with polymers via π–π stacking, one strategy is for the polymer chains to be synthesized with pyrene moieties as the termini of the polymer chains. RAFT polymerization can be a useful tool to achieve this aim. Polymers with pyrene end-groups have been made using RAFT mechanism in several recent papers [73, 80].

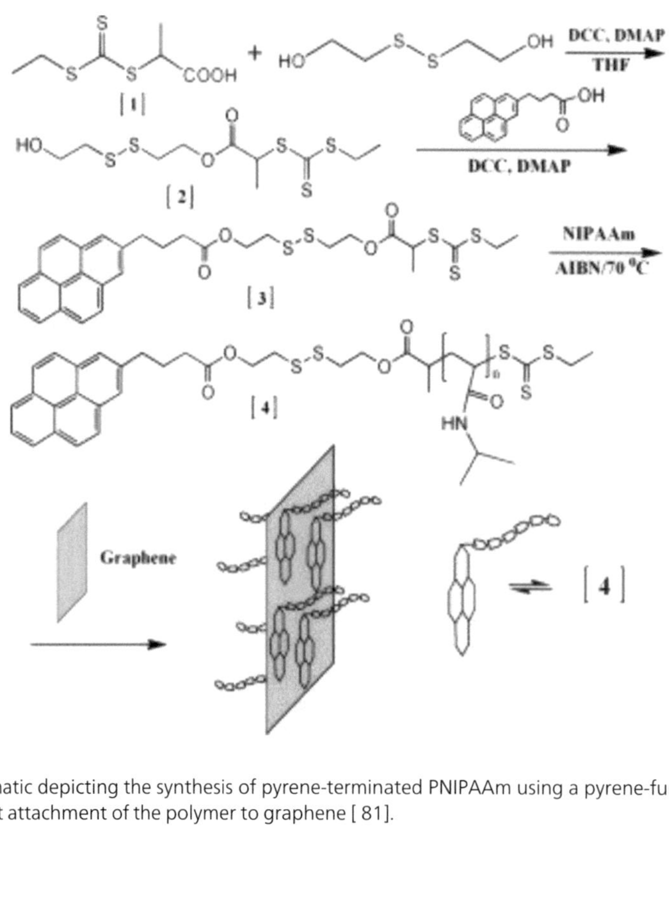

Figure 16. A schematic depicting the synthesis of pyrene-terminated PNIPAAm using a pyrene-functional RAFT agent and the subsequent attachment of the polymer to graphene [81].

Liu et al. synthesized thermoresponsive graphene-polymer nanocomposites. They first took advantage of RAFT polymerization to synthesize a well-defined thermoresponsive pyrene terminated poly(N-isopropylacrylamide) (PNIPAAm), followed by attachment onto the basal plane of graphene sheets via π-π stacking interactions (Figure 16) [81]. The lower critical solution temperature (LCST) of pyrene-terminated PNIPAAm was measured to be 33 ° C. However, after the pyrene-functional polymer functionalized with graphene sheets, the resultant graphene composites were also thermoresponsive in aqueous solutions, but with a lower LCST of 24 °C.

Similarly, Liu et al. also prepared pH sensitive graphene-polymer composites by functionalization of graphene with a pyrene-terminated positive charged polymer, poly(2-N,N′-(dimethyl amino ethyl acrylate)) (PDMAEA), and a negatively charged polymer, polyacrylic acid (PAA) [80]. During the process, a pyrene-terminated RAFT agent was used to prepare the pyrene-terminated PDMAEA and PAA. When manipulating the pH of the graphene −composite suspensions, phase transfer between the aqueous and organic phases was observed. Self-assembly of the two oppositely charged graphene-polymer composites afforded layer-by-layer (LbL) structures as evidenced by high-resolution scanning electron microscopy (SEM) and quartz crystal microbalance (QCM) measurements (Figure 17). In addition to RAFT mechanism, π-orbital rich polymers have also been synthesized by using of ATRP method for functionalization of r-GO to afford the fluorescent and water-soluble graphene composites via π-π stacking interactions [82].

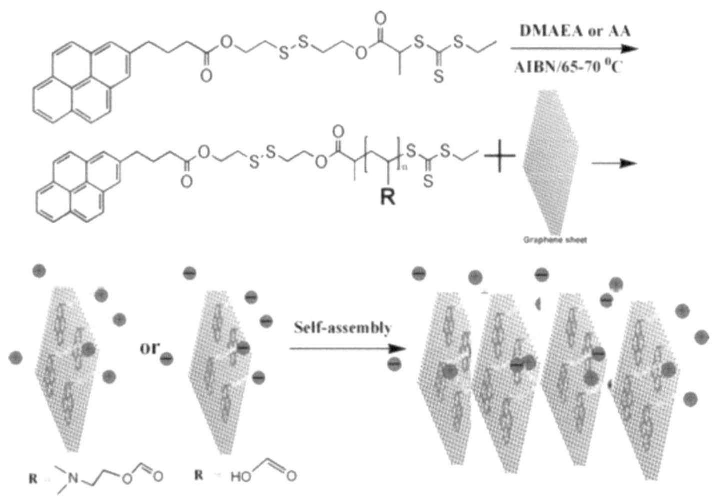

Figure 17. Synthesis of pH sensitive pyrene-polymer composites via π-π stacking interactions for the self-assembly of functionalized graphene into layered structures [80].

4.1.2. Conjugated polyelectrolytes

Moreover, conjugated polyelectrolytes with various functionalities have been used to modify r-GO nanosheets [83 - 85], in the hope to achieve good solubility in different kinds of solvents, and at the same time acquire added optoelectronic properties. Qi et al. has specially designed an amphiphilic coil–rod–coil conjugated triblock copolymer (PEG-OPE, chemical structure shown in Figure 18A) to improve the solubility of graphene-polymer nanocomposites in both high and low polar solvents [33, 83]. In the proposed configuration, the conjugated rigid-rod backbone of PEG-OPE can bind to the basal plane of the r-GO via the π-π stacking interaction (Figure 18B), whereas the lipophilic side chains and two hydrophilic coils of the backbone form an amphiphilic outer-layer surrounding the r-GO sheet. As a result, the obtained r-GO sheets with a uniformly coated polymer layer (Figure 18C) are soluble in both organic low polar (such as toluene and chloroform) and water-miscible high polar solvents (such as water and ethanol).

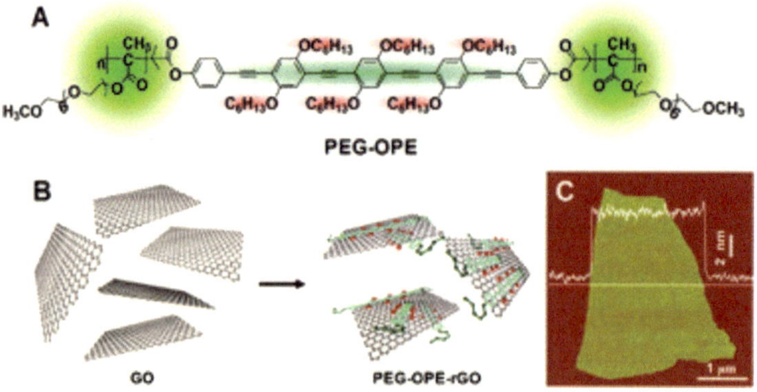

Figure 18. A) Chemical structure of PEG-OPE. (B) Schematic illustration of fabrication of PEG-OPE stabilized r-GO sheets. (C) Tapping-mode AFM image and cross-sectional analysis of PEG-OPE-r-GO on mica [67].

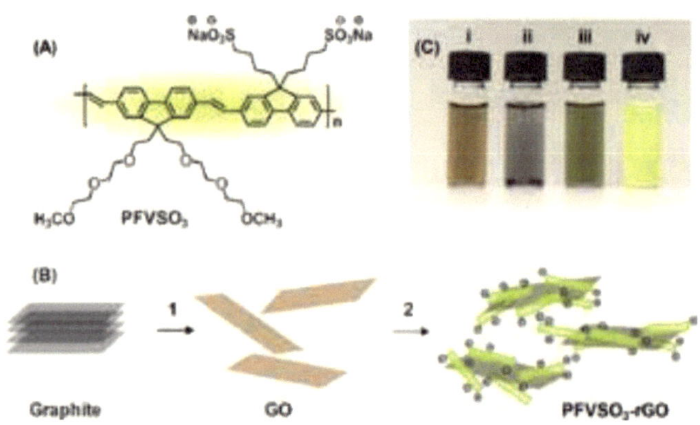

Figure 19. A) Chemical structure of the newly designed $PFVSO_3$. B) Schematic illustration of the synthesis of $PFVSO_3$-stabilized r-GO in H_2O: step 1, oxidative treatment of graphite (gray-black) yields single-layer GO sheets (brown); step 2, chemical reduction of GO with hydrazine in the presence of $PFVSO_3$ produces a stable aqueous suspension of $PFVSO_3$-functionalized r-GO sheets ($PFVSO_3$-r-GO). C) Photograph of aqueous dispersions of GO (i), r-GO (ii), $PFVSO_3$-r-GO (iii), and $PFVSO_3$ (iv) [84].

In another study, Qi et al. demonstrated the preparation of highly soluble r-GO hybrid material ($PFVSO_3$-r-GO) by taking advantage of strong π–π interactions between the anionic CPE and r-GO (Figure 19) [84]. The resulting CPE-functionalized r-GO (PFVSO3-r-GO) shows excellent solubility and stability in a variety of polar solvents, including water, ethanol, methanol, dimethyl sulfoxide, and dimethyl formamide. The morphology of $PFVSO_3$-r-GO is studied, which reveal a sandwich-like nanostructure. Within this nanostructure, the backbones of $PFVSO_3$ stack onto the basal plane of r-GO sheets via strong π–π interactions, while the charged hydrophilic side chains of $PFVSO_3$ prevent the rGO sheets from aggregating via electrostatic and steric repulsions, thus leading to the solubility and stability of $PFVSO_3$-rGO in polar solvents. Furthermore, the presence of $PFVSO_3$ within r-GO induces photoinduced charge transfer and p-doping of r-GO. As a result, the electrical conductivity of $PFVSO_3$-r-GO is not only much better than that of GO, but also than that of the unfunctionalized r-GO.

4.1.3. π–π stacking induced by melt blending

High temperature and strong shear forces are usually involved during the melt blending process, which tends to fracture the nanoparticle aggregates, and endow polymer chains with the ability to diffuse into the gaps of the nanoparticle interlayer. Furthermore, as suggested by theoretical and experimental studies [86, 87] chemical or physical interactions can be formed between the fillers and the polymer components. Zhang et al [88] found the melt blending led to enhanced interactions between PS and CNTs, which was indicated by increased amount of PS linked to CNTs and therefore dramatically increased solubility of CNTs in some solvents. Taking advantage of this method, Zhou et al [89] obtained PS-coated CNTs through simple melt mixing of PS with CNTs. Furthermore, Lu et al [90] studied the styrene-butadiene-styrene tri-block copolymer (SBS)/CNTs composite, and their results showed that there were interactions between CNTs and SBS occurred during melt mixing, leading to an improvement of the mechanical properties of SBS/CNTs composites, as well as the homogeneous dispersion of CNTs in SBS. The mechanism of melt blending on these enhanced interactions was mainly attributed to the formation of π-π stacking between the aromatic system of π-electrons of PS and the π-electrons system of CNTs during melt blending [10, 88].

Melt Blending can also graft PS chains onto the surface of graphene sheets via π-π interactions. The interaction between graphene and PS was significantly enhanced by melt blending, which led to an increased amount of PS-functional graphene (PSFG) exhibiting good solubility in some solvents [10]. The mechanism for the in-situ formation of π-π stacking was addressed, as illustrated in Figure 20. It was proposed that the strong shear action applied by extruder could stretch the PS chains and endow the polymer chains with possibility to diffuse into the interlayer gap of graphene sheets. Moreover, the PS chains could be pushed towards to graphene sheets to form the π-π stacking under high shear forces. The UV-vis absorption spectroscopy of PSFG presented an obvious red shift, suggested the presence of π-π stacking between PS and graphene.

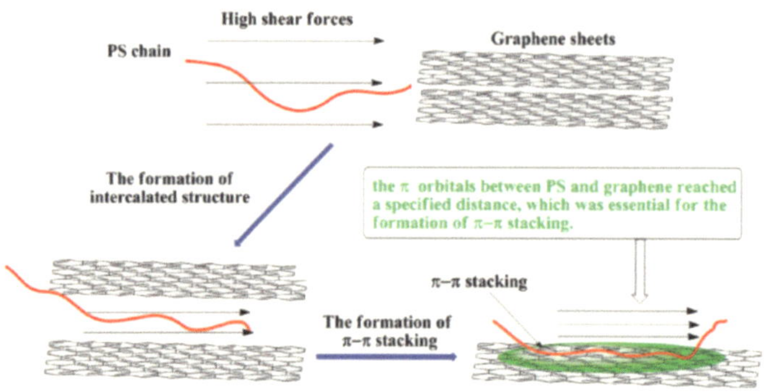

Figure 20. Schematic for the forming of π-π stacking in the process of melt blending [10].

4.2. Functionalizations via hydrogen bonding

Hydrogen bonding is very common in the biological world. An individual hydrogen bond is not very strong (2–8 kcal/mol); however, multiple hydrogen bonds will afford strong interactions as seen in DNA hybridization. Liang et al. prepared PVA nanocomposites with graphene by dispersing GO sheets into PVA matrix at molecular level [91]. The authors considered that the increased tensile strength and Young's modulus of the PVA/GO composite films were caused by the strong hydrogen bonding interactions between the residual oxygen-containing groups of GO sheets, such as epoxides and hydroxyls on their basal planes and carboxyls on the edges, and hydroxyl groups of the PVA chains. Polymer/graphene nanocomposites with other hydrophilic polymers, epoxy, poly(acrylonitrile) and polyaniline exhibited extraordinary high increase in modulus or glass transition temperature, attributed to hydrogen bonding interactions [92 - 94].

5. Applications of functionalized graphene

The use of graphene or functionalized graphene materials usually exploits properties such as the large surface area or high electrical conductivity. Currently, the applications of functionalized graphene are focused on clean energy devices and electronic devices, as well as on sensors, medical devices and catalysis. Two examples will be discussed below.

Non-covalent chemical modification between functionalized pyrene and graphene was adopted to achieve patterned arrays of glucose oxidase (GOD) for potential applications in glucose sensors, cell sensors and tissue engineering [96]. In research by Zeng et al. as illustrated in Figure 21, [95] chemically reduced GO was functionalized by pyrene-grafted poly(acrylic acid) (PAA) in aqueous solution on the basis of π-π stacking as well as van der Waals interactions. Then PAA-functionalized graphene (PAA-graphene) was LbL assembled with poly(ethyleneimine) (PEI). Graphene multilayer films facilitated the electron transfer, enhancing the electrochemical reactivity of H_2O_2. On the basis of this property, they fabricated a bienzyme biosensing system for the detection of maltose by successive LbL assembly of functionalized graphene, GOD, and glucoamylase (GA), which showed great promise in highly efficient sensors and advanced biosensing systems.

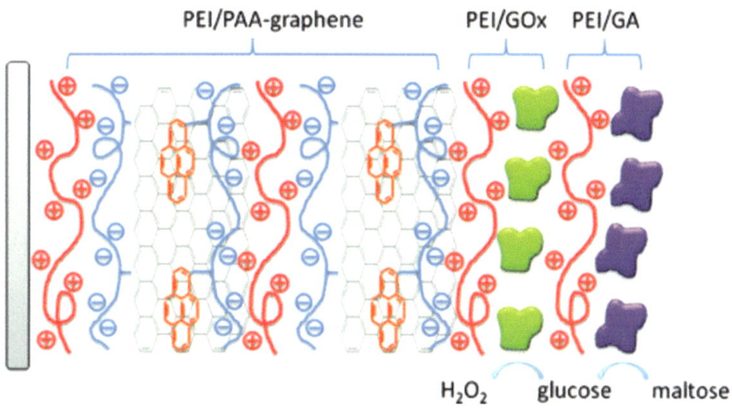

Figure 21. Schematic Illustration of the Strategies for layer-by-layer assembly of graphene multilayer films for enzyme-based glucose and maltose biosensing [95].

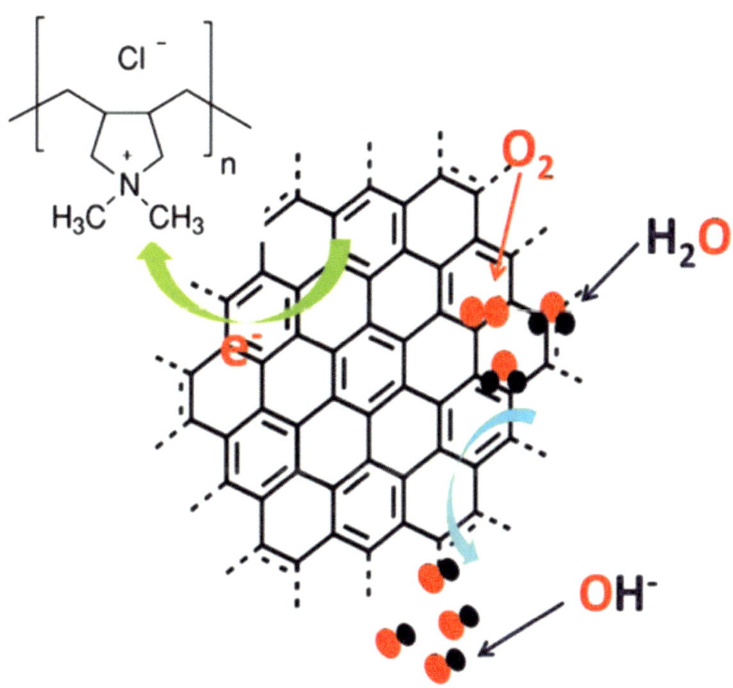

Figure 22. Schematic illustration of the electron-withdrawing from graphene by PDDA to facilitate the ORR process [97].

The planar structure and superb conductivity of graphene also provide an appropriate platform for novel electrochemical sensors. A metal-free electrocatalyst for the oxygen reduction reaction (ORR) was achieved with graphene sheets functionalized with an electron acceptor, poly(diallyldimethylammonium chloride) (PDDA) (Figure 22). The resultant positively charged graphene composite was demonstrated to show remarkable electrocatalytic activity toward ORR with better fuel selectivity, tolerance to CO poisoning, and long-term stability than that of the commercially available Pt/C electrode [97]. The observed ORR electrocatalytic activity induced by the intermolecular charge-transfer provides a general approach to various carbon-based metal-free ORR catalysts for oxygen reduction.

6. Conclusion

Generally speaking, functionalization of graphene with polymers can be achieved via either covalent or non-covalent interactions. Furthermore, the functionalization of graphene is always based on the graphene from previously prepared GO, which has multiple oxygen-containing functionalities, such as hydroxyl and carboxyl groups. In the covalent functionalizations, "grafting to" and "grafting from" techniques have been developed to graft the polymer chains onto the graphene surface. The "grafting from" method relies on the immobilization of initiators at the surface of graphene, followed by in situ surface-initiated polymerization to generate tethered polymer chains. The "grafting to" technique involves the bonding of preformed end functionalized polymer chains to the surface of graphene. Comparing both methods through the examples described above, it seems that "grafting-to" method allows the covalent bonding of a wider variety of polymers to graphene. The polymers "grafted from" graphene are those produced principally by some type of radical polymerization, such as ATRP and RAFT. However, the most relevant polymers grafted from graphene, such as PS and PMMA have also been attached to graphene by the "grafting-to" method. In principle, for polymers obtained by ATRP the "grafting-from" method might be the most appropriate, but this depends on the features desired in the final composite, because through "grafting to" method graphene forms part of the bulk polymer whereas in the "grafting from" it is limited to a terminal group. Moreover, except for a few exceptions, functional polymers and polymers synthesized by condensation reactions are principally bound to graphene by the grafting-to approach. Regarding to use PVA and PVC to functionalize graphene, we select the "grafting to" method due to the experimental procedures employed to obtain these polymers. In fact, it is well known that PVA is not prepared by polymerization of the corresponding monomer, and the majority of PVC is synthesized by free-radical polymerization through suspension or bulk processes. Finally, radiation-induced graft polymerization has been utilized in the "grafting to" and "grafting from" techniques, due to its advantages, including being a single-step chemical reaction, needing no additives or catalysts, being conducted at room temperature, cost-effective, and so on.

The covalent functionalization of polymers on graphene-based sheets holds versatile possibility due to the rich surface chemistry of GO/r-GO. Nevertheless, the non-covalent functionalization, which almost relies on hydrogen bonding or π-π stacking, is easier to carry out without altering the chemical structure of the capped rGO sheets, and provides effective means to tailor the electronic/optical property and solubility of the nanosheets. π-π stacking interactions usually occur between two relatively large non-polar aromatic rings having overlapping π orbitals. In order to functionalize graphene with polymers via π-π stacking, one strategy is for the polymer chains to be synthesized with pyrene moieties as the termini of the polymer chains. Moreover, conjugated polyelectrolytes with various functionalities have been used to functionalize graphene via π-π stacking. The hydrogen bonding interactions between the residual oxygen-containing groups of graphene sheets and hydroxyl groups of the hydrophilic polymer chains, such as PVA chains, have also been used to functionalize the graphene in order to obtain increased tensile strength and Young's modulus.

Author details

Wenge Zheng*, Bin Shen and Wentao Zhai

*Address all correspondence to: wgzheng@nimte.ac.cn

Ningbo Key Lab of Polymer Materials, Ningbo Institute of Material Technology and Engineering, Chinese Academy of Sciences, China

References

[1] Novoselov, K. S., Geim, A. K., Morozov, S. V., Jiang, D., Zhang, Y., Dubonos, S. V., Grigorieva, I. V., & Firsov, A. A. (2004). *Science*, 306, 666-669.

[2] Balandin, A. A., Ghosh, S., Bao, W., Calizo, I., Teweldebrhan, D., Miao, F., & Lau, C. N. (2008). *Nano Letters*, 8, 902-907.

[3] Lee, C., Wei, X., Kysar, J. W., & Hone, J. (2008). *Science*, 321, 385-388.

[4] Zhang, Y. W., Tan, W., Stormer, H. L., & Kim, P. (2005). *Nature*, 438, 201-204.

[5] Zhu, Y., Stoller, M. D., Cai, W., Velamakanni, A., Piner, R. D., Chen, D., & Ruoff, R. S. (2010). *ACS Nano*, 4, 1227-1233.

[6] Patil, A. J., Vickery, J. L., Scott, T. B., & Mann, S. (2009). *Advanced Materials*, 21, 3159-3164.

[7] Sudibya, H. G., He, Q., Zhang, H., & Chen, P. (2011). *ACS Nano*, 5, 1990-1994.

[8] Kuilla, T., Bhadra, S., Yao, D., Kim, N. H., Bose, S., & Lee, J. H. (2010). *Progress in Polymer Science*, 35, 1350-1375.

[9] Bai, H., Li, C., & Shi, G. (2011). *Advanced Materials*, 23, 1089-1115.

[10] Shen, B., Zhai, W., Chen, C., Lu, D., Wang, J., & Zheng, W. (2011). *ACS Applied Materials & Interfaces*, 3, 3103-3109.

[11] Bjo◎rk, J., Hanke, F., Palma, C. A., Samori, P., Cecchini, M., & Persson, M. (2010). *The Journal of Physical Chemistry Letters*, 1, 3407-3412.

[12] Stankovich, S., Piner, R. D., Chen, X., Wu, N., Nguyen, S. T., & Ruoff, R. S. (2006). *Journal of Materials Chemistry*, 16, 155-158.

[13] Huang, X., Qi, X., Boey, F., & Zhang, H. (2012). *Chemical Society Reviews*, 41, 666-686.

[14] Salavagione, H. J., Go′mez, M. n. A., & Marti′nez, G. (2009). *Macromolecules*, 42, 6331-6334.

[15] Liu, Z., Robinson, J. T., Sun, X., & Dai, H. (2008). *Journal of the American Chemical Society*, 130, 10876-10877.

[16] Park, S., & Ruoff, R. S. (2009). *Nat Nano*, 4, 217-224.

[17] Staudenmaier, L. (1898). *Berichte der deutschen chemischen Gesellschaft*, 31, 1481-1487.

[18] Hummers, W. S., & Offeman, R. E. (1958). *Journal of the American Chemical Society*, 1958(80), 1339.

[19] Dreyer, D. R., Park, S., Bielawski, C. W., & Ruoff, R. S. (2010). *Chemical Society Reviews*, 39.

[20] Zhu, Y., Murali, S., Cai, W., Li, X., Suk, J. W., Potts, J. R., & Ruoff, R. S. (2010). *Advanced Materials*, 22, 3906-3924.

[21] Park, S., An, J., Jung, I., Piner, R. D., An, S. J., Li, X., Velamakanni, A., & Ruoff, R. S. (2009). *Nano Letters*, 9, 1593-1597.

[22] Stankovich, S., Dikin, D. A., Piner, R. D., Kohlhaas, K. A., Kleinhammes, A., Jia, Y., Wu, Y., Nguyen, S. T., & Ruoff, R. S. (2007). *Carbon*, 45, 1558-1565.

[23] Lomeda, J. R., Doyle, C. D., Kosynkin, D. V., , W., Hwang, F., & Tour, J. M. (2008). *Journal of the American Chemical Society*, 130, 16201-16206.

[24] Wang, H., Robinson, J. T., Li, X., & Dai, H. (2009). *Journal of the American Chemical Society*, 131, 9910-9911.

[25] Stankovich, S., Dikin, D. A., Dommett, G. H. B., Kohlhaas, K. M., Zimney, E. J., Stach, E. A., Piner, R. D., Nguyen, S. T., & Ruoff, R. S. (2006). *Nature*, 442, 282-286.

[26] Si, Y., & Samulski, E. T. (2008). *Nano Letters*, 8, 1679-1682.

[27] Wang, G., Yang, J., Park, J., Gou, X., Wang, B., Liu, H., & Yao, J. (2008). *The Journal of Physical Chemistry C*, 112, 8192-8195.

[28] Ferna′ndez-Merino, M. J., Guardia, L., Paredes, J. I., Villar-Rodil, S., Soli′s-Ferna′ndez, P., Marti′nez-Alonso, A., & Tasco′n, J. M. D. (2010). *The Journal of Physical Chemistry C*, 114, 6426-6432.

[29] Schniepp, H. C., Li, J. L., Mc Allister, M. J., Sai, H., Herrera-Alonso, M., Adamson, D. H., Prud'homme, R. K., Car, R., Saville, D. A., & Aksay, I. A. (2006). *The Journal of Physical Chemistry B*, 110, 8535-8539.

[30] Mc Allister, M. J., Li, J. L., Adamson, D. H., Schniepp, H. C., Abdala, A. A., Liu, J., Herrera-Alonso, M., Milius, D. L., Car, R., Prud'homme, R. K., & Aksay, I. A. (2007). *Chemistry of Materials*, 19, 4396-4404.

[31] Steurer, P., Wissert, R., Thomann, R., & Mülhaupt, R. (2009). *Macromolecular Rapid Communications*, 30, 316-327.

[32] Kudin, K. N., Ozbas, B., Schniepp, H. C., Prud'homme, R. K., Aksay, I. A., & Car, R. (2007). *Nano Letters*, 8, 36-41.

[33] Wang, X., Hu, Y., Song, L., Yang, H., Xing, W., & Lu, H. (2011). *Journal of Materials Chemistry*, 21, 4222-4227.

[34] Kang, S. M., Park, S., Kim, D., Park, S. Y., Ruoff, R. S., & Lee, H. (2011). *Advanced Functional Materials*, 21, 108-112.

[35] Etmimi, H. M., Tonge, M. P., & Sanderson, R. D. (2011). *Journal of Polymer Science Part A: Polymer Chemistry*, 49, 1621-1632.

[36] Choi, E. K., Jeon, I. Y., Oh, S. J., & Baek, J. B. (2010). *Journal of Materials Chemistry*, 20, 10936-10942.

[37] Huang, Y., Qin, Y., Zhou, Y., Niu, H., Yu, Z. Z., & Dong, J. Y. (2010). *Chemistry of Materials*, 22, 4096-4102.

[38] Lee, S. H., Dreyer, D. R., An, J., Velamakanni, A., Piner, R. D., Park, S., Zhu, Y., Kim, S. O., Bielawski, C. W., & Ruoff, R. S. (2010). *Macromolecular Rapid Communications*, 31, 281-288.

[39] Lee, S. H., Kim, H. W., Hwang, J. O., Lee, W. J., Kwon, J., Bielawski, C. W., Ruoff, R. S., & Kim, S. O. (2010). *Angewandte Chemie International Edition*, 49, 10084-10088.

[40] Goncalves, G., Marques, P. A. A. P., Barros-Timmons, A., Bdkin, I., Singh, M. K., Emami, N., & Gracio, J. (2010). *Journal of Materials Chemistry*, 20, 9927-9934.

[41] Li, G. L., Liu, G., Li, M., Wan, D., Neoh, K. G., & Kang, E. T. (2010). *The Journal of Physical Chemistry C*, 114, 12742-12748.

[42] Yang, Y., Wang, J., Zhang, J., Liu, J., Yang, X., & Zhao, H. (2009). *Langmuir*, 25, 11808-11814.

[43] Fang, M., Wang, K., Lu, H., Yang, Y., & Nutt, S. (2009). *Journal of Materials Chemistry*, 19, 7098-7105.

[44] Fang, . M., Wang, K., Lu, H., Yang, Y., & Nutt, S. (2010). *Journal of Materials Chemistry*, 20, 1982-1992.

[45] Wang, D., Ye, G., Wang, X., & Wang, X. (2011). *Advanced Materials*, 23, 1122-1125.

[46] Pramoda, K. P., Hussain, H., Koh, H. M., Tan, H. R., & He, C. B. (2010). *Journal of Polymer Science Part A: Polymer Chemistry*, 48, 4262-4267.

[47] Coessens, V., Pintauer, T., & Matyjaszewski, K. (2001). *Progress in Polymer Science*, 26, 337-377.

[48] Bang, J. H., & Suslick, K. S. (2010). *Adv. Mater.*, 22, 1039-1059.

[49] Suslick, K. S., & Price, G. J. (1999). *Annu. Rev. Mater. Sci.*, 29, 295-326.

[50] Zhai, W., Yu, J., & He, J. (2008). *Polymer*, 49, 2430-2434.

[51] Cravotto, G., & Cintas, P. (2010). *Chemistry- A European Journal*, 16, 5246-5259.

[52] Suslick, K. S., & Flannigan, D. J. (2008). *Annual Review of Physical Chemistry*, 59, 659-683.

[53] Xu, H., & Suslick, K. S. (2011). *Journal of the American Chemical Society*, 133, 9148-9151.

[54] Steenackers, M., Gigler, A. M., Zhang, N., Deubel, F., Seifert, M., Hess, L. H., Lim, C. H. Y. X., Loh, K. P., Garrido, J. A., Jordan, R., Stutzmann, M., & Sharp, I. D. (2011). *Journal of the American Chemical Society*, 2011(133), 10490-10498.

[55] Zhang, B., Zhang, Y., Peng, C., Yu, M., Li, L., Deng, B., Hu, P., Fan, C., Li, J., & Huang, Q. (2012). *Nanoscale*, 2012(4), 1742-1748.

[56] Lee, W. H., Suk, J. W., Chou, H., Lee, J., Hao, Y., Wu, Y., Piner, R., Akinwande, D., Kim, K. S., & Ruoff, R. S. (2012). *Nano Letters*.

[57] Veca, L. M., Lu, F., Meziani, M. J., Cao, L., Zhang, P., Qi, G., Qu, L., Shrestha, M., & Sun, Y. P. (2009). *Chemical Communications*, 2565-2567.

[58] Yu, D., & Dai, L. (2009). *The Journal of Physical Chemistry Letters*, 1, 467-470.

[59] Salavagione, H. J., & Martı́nez, G. (2011). *Macromolecules*, 44, 2685-2692.

[60] Zhuang, X. D., Chen, Y., Liu, G., Li, P. P., Zhu, C. X., Kang, E. T., Noeh, K. G., Zhang, B., Zhu, J. H., & Li, Y. X. (2010). *Advanced Materials*, 22, 1731-1735.

[61] Martínez, G., & Millán, J. L. (2004). *Journal of Polymer Science Part A: Polymer Chemistry*, 42, 6052-6060.

[62] Martínez, G. (2006). *Journal of Polymer Science Part A: Polymer Chemistry*, 44, 2476-2486.

[63] Yu, D., Yang, Y., Durstock, M., Baek, J. B., & Dai, L. (2010). *ACS Nano*, 4, 5633-5640.

[64] Choi, . J., Kim, K. j., Kim, B., Lee, H., & Kim, S. (2009). *The Journal of Physical Chemistry C*, 113, 9433-9435.

[65] Strom, T. A., Dillon, E. P., Hamilton, C. E., & Barron, A. R. (2010). *Chemical Communications*, 46, 4097-4099.

[66] Xu, X., Luo, Q., Lv, W., Dong, Y., Lin, Y., Yang, Q., Shen, A., Pang, D., Hu, J., Qin, J., & Li, Z. (2011). *Macromolecular Chemistry and Physics*, 212, 768-773.

[67] He, H., & Gao, C. (2010). *Chemistry of Materials*, 22, 5054-5064.

[68] Lin, Y., Jin, J., & Song, M. (2011). *Journal of Materials Chemistry*, 21, 3455-3461.

[69] Shan, C., Yang, H., Han, D., Zhang, Q., Ivaska, A., & Niu, L. (2009). *Langmuir*, 25, 12030-12033.

[70] Deng, Y., Li, Y., Dai, J., Lang, M., & Huang, X. (2011). *Journal of Polymer Science Part A: Polymer Chemistry*, 49, 1582-1590.

[71] Vuluga, D., Thomassin, J. M., Molenberg, I., Huynen, I., Gilbert, B., Jerome, C., Alexandre, M., & Detrembleur, C. (2011). *Chemical Communications*, 47, 2544-2546.

[72] Shen, B., Zhai, W., Lu, D., Wang, J., & Zheng, W. (2012). *RSC Advances*, 2, 4713-4719.

[73] Duan, Q., Miura, Y., Narumi, A., Shen, X., Sato, S. I., Satoh, T., & Kakuchi, T. (2006). *Journal of Polymer Science Part A: Polymer Chemistry*, 44, 1117-1124.

[74] Scales, C. W., Convertine, A. J., & Mc Cormick, C. L. (2006). *Biomacromolecules*, 7, 1389-1392.

[75] Zhou, N., Lu, L., Zhu, J., Yang, X., Wang, X., Zhu, X., & Zhang, Z. (2007). *Polymer*, 48, 1255-1260.

[76] Segui, F., Qiu, X. P., & Winnik, F. M. (2008). Journal of Polymer Science Part A: Polymer Chemistry ., 46, 314-326.

[77] Meuer, S., Braun, L., & Zentel, R. (2008). *Chemical Communications*, 3166-3168.

[78] Meuer, S., Braun, L., Schilling, T., & Zentel, R. (2009). *Polymer*, 50, 154-160.

[79] Xu, J., Tao, L., Boyer, C., Lowe, A. B., & Davis, T. P. (2010). *Macromolecules*, 44, 299-312.

[80] Liu, J., Tao, L., Yang, W., Li, D., Boyer, C., Wuhrer, R., Braet, F., & Davis, T. P. (2010). *Langmuir*, 26, 10068-10075.

[81] Liu, J., Yang, W., Tao, L., Li, D., Boyer, C., & Davis, T. P. (2010). *Journal of Polymer Science Part A: Polymer Chemistry*, 48, 425-433.

[82] Xu, L. Q., Wang, L., Zhang, B., Lim, C. H., Chen, Y., Neoh, K. G., Kang, E. T., & Fu, G. D. (2011). *Polymer*, 52, 2376-2383.

[83] Qi, X., Pu, K. Y., Li, H., Zhou, X., Wu, S., Fan, Q. L., Liu, B., Boey, F., Huang, W., & Zhang, H. (2010). *Angewandte Chemie International Edition*, 49, 9426-9429.

[84] Qi, X., Pu, K. Y., Zhou, X., Li, H., Liu, B., Boey, F., Huang, W., & Zhang, H. (2010). *Small*, 6, 663-669.

[85] Yang, H., Zhang, Q., Shan, C., Li, F., Han, D., & Niu, L. (2010). *Langmuir*, 26, 6708-6712.

[86] Cheah, K., Simon, G. P., & Forsyth, M. (2001). *Polym. Int.*, 50, 27-36.

[87] Ginzburg, V. V., Gendelman, O. V., & Manevitch, L. I. (2001). *Phys. Rev. Lett.*, 86, 5073.

[88] Zhang, Z., Zhang, J., Chen, P., Zhang, B., He, J., & Hu, G. H. (2006). *Carbon*, 44, 692-698.

[89] Zhou, T. N., Hou, Z. C., Wang, K., Zhang, Q., & Fu, Q. (2011). *Polymers for Advanced Technologies*, 22, 1359-1365.

[90] Lu, L., Zhou, Z., Zhang, Y., Wang, S., & Zhang, Y. (2007). *Carbon*, 45, 2621-2627.

[91] Liang, J., Huang, Y., Zhang, L., Wang, Y., Ma, Y., Guo, T., & Chen, Y. (2009). *Advanced Functional Materials*, 19, 2297-2302.

[92] Rafiee, M. A., Rafiee, J., Wang, Z., Song, H., Yu, Z. Z., & Koratkar, N. (2009). *ACS Nano*, 3, 3884-3890.

[93] Ramanathan, T. A. A., Abdala, Stankovich. S., Dikin, D. A., Herrera, M., Alonso, R. D., Piner, D. H., Adamson, H. C., Schniepp, Chen. X., Ruoff, R. S., Nguyen, S. T., Aksay, I. A., Prud'Homme, R. K., & Brinson, L. C. (2008). *Nat Nano*, 3, 327-331.

[94] Wang, H., Hao, Q., Yang, X., Lu, L., & Wang, X. (2010). ACS Applied Materials & Interfaces,. 2, 821-828.

[95] Zeng, G., Xing, Y., Gao, J., Wang, Z., & Zhang, X. (2010). *Langmuir*, 26, 15022-15026.

[96] Cardinali, M., Valentini, L., & Kenny, J. M. (2011). *The Journal of Physical Chemistry C*, 115, 16652-16656.

[97] Wang, S., Yu, D., Dai, L., Chang, D. W., & Baek, J. B. (2011). ACS Nano, , 5, 6202-6209.

Chapter 9

Graphene Nanowalls

Mineo Hiramatsu, Hiroki Kondo and Masaru Hori

Additional information is available at the end of the chapter

http://dx.doi.org/10.5772/51528

1. Introduction

Graphene, hexagonal arrangement of carbon atoms forming one-atom thick planar sheet, is a promising material for future electronic applications due to their high electrical conductivity as well as chemical and physical stability [1]. Planar graphene films with respect to the substrate have been synthesized using various methods including mechanical exfoliation from highly oriented pyrolytic graphite, chemical exfoliation from bulk graphite, thermal decomposition of carbon-terminated silicon carbide, and chemical vapor deposition (CVD) on metals such as nickel and copper substrates [2–6].

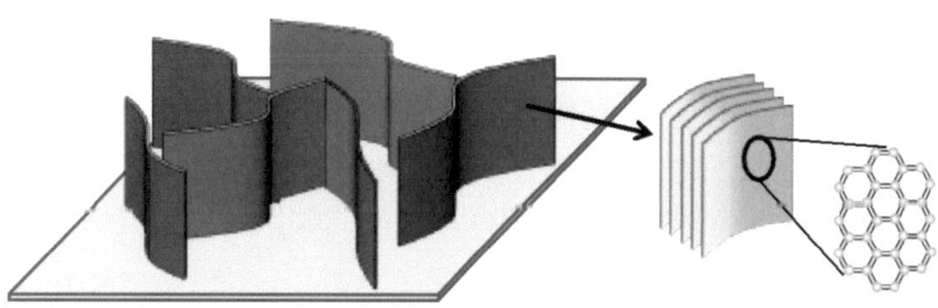

Figure 1. Schematic illustration of graphene nanowalls

On the other hand, plasma-enhanced CVD (PECVD) is among the early methods to synthesize of vertically standing few layer graphenes or graphene nanowalls (GNWs) [7–12]. GNWs can be described as self-assembled, vertically standing, few-layered graphene sheet nanostructures, which are also called as carbon nanowalls, carbon nanosheets, and carbon nanoflakes. As illustrated in Fig. 1, the sheets form a self-supported network of wall structures with thicknesses ranging from a few nanometers to a few tens of nanometers. GNWs have a high density of atomic scale graphitic edges that are potential sites for electron field emission, which might lead to the application in flat panel displays and light sources [13,14].

The large surface area of GNWs is useful as templates for the fabrication of other types of nanostructured materials, electrodes for energy storage devices and biosensors [15–20]. These graphene-based materials are applied in the field of electrochemistry including electrode for fuel cell and electrochemical sensors. In these applications, graphene-based materials are often decorated with metal nanoparticles and other materials.

In this chapter, basic properties of GNWs and their prospective applications are described. First of all, synthesis and characterization of GNWs are outlined. PECVD is becoming one of the most promising techniques for the production of carbon materials including diamond, aligned carbon nanotube films and GNWs, due to its feasibility and potentiality for large-area production with reasonable growth rates at relatively low temperatures. In the present study, GNW growth using inductively coupled plasma (ICP) enhanced CVD is featured, since the ICP CVD system has advantages of simple design and scalability to large area growth. The growth mechanism, characterization of GNWs, and several decoration techniques of GNW surface are described.

Due to the large surface area of GNWs, we can expect a variety of electrochemical applications using GNWs such as batteries, capacitors, and sensors. To these ends, GNWs are often decorated with nanoparticles or films. In the latter half of this chapter, application of GNWs as electrode for fuel cell is described. GNWs were grown on the carbon fiber paper. Then metal organic chemical fluid deposition (MOCFD) using supercritical fluid (SCF) was applied to form platinum (Pt) nanoparticles on the surface of GNWs. Using this method, highly dispersed Pt nanoparticles of approximately 2 nm in diameter were formed on the surface of GNWs grown on the carbon fiber paper. Furthermore, the application as a biosensor using GNWs is described. As another example, the electrocatalytic activity of GNWs for determining dopamine, ascorbic acid and uric acid in phosphate buffer solution was investigated. The ability of GNWs as a platform to create graphene-based hybrid materials is demonstrated.

2. Growth of graphene nanowalls and their growth mechanism

2.1. Growth of graphene nanowalls using inductively coupled plasma CVD

Synthesis techniques for GNWs and related vertical graphene structures are similar to those used for diamond films and carbon nanotubes (CNTs). In general, a mixture of hydrocarbon and hydrogen or argon gases, typically CH_4 and H_2, is used as source gases for the synthesis of GNWs. Unlike the CNT growth, GNWs can be fabricated on a variety of substrates, including Si, SiO_2, Al_2O_3, Ni, Ti, and stainless steel, at substrate temperatures of 500–750 °C without the use of catalysts [8]. To date, GNWs have been grown using various PECVD methods employing microwave plasma, inductively coupled plasma (ICP), capacitively coupled plasma (CCP) with H radical injection, very high frequency (VHF) plasma with H radical injection, electron beam excited plasma, and DC plasma [7–12,19].

Radio frequency (RF: 13.56 MHz) ICP is one of high-density plasmas, and has been used to etch several materials including Si, SiO_x, SiN_x, and metal films in the LSI fabrication process. The ICP is operated at relatively low pressures below 100 mTorr (13.3 Pa). In the case of pla-

http://dx.doi.org/10.5772/51528

nar geometry, RF power is inductively coupled into the process chamber with a planar-coil antenna through a quartz (fused silica) window, and plasma is generated in the chamber. Plasma density of ICP discharge is on the order of 10^{12} cm^{-3}. Figure 2 shows a schematic of ICP reactor with planar geometry used for the growth of GNWs. The ICP reactor was 16 cm in diameter and 30 cm in height. A one-turn coil antenna with a diameter of 10 cm was set on a quartz window at the top of reactor. Si or SiO_2-coated Si substrates were set on the middle of the substrate holder at 10 cm below the quartz window.

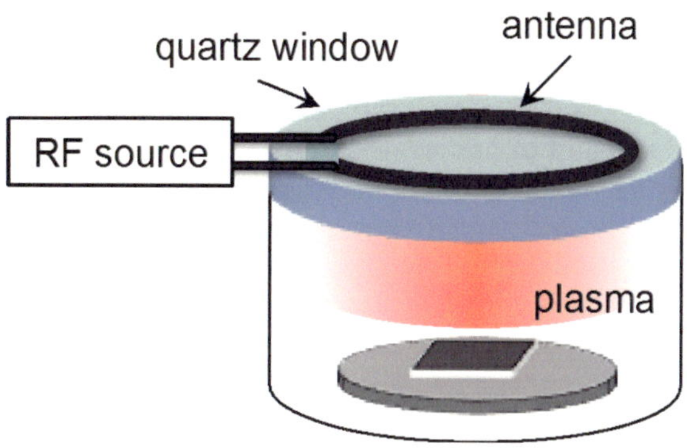

Figure 2. Schematic of inductively coupled plasma reactor with planar geometry used for the growth of GNWs.

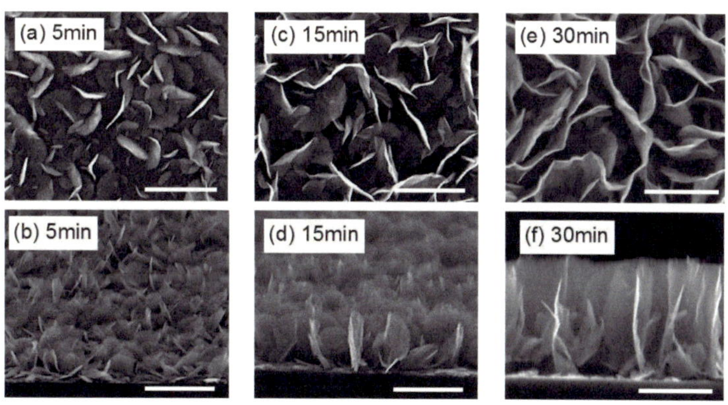

Figure 3. SEM images of the GNW films grown by ICP-CVD employing a mixture of CH_4 and Ar for (a–b) 5 min, (c–d) 15 min, and (e–f) 30 min. Scale bar: 1 μm [21].

Figures 3(a) –3(f) show scanning electron microscopy (SEM) images of GNW films grown by ICP-CVD employing a mixture of CH_4 and Ar for different growth periods. After the nucleation stage of GNWs, growth of less-aligned, isolated carbon sheets with a semicircular shape standing on the substrate is confirmed as shown in Figs. 3(a) and 3(b). As the growth period increased, density of isolated nanosheets increased and those standing almost vertically on the substrate continued preferably to spread faster. Then, spreading nanosheets met one another; eventually resulting in the formation of linked nanowalls as shown in Figs. 3(c) and 3(d). During the early growth stage after the nucleation up to the steady-state growth, the growth for the inclined smaller nanowalls was terminated, while the vertical nanowalls preferentially continued to grow. Therefore, with the increase of growth period in the early stage, the spacing between nanowalls at their top increased gradually, and then became almost saturated, resulting in the formation of two-dimensional carbon sheets standing vertically on the substrate with high aspect ratio. As shown in Figs. 3(e) and 3(f), with the further increase of growth period during the steady-state growth, the height of vertical aligned GNWs increased, while the thickness of nanowalls and the spacing between nanowalls became almost saturated with keeping the morphology of GNWs.

Growth rate curve for the GNWs fabricated by ICP-CVD employing CH_4/Ar system was obtained by measuring the height of the nanowalls for differing period of growth (0–120 min). Figure 4 shows the average height of GNW films as a function of growth period. As shown in Fig. 4, the height of GNWs almost linearly increased with the increase of growth period in the range from 10 to 120 min, while it took approximately 5 min for the nucleation. The growth rate of GNWs in the steady-state condition was constant at approximately 60 nm/min.

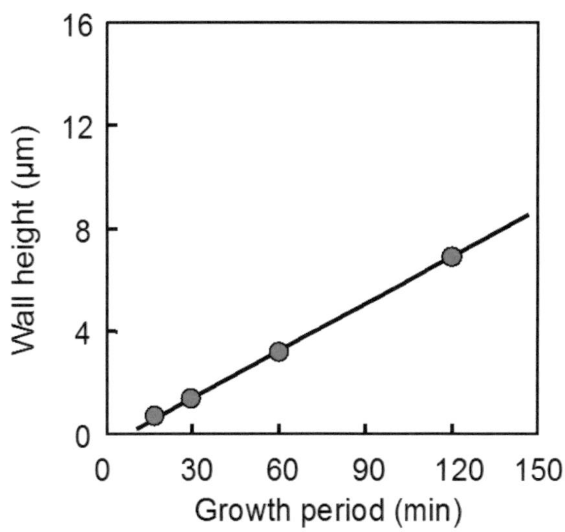

Figure 4. Wall height of GNWs as a function of growth period. The growth rate data were obtained from the samples grown for different period on Si substrates by ICP-CVD employing CH_4/Ar system [21].

2.2. Characterization of graphene nanowalls

As was illustrated in Fig. 1, GNWs can be described as graphite sheet nanostructures with edges that are composed of stacks of graphene sheets standing almost vertically on the substrate. The sheets form a self-supported network of wall structures with thicknesses in the range from a few nanometers to a few tens of nanometers, and with a high aspect ratio. In this section, typical GNWs grown using PECVD are characterized using SEM, transmission electron microscopy (TEM), in-plane synchrotron X-ray diffraction, and Raman spectroscopy using a 514.5 nm line of argon laser.

Figure 5(a) shows a typical SEM image of GNW film, indicating the vertical growth of the two-dimensional carbon sheets with honeycomb structure on the substrate. Actually, the morphology and structure of GNW film depend on the source gases, pressure, process temperature, as well as the type of plasma used for the growth of GNWs. In addition to the vertically standing maze-like structure, isolated very thin nanosheets, less aligned petal-like, highly branched type, and a kind of porous film have been fabricated so far.

Figure 5(b) shows a low-magnification TEM image of a piece of typical GNW of a micrometer-high planar nanosheet structure with a relatively smooth surface. The GNW is composed of nano-domains of a few tens of nanometers in size. The high-resolution TEM image of the GNW shown in Fig. 5(c) reveals the graphene layers, which indicates the graphitized structure of the GNWs. The spacing between neighboring graphene layers was measured as approximately 0.34 nm.

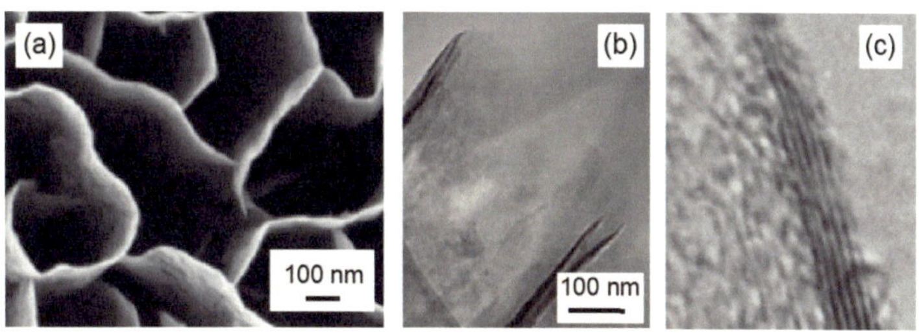

Figure 5. a) Typical SEM image of GNW film, (b) low-resolution TEM image of GNW on a microgrid, and (c) high-resolution TEM image of GNW, showing graphene layers at the fold of GNWs.

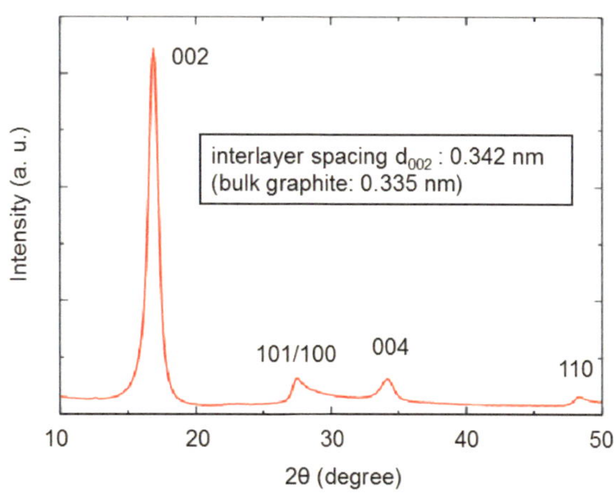

Figure 6. SR X-ray diffraction pattern of GNW film measured at beam line BL13XU of SPring-8 [22].

The crystallinity of GNWs was analyzed using synchrotron X-ray surface diffraction at grazing incidence and exit at the beamline BL13XU of SPring-8 [22]. The in-plane diffraction technique measures diffracted beams, which are scattered nearly parallel to the sample surface and hence measures lattice planes that are perpendicular to the sample surface. The X-ray beam was incident on the GNW film sample at grazing angle of 0.3° relative to the substrate surface. Figure 6 shows the grazing incidence in-plane X-ray diffraction pattern of GNW film sample. An intense 002 Bragg peak, the plane of which is normal to the substrate, is at $2\theta=16.9°$and there are also weak 100/101, 004, and 110 Bragg peaks. The interlayer spacing d_{002} was determined from the 002 peak by applying Bragg's law with a wavelength of 0.1003 nm. It was found to be 0.342 nm for all samples, which is slightly larger than that of bulk graphite (0.335 nm).

Raman spectrum for GNWs grown on Si substrate is shown in Fig. 7. Typical Raman spectrum for the GNWs has G band peak at 1590 cm^{-1} indicating the formation of a graphitized structure and D band peak at 1350 cm^{-1} corresponding to the disorder-induced phonon mode. The peak intensity of D band is twice as high as that of G band. The G band peak is accompanied by a shoulder peak at 1620 cm^{-1} (D' band), which is associated with finite-size graphite crystals and graphene edges [23,24]. The 2D band peak at 2690 cm^{-1} is used to confirm the presence of graphene and it originates from a double resonance process that links phonons to the electronic band structure [25,26]. The strong and sharp D band peak and D' band peak suggest a more nanocrystalline structure and the presence of graphene edges and defects, which are prevalent features of GNWs [8,10,12].

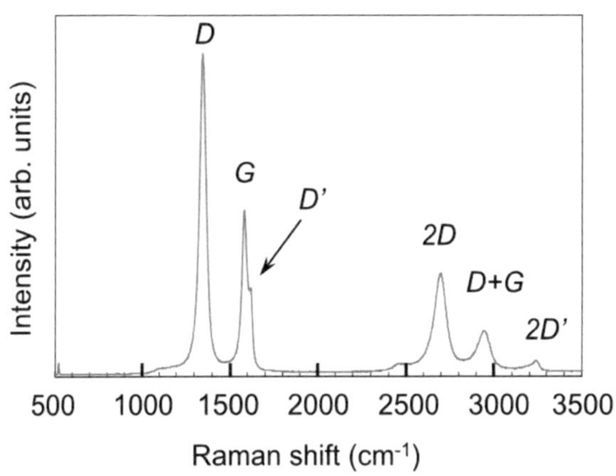

Figure 7. Typical Raman spectrum of GNWs measured using the 514.5 nm line of an Ar laser [21]

2.3. Nucleation of vertical nanographenes

From the temporal dependence of nanowall height shown in Fig. 3, nucleation of GNWs is considered to occur before the commencement of steady-state growth. Significant interest exists in clarifying the nucleation mechanism of GNWs at the very early stage. Figure 8(a) shows a top view SEM image of SiO_2 substrate surface after 30 sec growth. First the surface of substrate was

http://dx.doi.org/10.5772/51528

covered with amorphous carbon layer, which is later confirmed using Raman spectroscopy. In 1 min, the onset of nano-graphene growth was observed in the top view SEM image of deposits formed for 1 min as shown Fig. 8(b). Figures 8(c) and 8(d) show tilted SEM image and cross-sectional TEM image of deposits, respectively, formed for 2 min. For the cross-sectional TEM observation, sample surface was coated with the epoxy resin, the deposits embedded in the epoxy resin were peeled off from the substrate, and then the substrate side (interface side) of resin embedding the deposits was coated again with the epoxy resin. In Fig. 8(d), the red line corresponds to the interface to the substrate surface and the red arrow indicates the growth direction. In 2 min, the formation of isolated graphene sheets was observed on the amorphous carbon layer as shown in Fig. 8(c). The thickness of amorphous carbon layer was estimated to be 30 nm from the TEM observation shown in Fig. 8(d).

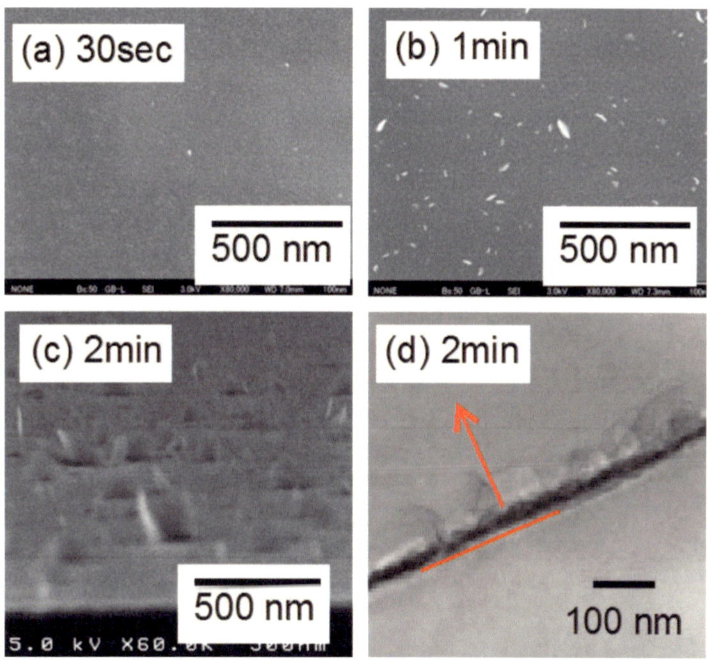

Figure 8. SEM and TEM images of deposits formed at the nucleation stage. (a) Top view SEM image of substrate surface after 30 sec. (b) Top view SEM image of deposits formed for 1 min, indicating the commencement of nano-graphene growth. (c) Tilted SEM image of deposits formed for 2 min. (d) Cross-sectional TEM image of deposits formed for 2 min. Deposits were detached from the Si substrate. Red line in (d) corresponds to the interface to the substrate surface and the red arrow indicates the growth direction [21].

Figure 9 shows Raman spectra of deposits formed on SiO_2-coated Si substrate in the nucleation stage. The peak around 950 cm^{-1} originates from the Si wafer [27]. The broad peak at 1340 cm^{-1} in the Raman spectrum at 30 sec indicates that the deposits are amorphous carbon or diamond-like carbon. Namely, during the nucleation period, graphene component was scarcely contained in the underlying interface layer shown in Fig. 9.

So far, several papers have been published on the observation of GNW growth in the early growth stage and the nucleation mechanism for the formation of vertical layered-graphenes using various CVD methods [8,28-35]. The things in common in previous observations are that there is an induction period of 1–5 min before the onset of vertical nano-graphene growth and an interface layer exists between vertical nano-graphenes and the surface of Si and SiO_2 substrates. Zhu, et al. [31,32] reported the presence of graphenes parallel to the substrate surface before the onset of vertical nanosheet growth, although neither Raman spectrum nor TEM image of base layer was attached. In their model, these few-layer graphenes would grow parallel to the substrate surface until a sufficient level of force develops at the grain boundaries to curl the leading edge of the top layers upward. The vertical orientation of these sheets would result from the interaction of the plasma electric field with their anisotropic polarizability [32]. In the case of GNW growth in the present work, on the other hand, the interface layer under the CNWs was an amorphous or diamond-like carbon, which is similar to the cases using radical injection PECVD, multi-beam CVD, and DC PECVD [8,28-30,33,34]. Moreover, ion bombardment on the surface will play an important role for nucleation by creating active sites for neutral radical bonding [30]. The existence of amorphous or diamond-like carbon interface layer will enable us to grow GNWs and similar structures on a variety of substrates without catalyst.

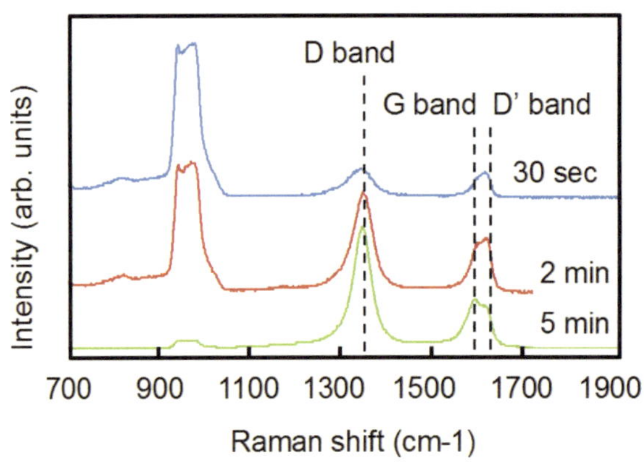

Figure 9. Raman spectra of deposits formed in the nucleation stage on SiO_2-coated Si substrate. Peak around 950 cm^{-1} originates from the Si wafer [8].

2.4. Growth mechanism of graphene nanowalls

Figure 10 depicts the initial growth process of GNWs. A model for the initial growth mechanism is as follows. (1) In the beginning, hydrocarbon radicals such as CH_3 are adsorbed on the substrate, forming a very thin amorphous carbon layer. Ion irradiation induces the formation of dangling bonds on the growing surface, resulting in the formation of nucleation sites. (2) Adsorbed carbon species are migrating on the surface and condensed to form nano-islands with dangling bonds. (3) Ion irradiation would also enhance the adsorption of CH_x radicals on the surface. (4) Small, disordered graphene nanosheets are nucleated at these dangling bonds, followed by two-dimensional growth and subsequent formation of nano-graphene sheets with random orientation. (5) Among the nucleated graphene sheets with random orientations, those standing almost vertically on the substrate continued preferably to grow up faster to vertically standing nanosheets owing to the difference in the growth rates along the strongly bonded planes of graphene sheets expanding and in the weakly bonded stacking direction. Reactive carbon species arriving at the edge of the graphene layer are easily bonded to the edge, and eventually the graphene layer would expand preferably along the direction of radical diffusion, perpendicular to the electrode plane. On the other hand, low-lying inclined graphene sheets were shadowed by the high-grown vertical graphene sheets. As a result, the amounts of reactive carbon species arriving at the low-lying inclined graphene sheets decreased, resulting in the termination of growth for the inclined smaller nanowalls, while the vertical nanowalls preferentially continued to grow. As growth period increased, spreading vertical nanowalls met one another, eventually resulting in the formation of linked nanowalls similar to a maze. With further increase of growth period, the spacing between nanowalls at their top increased gradually, and then became almost saturated, resulting in the formation of two-dimensional graphene sheets standing vertically on the substrate with high aspect ratio. In the steady-state growth condition, the height of nanowalls increased almost linearly with keeping their morphology.

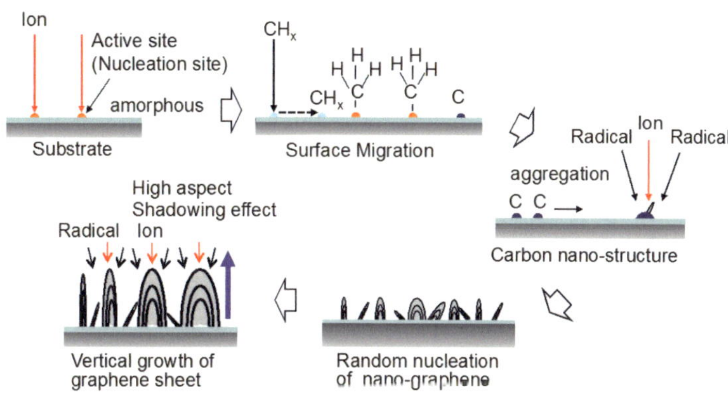

Figure 10. Illustration of the initial growth model of GNWs

3. Functionalization of graphene nanowalls

3.1. Decoration of graphene nanowall surface

Due to the large surface area (high surface-to-volume ratio) of GNWs, we can expect a variety of electrochemical applications using GNWs such as batteries, capacitors, and gas sensors. To these ends, GNWs are decorated with nanoparticles or films of metals, semiconductors, and insulators, by using several techniques including vacuum evaporation, sputtering, CVD, and plating. Figure 11 shows schematic illustrations of decorated

GNWs with different morphologies. GNWs are used as the templates to fabricate other types of nanostructures. The morphology of GNWs decorated with nanoparticles or film depends on the deposition methods of materials. Conformal deposition (Fig. 11(a)) and gap filling (Fig. 11(b)) will be achieved using metal-organic chemical vapor deposition (MOCVD), sputtering, atomic layer deposition, and plating in liquid phase. In Fig. 11(c), thin film or aggregation of nanoparticles would be formed on the top edges of GNWs. Diamond surface is modified with several types of surface termination, e.g. C–NH_2, C–OH, and C–COOH, and DNA and proteins were immobilized on the surface of diamond and nanodiamond films for bio-sensing application [35–38]. As is the case with the diamond surface, the edges of GNWs can also be modified with similar surface termination, and covalent immobilization of DNA and proteins on the GNWs will be realized. Previously, Wu et al. used GNWs as templates to fabricate large surface area materials, including conformal deposition of Au and Cu by electron beam evaporation; conformal deposition of ZnO, TiO_2, SiO_x, SiN_x, and AlO_x by atomic layer deposition; conformal deposition of Ni, NiFe and CoNiFe nanoparticles by electrochemical deposition; gap filling with dispersed Fe nanoparticles by the immersion of GNWs into a mixed solution of Fe particles and isopropanol in an ultrasonic bath; Se deposition on the top of edges of GNWs by electrochemical deposition [39–41]. In Fig. 11(d), on the other hand, metal nanoparticles are dispersed on the surface of GNWs, which is a kind of nanocomposite. This morphology can be moderately achieved by plating, sputtering, and laser ablation.

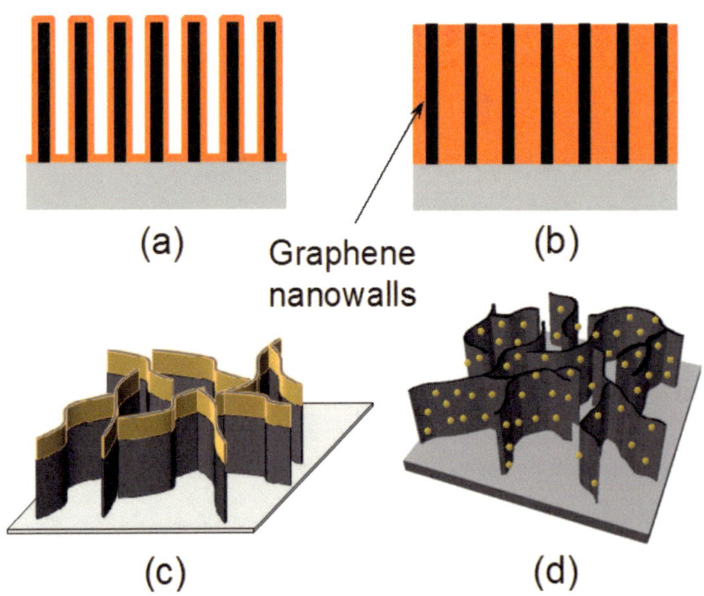

Figure 11. Schematic illustrations of decorated GNWs with different morphologies: (a) conformal deposition, (b) gap filling, (c) deposition on the top edges of GNWs, and (d) dispersed nanoparticle deposition on the surface of GNWs

Because of the unique structure of GNWs with high surface-to-volume ratio, GNWs can be potentially used as catalyst support materials for the electrodes of fuel cells. In this application, it is required to support platinum (Pt) nanoparticles as catalysts on the GNW surface. It is well known that the specific activity of catalysts is strongly related to their size, dispersion, and compatibility with supporting materials. Highly dispersed catalyst nanoparticles with small size and narrow size distribution supported on the surface of carbon nanostructures are ideal for high electrocatalyst activity due to their large surface-to-volume ratio. To support metal nanoparticles on the surface of carbon nanostructures, metal compounds in the form of liquids are generally employed. A few papers have been published on the preparation of Pt nanoparticles on CNT surfaces by the reduction of Pt salt precursors such as H_2PtCl_6 in solution [42,43]. However, it is difficult to treat the entire surface of GNWs with a metal compound in a liquid phase, because of the high surface tension of GNWs due to their high aspect ratio with narrow interspaces. On the other hand, in gas phase deposition such as sputtering and CVD, metal nanoparticles are deposited only around the tops of GNWs and tend to easily clump together, resulting in the formation of larger particles or films on the top of carbon nanostructures [44].

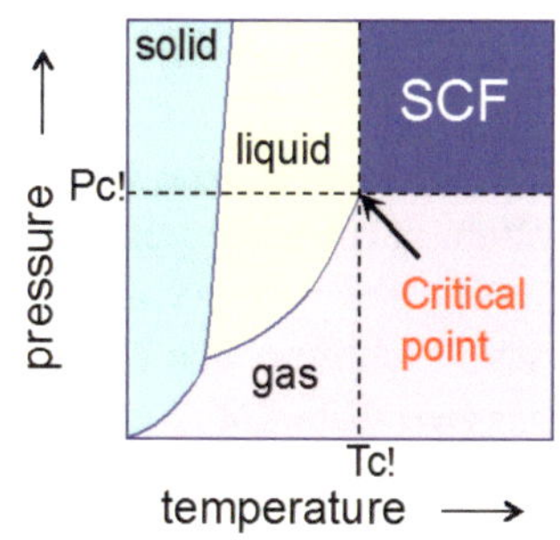

Properties of liquid phase, gas phase, and supercritical fluid (SCF)

	Liquid	SCF	Gas
Density (g/cm^3)	1	0.1 – 1	10^{-3}
Viscosity (Pa-s)	10^{-3}	$10^{-4} – 10^{-5}$	10^{-5}
Diffusivity (cm^2/s)	10^{-5}	10^{-3}	10^{-1}

Figure 12. Phase diagram of substance (SCF supercritical fluid, Pc critical pressure, Tc critical temperature) and properties of liquid phase, gas phase, and supercritical fluid

3.2. Supercritical fluids

As an alternative approach to support metal nanoparticles on the surfaces of dense, aligned CNTs and GNWs with narrow interspaces, we have demonstrated a method employing metal-organic chemical fluid deposition (MOCFD), where supercritical carbon dioxide (sc-CO_2) is used as a solvent of metal-organic compounds [45,46]. The phase diagram of substance is shown in Fig. 12, together with the properties of liquid phase, gas phase, and supercritical fluid (SCF). Generally, materials can exist in three phases depending on the temperature and pressure, namely, solid, liquid, and gas. The SCF possesses attractive properties of both the gas and the liquid phases. Rapid diffusion and permeation are realized by its gas-like diffusivity and viscosity, while its liquid-like density enables dissolution of a wide range of materials. To produce an SCF phase, the temperature and pressure of the material are required to exceed the critical point. The critical point of sc-CO_2 exists at 7.38 MPa

(72.8 atm) and 31.1 °C. Among SCFs, sc-CO_2 is particularly attractive since it is environmentally friendly and safe due to its low toxicity, low reactivity and nonflammability.

In the case of Pt deposition, the SCF using sc-CO_2 was first applied to the preparation of polymer-supported Pt nanoparticles using dimethyl (1,5-cyclooctadiene) platinum(II), ($PtMe_2$(cod)) as a precursor [47]. Erkey's group has demonstrated the preparation of Pt nanoparticles on a wide range of materials, including carbon aerogel, carbon black, silica aerogel, alumina, and Nafion [47–52]. In their method, $PtMe_2$(cod) was dissolved in sc-CO_2 and impregnated into the supporting materials, and after depressurization the impregnated $PtMe_2$(cod) molecules were then reduced to metallic Pt nanoparticles by heat treatment or by chemical reduction with hydrogen, resulting in the formation of uniformly dispersed nanoparticles with narrow size distributions. However, it took almost 10 hours to complete this process. In our case, in contrast, the supporting carbon nanostructures such as CNTs and GNWs were selectively heated in the sc-CO_2 with Pt precursors during the process. Therefore, at the heated surface of the carbon nanostructures during *in situ* thermal reduction under the SCF environment, decomposition of the adsorbed precursor molecules and growth of the particles would occur without reduction process.

3.3. Experimental procedure of metal-organic chemical fluid deposition using supercritical carbon dioxide

Figure 13 shows the SCF-MOCFD system employing sc-CO_2 used for the deposition of Pt nanoparticles on the surface of GNWs. The MOCFD system is composed of two high-pressure stainless steel vessels with a compressor, heating units, and a reservoir for the metal-organic compound. The preliminary vessel contains a screw agitator. The temperature and pressure in each vessel can be set independently, so that two different supercritical conditions employing CO_2 can be produced in these two vessels. As the precursor, (methylcyclopentadienyl) trimethyl platinum ($(CH_3C_5H_4)Pt(CH_3)_3$: $MeCpPtMe_3$) dissolved in hexane was used. In the preliminary vessel, the precursor was stirred with the sc-CO_2 for about 30 min. In the impregnation vessel, the selective heating of GNW samples during the MOCFD process facilitated selective metal deposition on the surface of the carbon nanostructures. In the preliminary vessel, the pressure and temperature of sc-CO_2 were maintained at 11 MPa and 50°C, respectively, and $MeCpPtMe_3$ was dissolved in the sc-CO_2. In the impregnation vessel, the pressure and temperature of sc-CO_2 were maintained at 9 MPa and 70°C, respectively, and the temperature of GNW samples was controlled in the range of 70–170°C. The solutions were mixed and Pt nanoparticles formation was carried out for 30 min; the vessel was then depressurized slowly in 30 min to atmospheric conditions. After the depressurization, additional heat treatment was not carried out in the present work.

3.4. Characterization of platinum nanoparticles formed by metal-organic chemical fluid deposition using supercritical carbon dioxide

Figures 14(a)–14(c) show TEM images of the Pt-supported GNW surface after the SCF-MOCFD at sample temperatures of 120, 150, and 170°C, respectively. In this experiment, GNW samples were fabricated on the Si substrate by fluorocarbon (C_2F_6) PECVD assisted by

http://dx.doi.org/10.5772/51528

hydrogen radical injection, which comprises a parallel-plate very high frequency (VHF: 100 MHz), capacitively coupled plasma region, and a hydrogen radical injection source that employs a surface-wave-excited microwave (2.45 GHz) H_2 plasma [11]. As can be seen from these TEM images, the spatial density of the Pt nanoparticles (particle numbers/area) supported on the GNW surface strongly depended on the sample temperature during the SCF-MOCFD, while the average size of the Pt nanoparticles increased from 1.5 to 3 nm with an increase of the sample temperature from 120 to 170°C.

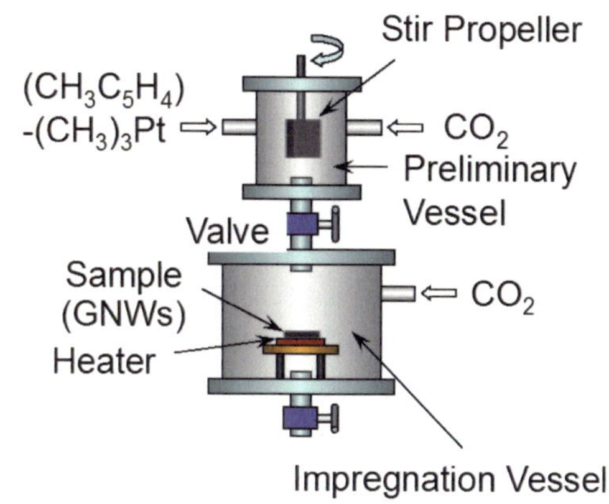

Figure 13. Schematic of the supercritical fluid, metal-organic chemical fluid deposition (SCF-MOCFD) system [53]

Ex situ X-ray photoelectron spectroscopy (XPS) analysis was conducted to gain an insight into the state of the platinum for the supported Pt surface fabricated by the SCF-MOCFD. Figure 14(d) shows an XPS spectrum of the Pt(*4f*) region of the Pt-supported GNW film after the SCF-MOCFD at a sample temperature of 150°C. The presence of two prominent sets of Pt (*4f*) peaks, corresponding to the $4f_{7/2}$ and $4f_{5/2}$ orbital states, is further confirmation of platinum being present on the GNW surface. The peak regions in Fig. 14(d) can be fitted with two sets of peaks at 71.4 eV ($4f_{7/2}$) and 74.6 eV ($4f_{5/2}$) [54]. These correspond to platinum in the metallic state, indicating that only pure Pt exists without being oxidized on the surface of the GNWs after the SCF-MOCFD.

The Pt/C ratio of the GNW film surface was obtained from the ratio of the intensities of the XPS C(*1s*) and Pt(*4f*) peaks. Figure 14(e) shows the relative Pt/C ratio of the surface of the GNW film as a function of the temperature of the GNW sample during the SCF-MOCFD process. As the sample temperature during the SCF-MOCFD increased up to 120◎?°C, the relative Pt/C ratio of the surface of the GNW film increased gradually. By further increasing the sample temperature above 120°C, the relative Pt/C ratio increased rapidly.

3.5. Mechanism of platinum nanoparticle formation by metal-organic chemical fluid deposition using supercritical carbon dioxide

In the case of Pt deposition by the conventional process including impregnation, depressurization, and reduction, $PtMe_2$(cod) was dissolved in sc-CO_2 and impregnated into the supporting materials, and after depressurization the impregnated $PtMe_2$(cod) molecules were then reduced to metallic Pt nanoparticles by heat treatment or by chemical reduction with hydrogen. It was proposed that the precursor molecules in the sc-CO_2 phase are adsorbed on the surface during the impregnation period, and after depressurization these adsorbed molecules are in turn reduced to elemental platinum and the resulting particles at the surface continue to grow until all the adsorbed precursor molecules are converted to the metal.

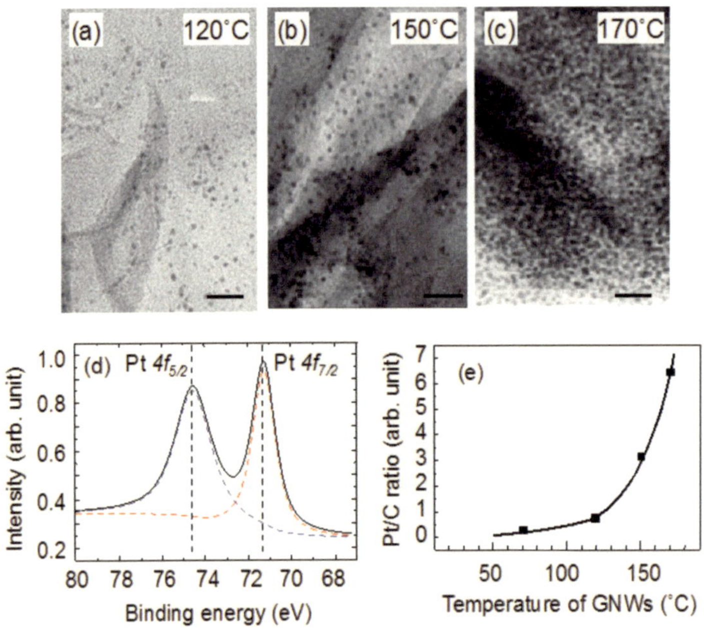

Figure 14. a)–(c) High-resolution TEM images of the surface of the GNWs supporting Pt nanoparticles after the SCF-MOCFD at sample temperatures of 120, 150, and 170�°C, respectively. Scale bar: 20 nm. (d) X-ray photoelectron spectroscopy spectrum of the Pt-supported GNW film after SCF-MOCFD. (e) Relative Pt/C ratio of the surface of the GNW film as a function of temperature of the GNW sample during SCF-MOCFD [53]

In our case, in contrast, the supporting carbon nanostructures were selectively heated in the sc-CO_2 with Pt precursors during the process. Figure 15 depicts the model of Pt nanoparticle formation on the surface of carbon nanostructures using metal-organic chemical fluid deposition in the supercritical CO_2. At the heated surface of the carbon nanostructures during *in situ* thermal reduction under the SCF environment, decomposition of the adsorbed precursor molecules and growth of the particles would occur. GNWs have been reported to consist of nano-domains a few tens of nanometers in size, and individual GNWs were found to have many defects [24,55]. It is suggested that the surface-migrating Pt adatoms, produced by the decomposition of $MeCpPtMe_3$ precursors, merge to form Pt clusters from several Pt atoms preferentially at chemically active sites such as defects and grain boundaries on the surface of the carbon nanostructures, resulting in the nucleation of Pt nanoparticles. It has been reported recently that the density of Pt nanoparticles formed by SCF-MOCFD method increased with increase in the surface defect density

[56]. The reaction temperature at the surface would be a significant factor influencing the particle number density and particle size. When the temperature of GNWs is increased, both reduction of metal-organic precursors and surface migration of Pt atoms would be enhanced, which may lead to an increase in the particle number density and particle size. As can be seen from the TEM images in Figs. 14(a)–14(c), the average size of Pt nanoparticles increased from 1.5 to 3 nm with an increase in the sample temperature from 120 to 170°C, while the Pt particle number density increased drastically.

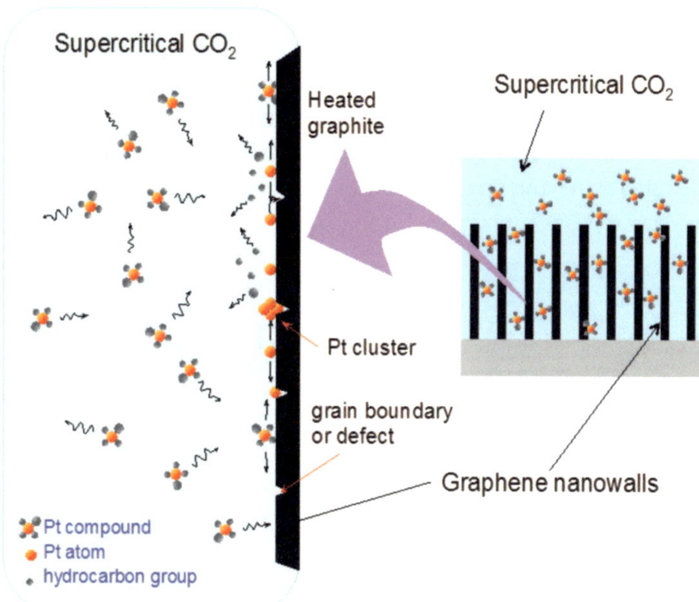

Figure 15. Illustration of the formation model of Pt nanoparticles on the surface of carbon nanostructures using metal-organic chemical fluid deposition in the supercritical CO_2

4. Fabrication of graphene nanowalls on carbon fiber paper for fuel cell application

Carbon fiber paper (CFP) or carbon fiber cloth, which are composed of an open mesh of carbon fibers, have been used as gas diffusion layer in proton exchange membrane (PEM) fuel cell application [57–59]. PEM fuel cells have been widely recognized as the most promising candidates for future power generating devices in the automotive, distributed power generation and portable electronic applications. Waje, et al. demonstrated the preparation of Pt nanoparticles 2–2.5 nm in size on organically functionalized CNTs grown on CFP [60]. Very recently Lisi, et al. demonstrated the growth of GNWs on CFP by hot-filament CVD and investigated the microstructure of the GNWs both at the tip and at the fiber–nanowall base interface [61]. Our current interest in GNWs is to use them as catalyst supports for Pt in

PEM fuel cells. In this section, in order to demonstrate the usefulness of GNWs in the fuel cell application, GNWs were directly grown on the CFP using PECVD. Subsequently, Pt nanoparticles were formed on the surface of GNWs using the SCF-MOCFD method. This configuration ensures that all the supported Pt nanoparticles are in electrical contact with the external electrical circuit. Such a design would improve Pt utilization and potentially decrease Pt usage.

4.1. Growth of graphene nanowalls on carbon fiber paper

Commercially available CFP (Engineered Fibers Technology, Spectracarb 2050A porous carbon-carbon paper, 200 μm thick) was decorated with GNWs using RF-ICP employing Ar/CH_4 mixture. The growth experiments were carried out on CFP and Si substrates for 30 min at RF power of 500 W, total gas pressure of 20 mTorr, substrate (CFP) temperature of 720 °C, and flow rates of Ar and CH_4 of 100 and 50 sccm, respectively.

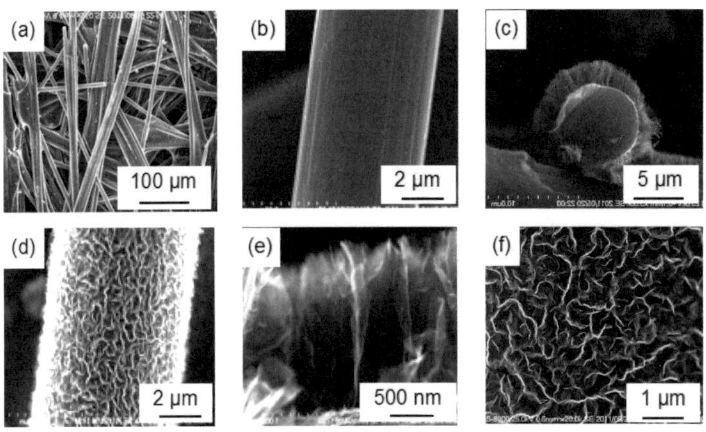

Figure 16. a), (b) SEM images of carbon fiber used in this study. (c)–(f) SEM images of GNWs grown on carbon fiber by ICP-CVD for 30 min at different magnifications [62].

Figures 16(a) and 16(b) show SEM images of carbon fiber used in this study. SEM images of GNWs grown on carbon fiber by ICP-CVD for 30 min at different magnifications are shown in Figs. 16(c)–16(f), indicating that GNWs were successfully grown on the CFP using ICP-CVD. As shown in Fig. 16(e), GNWs were grown almost vertically on the surface of carbon fibers forming paper structure. The height of GNWs grown on the CFP at the surface facing the plasma was about 1.5 μm. Interestingly the GNW growth proceeds in a conformal manner all the way around each carbon fiber into the CFP to a depth of a few tens of micrometers.

4.2. Pt nanoparticle formation on GNW-decorated carbon fiber paper

Pt nanoparticles were prepared on the GNWs grown on the CFP by the SCF-MOCFD method. The pressure and temperature of $sc\text{-}CO_2$ were 10 MPa and 130 °C, respectively, and the temperature of GNW-decorated CFP was maintained at 180 °C. Pt nanoparticle formation was carried out for 30 min. Compared to the SEM images of typical as-grown GNW films without the SCF treatment, no change in the surface morphology was observed after the SCF treatment. It was confirmed that the unique nanostructure of the GNWs was maintained, even after being exposed to the high-pressure fluid. Figure 17(a) shows SEM image of the surface of the GNW supporting Pt nanoparticles after the SCF-MOCFD for 30 min. It was found that dispersed Pt nanoparticles of approximately 2 nm in diameter were supported on the surfaces of GNWs grown on CFP. The area density of Pt nanoparticles on GNW surface was approximately 3×10^{12} cm^{-2}. Figure 17(b) shows TEM image of the surface of the GNW supporting Pt nanoparticles after the SCF-MOCFD for 30 min. The GNWs consist of nano-domains and individual GNWs were found to have many defects. As shown in Fig. 17(b), Pt nanoparticles were formed preferentially at the domain boundaries on the surface of the GNWs.

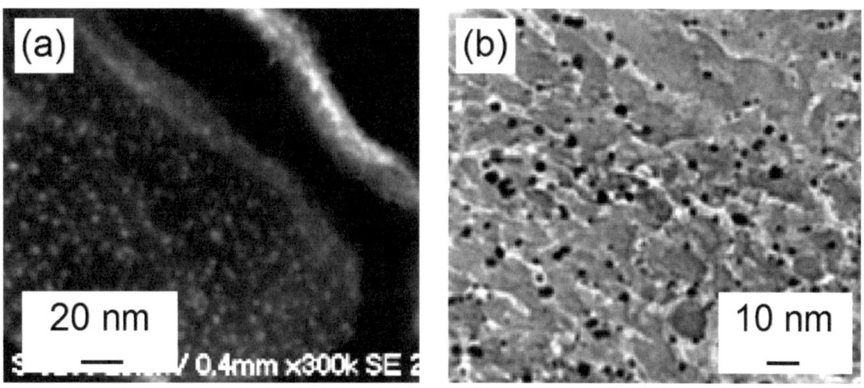

Figure 17. a) SEM image of the surface of the GNW supporting Pt nanoparticles after the SCF-MOCFD for 30 min. (b) TEM image of the surface of the GNW supporting Pt nanoparticles after the SCF-MOCFD for 30 min

4.3. Fuel cell unit using Pt-supported GNW/CFP electrode

A test cell unit using Pt-supported GNW/CFP electrode was constructed experimentally. A schematic of a single PEM fuel cell with active surface area of 1×1 cm^2 is shown in Fig. 18. The membrane electrode assembly (MEA) consists of a Nafion 115 membrane in combination with Pt- supported GNWs on CFPs. At present, the voltage-current curve for the test PEM fuel cell exhibited unexpectedly poor performance due to the high ohmic resistance. Very recently, Shin, et al. demonstrated the preparation of Pt nanoparticles with an average diameter of 3.5 nm on GNWs by a solution-reduction method, and investigated the electrocatalytic activity of powdered Pt/ GNWs peeled off from the substrate [63]. They suggested that the domain structure of GNWs is useful as catalytic support for fuel cells, although they did not demonstrate an actual PEM fuel cell using Pt/GNWs. For the evaluation of the Pt-supported GNWs on CFPs as catalyst layer for the practical use in our case, it is necessary to realize optimum operating conditions including compression, temperature, pressure, and ionomer loading.

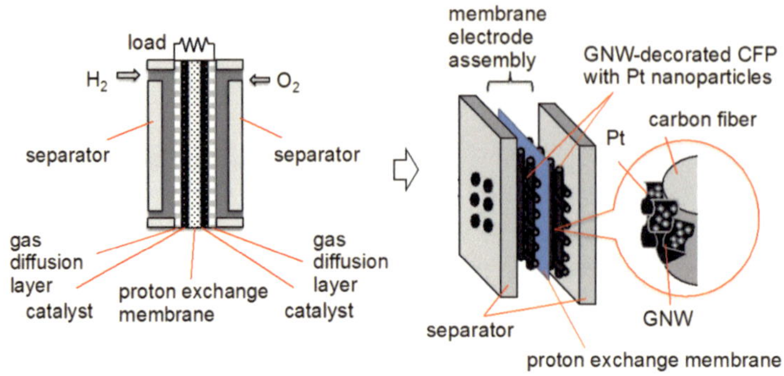

Figure 18. A schematic of a test single proton exchange membrane (PEM) fuel cell. The membrane electrode assembly (MEA) consists of a Nafion 115 membrane in combination with Pt-loaded GNWs on CFPs [62].

5. Toward the application to electrochemical sensors and biosensors

There has been an increase in research into the biological applications of CVD diamond. Diamond has attractive characteristics for some biological applications, such as its wide potential window, chemical–physical stability, and biocompatibility. DNA and proteins were immobilized on the surface of diamond and nanodiamond films for bio-sensing application [35–38]. Covalent modification of diamond surfaces with molecular monolayers serves as a starting point for linking biomolecules such as DNA and proteins to surfaces. In these cases, diamond surface is modified with several types of surface termination, e.g., $C–NH_2$, C–OH, and C–COOH. It is considered that the surfaces and edges of GNWs can also be modified with similar surface termination. Therefore, covalent immobilization of DNA and proteins on the GNWs will be realized. In the near future, GNWs will be used as a stable, highly selective platform in subsequent surface hybridization processes.

Recently, graphene has proved to be an excellent nanomaterial for applications in electrochemistry. Graphene-based materials with large surface area are useful as electrodes for electrochemical sensors and biosensors [62–64]. Especially, GNWs and related carbon nanostructures can be one of the best electrode materials to investigate basic electrochemical phenomena, due to their edge defects and graphene structures.

It is interesting to investigate the electrochemical properties of GNWs as electrochemical electrodes. In order to evaluate the potential window of GNW electrode, electrochemical measurements were carried out with a conventional three-electrode arrangement controlled by a commercial potentiostat, with a GNW (boron (B)-doped diamond for comparison) working electrode, a Pt coil counter electrode and an Ag/AgCl reference electrode. The back face of the substrate and the electrical contact were protected from the electrolyte solution by insulating epoxy adhesive, so that only an active GNW electrode area was exposed to the electrolyte solution. Figure 19 shows the cyclic voltammograms for the GNW and B-doped diamond electrodes in HNO_3 (0.2 mol/l) obtained in the potential range between –5 and 5 V at a scan rate of 100 mV/s. As shown in Fig. 19, a work potential window nearly 3 V was obtained for the GNW electrode, which was comparable to that for B-doped diamond electrode. The oxygen evolution for the GNW electrodes occurs at about 1 V and the hydrogen evolution at –2 V. Reported potential windows for glassy carbon and highly oriented pyrolytic graphite electrodes are 1.5 and 2.0 V, respectively [67].

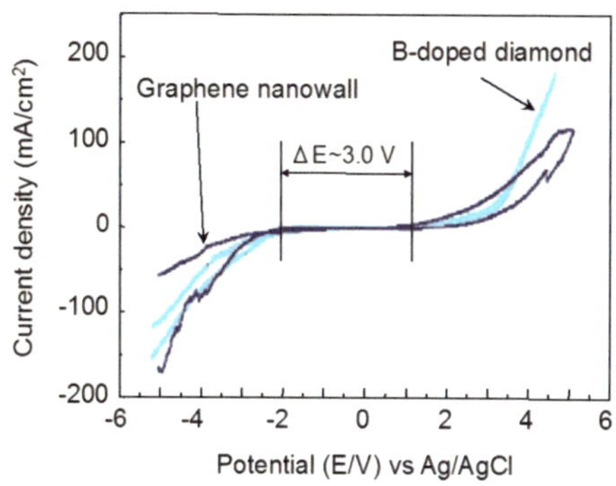

Figure 19. Cyclic voltammograms for the GNWs and boron-doped diamond electrodes in HNO_3 (0.2 Mol/l); at 100 mV/s scan rate

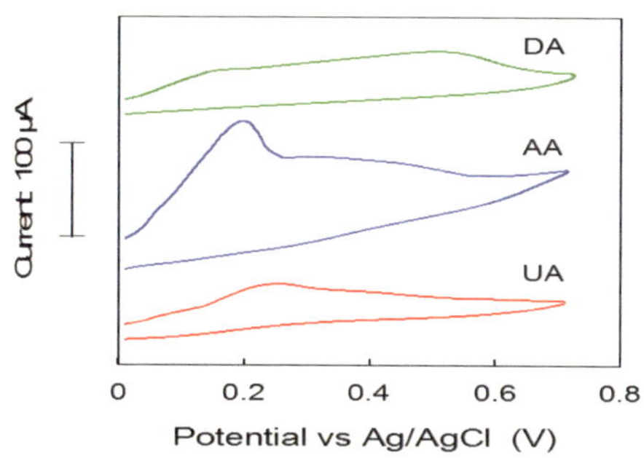

Figure 20. Cyclic voltammograms of GNW electrode in the solution of PBS with DA, AA, and UA, at 100 mV/s scan rate

Electrochemical activity of GNW electrode has been investigated by cyclic voltammetry measurements in an aqueous solution of ferrocyanide and a faster electron transfer between the electrolyte and the nanowall (or nanosheet) surface has been demonstrated [17–20]. Furthermore, biosensing with GNWs is a promising application. Dopamine (DA), ascorbic acid (AA), and uric acid (UA) are compounds of great biomedical interest, which all are essential biomolecules in our body fluids. Figure 20 shows cyclic voltammogram curves obtained from GNW electrode in the phosphate buffer solution (PBS) with DA, AA, and UA, at 100 mV/s scan rate. At present, researches on the sensing of biological molecules became popular. Shang, et al. demonstrated the excellent electrocatalytic activity of multilayer graphene nanoflake films for simultaneously determining DA, AA and UA in PBS [17]. The GNW-based electrochemical platform, which possesses large surface area with edges and electrochemical activity, offers great promise for providing a new class of nanostructured electrodes for biosensing and energy-conversion applications.

6. Conclusion

Self-organized graphite sheet nanostructures composed of graphene have been studied intensively. Graphene nanowalls and related sheet nanostructures are layered graphenes with open boundaries. The sheets form a self-supported network of wall structures with thicknesses in the range from a few nanometers to a few tens of nanometers, and with a high aspect ratio. The large surface area and sharp edges of graphene nanowalls could prove useful for a number of different applications.

Using graphene nanowalls as templates would be the most promising and important application. Graphene nanowalls can be used as templates for fabricating a variety of nanostructured materials based on the surface morphology of the graphene nanowalls and nanocomposites of carbon and nanoparticles of other materials. These structures could prove useful in batteries, sensors, solar cells, electrodes, and biomedical devices. For this purpose, it is necessary to establish decorating methods of graphene nanowall surface with a variety of materials. Furthermore, it is important to evaluate electrochemical characteristics of nanocomposites of carbon and other materials systematically.

In order to demonstrate the usefulness of graphene nanowalls in the fuel cell application, graphene nanowalls were directly grown on the carbon fiber paper using the inductively coupled plasma-enhanced chemical vapor deposition method. Subsequently, highly dispersed Pt nanoparticles 2 nm in size were formed on the surface of graphene nanowalls using metal-organic chemical deposition employing supercritical CO_2. This configuration ensures that all the supported Pt nanoparticles are in electrical contact with the external electrical circuit. Such a design would improve Pt utilization and potentially decrease Pt usage. Pt-supported graphene nanowalls grown on the carbon fiber paper will be well suited to the application for the electrodes of fuel cells.

Furthermore, the application as a biosensor using GNWs was briefly described. The GNW-based electrochemical platform offers great promise for providing a new class of nanostructured electrodes for electrochemical sensing, biosensing and energy-conversion applications.

Author details

Mineo Hiramatsu[1*], Hiroki Kondo[2] and Masaru Hori[2]

1 Meijo University, Japan

2 Nagoya University, Japan

References

[1] Geim, A. K., & Novoselov, K. S. (2007). The rise of graphene. *Nature Materials*, 6, 183-191, nmat1849.

[2] Novoselov, K. S., Geim, A. K., Morozov, S. V., Jiang, D., Zhang, Y., Dubonos, S. V., Grigorieva, I. V., & Firsov, A. A. (2004). Electric Field Effect in Atomically Thin Carbon Films. *Science*, 306(5696), 666-669, science.1102896.

[3] Eda, G., Fanchini, G., & Chhowalla, M. (2008). Large-area ultrathin films of reduced graphene oxide as a transparent and flexible electronic material. *Nature Nanotechnology*, 3, 270-274, nnano.2008.83.

[4] Berger, C., Song, Z. M., Li, X. B., Wu, X. S., Brown, N., Naud, C., Mayo, D., Li, T. B., Hass, J., Marchenkov, A. N., Conrad, E. H., First, P. N., & de Heer, W. A. (2006). Electronic Confinement and Coherence in Patterned Epitaxial Graphene. *Science*, 312(5777), 1191-1196, science.1125925.

[5] Yu, Q., Lian, J., Siriponglert, S., Li, H., Chen, Y. P., & Pei, S. S. (2008). Graphene segregated on Ni surfaces and transferred to insulators. *Applied Physics Letters*, 93(11).

[6] Li, X., Cai, W., An, J., Kim, S., Nah, J., Yang, D., Piner, R., Velamakanni, A., Jung, I., Tutuc, E., Banerjee, S. K., Colombo, L., & Ruoff, R. S. (2009). Large-Area Synthesis of High-Quality and Uniform Graphene Films on Copper Foils. *Science*, 324(5932), 1312-1314.

[7] Wu, Y. H., Qiao, P. W., Chong, T. C., & Shen, Z. X. (2002). Carbon Nanowalls Grown by Microwave Plasma Enhanced Chemical Vapor Deposition. *Advanced Materials*, 14(1), 64-67, AID- ADMA64>3.0.CO;2-G.

[8] Hiramatsu, M., Shiji, K., Amano, H., & Hori, M. (2004). Fabrication of vertically aligned carbon nanowalls using capacitively coupled plasma-enhanced chemical vapor deposition assisted by hydrogen radical injection. *Applied Physics Letters*, 84(23), 4708 -4710.

[9] Wang, J. J., Zhu, M. Y., Outlaw, R. A., Zhao, X., Manos, D. M., Holloway, B. C., & Mammana, V. P. (2004). Free-standing subnanometer graphite sheets. *Applied Physics Letters*, 85(7), 1265, 10.1063/1.1782253.

[10] Hiramatsu, M., & Hori, M. (2006). Fabrication of Carbon Nanowalls Using Novel Plasma Processing. *Japanese Journal of Applied Physics*, 45, 5522-5527, JJAP.45.5522.

[11] Kondo, S., Hori, M., Yamakawa, K., Den, S., Kano, H., & Hiramatsu, M. (2008). Highly reliable growth process of carbon nanowalls using radical injection plasma-enhanced chemical vapor deposition. *Journal of Vacuum Science & Technology B*, 26(4), 1294-1300.

[12] Mori, T., Hiramatsu, M., Yamakawa, K., Takeda, K., & Hori, M. (2008). Fabrication of carbon nanowalls using electron beam excited plasma-enhanced chemical vapor deposition. *Diamond and Related Materials*, 17(7-10), 1513-1517, j.diamond.2008.01.070.

[13] Chuang, A. T. H., Robertson, J., Boskovic, B. O., & Koziol, K. K. K. (2007). Three-dimensional carbon nanowall structures. *Applied Physics Letters*, 90(12), 123107 .

[14] Hou, K., Outlaw, R. A., Wang, S., Zhu, M., Quinlan, R. A., Manos, D. M., Kordesch, M. E., Arp, U., & Holloway, B. C. (2008). Uniform and enhanced field emission from chromium oxide coated carbon nanosheets. *Applied Physics Letters*, 92(13).

[15] Yang, B. J., Wu, Y. H., Zong, B. Y., & Shen, Z. X. (2002). Electrochemical Synthesis and Characterization of Magnetic Nanoparticles on Carbon Nanowall Templates. *Nano Letters*, 2(7), 751-754, nl025572r.

[16] Giorgi, L., Makris, T. D., Giorgi, R., Lisi, N., & Salernitano, E. (2007). Electrochemical properties of carbon nanowalls synthesized by HF-CVD. *Sensors and Actuators B: Chemical*, 126(1), 144-152, j.snb.2006.11.018.

[17] Shang, N. G., Papakonstantinou, P., Mc Mullan, M., Chu, M., Stamboulis, A., Potenza, A., Dhesi, S. S., & Marchetto, H. (2008). Catalyst-Free Efficient Growth, Orientation and Biosensing Properties of Multilayer Graphene Nanoflake Films with Sharp Edge Planes. *Advanced Functional Materials*, 18(21), 3506-3514, adfm.200800951.

[18] Luais, E., Boujtia, M., Gohier, A., Tailleur, A., Casimirius, S., Djouadi, M. A., Granier, A., & Tessier, P. Y. (2009). Carbon nanowalls as material for electrochemical transducers. *Applied Physics Letters*, 95(1), 014104.

[19] Tanaike, O., Kitada, N., Yoshimura, H., Hatori, H., Kojima, K., & Tachibana, M. (2009). Lithium insertion behavior of carbon nanowalls by dc plasma CVD and its heat-treatment effect. *Solid State Ionics*, 180(4-5), 381-385, j.ssi.2009.01.012.

[20] Wang, Z., Shoji, M., & Ogata, H. (2011). Carbon nanosheets by microwave plasma enhanced chemical vapor deposition in CH4-Ar system. *Applied Surface Science*, 257(21), 9082-9085, j.apsusc.2011.05.104.

[21] Hiramatsu, M., Nihashi, Y., Kondo, H., & Hori, M. (2012). Nucleation control of self-organized vertical nano-graphenes using inductively coupled plasma enhanced chemical vapor deposition. *Japanese Journal of Applied Physics, 51 (to be published)*, 51.

[22] Nemanich, R. J., & Solin, S. A. (1979). First- and second-order Raman scattering from finite-size crystals of graphite. *Physical Review B*, 20(2), 392-401, PhysRevB.20.392.

[23] Kurita, S., Yoshimura, A., Kawamoto, H., Uchida, T., Kojima, K., Tachibana, M., Molina-Morales, P., & Nakai, H. (2005). Raman spectra of carbon nanowalls grown by plasma-enhanced chemical vapor deposition. *Journal of Applied Physics*, 97(10).

[24] Ferrari, A. C., Meyer, J. C., Scardaci, V., Casiraghi, C., Lazzeri, M., Mauri, F., Piscanec, S., Jiang, D., Novoselov, K. S., Roth, S., & Geim, A. K. (2006). Raman Spectrum of Graphene and Graphene Layers. *Physical Review Letters*, 97, PhysRevLett.97.187401.

[25] Ferrari, A. C. (2007). Raman spectroscopy of graphene and graphite: Disorder, electron-phonon coupling, doping and nonadiabatic effects. *Solid State Communications*, 143(1-2), 47-57, j.ssc.2007.03.052.

[26] Kawata, M., Nadahara, S., Shinozawa, J., Watanabe, M., & Katoda, T. (1990). Characterization of stress in doped and undoped polycrystalline silicon before and after annealing or oxidation with laser raman spectroscopy. *Journal of Electronic Materials*, 19(5), 407-411, BF02657998.

[27] Kondo, S., Kawai, S., Takeuchi, W., Yamakawa, K., Den, S., Kano, H., Hiramatsu, M., & Hori, M. (2009). Initial growth process of carbon nanowalls synthesized by radical injection plasma-enhanced chemical vapor deposition. *Journal of Applied Physics*, 106(9), 10.1063/1.3253734.

[28] Kawai, S., Kondo, S., Takeuchi, W., Kondo, H., Hiramatsu, M., & Hori, M. (2010). Optical Properties of Evolutionary Grown Layers of Carbon Nanowalls Analyzed by Spectroscopic Ellipsometry. *Japanese Journal of Applied Physics*, 49, JJAP.49.060220.

[29] Kondo, S., Kondo, H., Hiramatsu, M., Sekine, M., & Hori, M. (2010). Critical Factors for Nucleation and Vertical Growth of Two Dimensional Nano-Graphene Sheets Employing a Novel Ar+ Beam with Hydrogen and Fluorocarbon Radical Injection. *Applied Physics Express*, 3, APEX.3.045102.

[30] Zhao, X., Outlaw, R. A., Wang, J. J., Zhu, M. Y., Smith, G. D., & Holloway, B. C. (2006). Thermal desorption of hydrogen from carbon nanosheets. *Journal of Chemical Physics*, 124(19).

[31] Zhu, M., Wang, J., Holloway, B. C., Outlaw, R. A., Zhao, X., Hou, K., Shutthanandan, V., & Manos, D. M. (2007). A mechanism for carbon nanosheet formation. *Carbon*, 45(11), 2229-2234, j.carbon.2007.06.017.

[32] Seo, D. H., Kumar, S., & Ostrikov, K. (2011). Control of morphology and electrical properties of self-organized graphenes in a plasma. *Carbon*, 49(13), 4331-4339, j.carbon.2011.06.004.

[33] Krivchenko, V. A., Dvorkin, V. V., Dzbanovsky, N. N., Timofeyev, M. A., Stepanov, A. S., Rakhimov, A. T., Suetin, N. V., Vilkov, O. Y., & Yashina, L. V. (2012). Evolution of carbon film structure during its catalyst-free growth in the plasma of direct current glow discharge. *Carbon*, 50(4), 1477-1487, j.carbon.2011.11.018.

[34] Ushizawa, K., Sato, Y., Mitsumori, T., Machinami, T., Ueda, T., & Ando, T. (2002). Covalent immobilization of DNA on diamond and its verification by diffuse reflectance infrared spectroscopy. *Chemical Physics Letters*, 351(1-2), 105-108, S0009-2614(01)01362-8.

[35] Yang, W., Auciello, O., Butler, J. E., Cai, W., Carlisle, J. A., Gerbi, J. E., Gruen, D. M., Knickerbocker, T., Lasseter, T. L., Russell, J. N., Smith, L. M., & Hamers, R. J. (2002). DNA-modified nanocrystalline diamond thin-films as stable, biologically active substrates. *Nature Materials*, 1(4), 253-257, nmat779.

[36] Wenmackers, S., Haenen, K., Nesladek, M., Wagner, P., Michiels, L., van de Ven, M, & Ameloot, M. (2003). Covalent immobilization of DNA on CVD diamond films. *Physica Status Solidi (a)*, 199(1), 44-48, pssa.200303822.

[37] Takahashi, K., Tanga, M., Takai, O., & Okamura, H. (2003). DNA preservation using diamond chips. *Diamond and Related Materials*, 12(3-7), 572-576, S0925-9635(03)00070-0.

[38] Wu, Y. H., Yang, B. J., Zong, B. Y., Sun, H., Shen, Z. X., & Feng, Y. P. (2004). Carbon nanowalls and related materials. *Journal of Materials Chemistry*, 14(4), 469-477, B311682D.

[39] Yang, B. J., Wu, Y. H., Zong, B. Y., & Shen, Z. X. (2002). Electrochemical synthesis and characterization of magnetic nanoparticles on carbon nanowall templates. *Nano Letters*, 2(7), 751-754, nl025572r.

[40] Wu, Y. H., Yang, B. J., Han, G. C., Zong, B. Y., Ni, H. Q., Luo, P., Chong, T. C., Low, T. S., & Shen, Z. X. (2002). Fabrication of a class of nanostructured materials using carbon nanowalls as the templates. *Advanced Functional Materials*, 12(8), 489-494, AID-ADFM489>3.0.CO;2-X.

[41] Huang, J. E., Guo, D. J., Yao, Y. G., & Li, H. L. (2005). High dispersion and electrocatalytic properties of platinum nanoparticles on surface-oxidized single- walled carbon nanotubes. *Journal of Electroanalytical Chemistry*, 577(1), 93-97, j.jelechem. 2004.11.019.

[42] Mu, Y., Liang, H., Hu, J., Jiang, L., & Wan, L. (2005). Controllable Pt Nanoparticle Deposition on Carbon Nanotubes as an Anode Catalyst for Direct Methanol Fuel Cells. *Journal of Physical Chemistry B*, 109(47), 22212-22216, jp0555448.

[43] Rabat, H., Andreazza, C., Brault, P., Caillard, A., Béguin, F., Charles, C., & Boswell, R. (2009). Carbon/platinum nanotextured films produced by plasma sputtering. *Carbon*, 47(1), 209-214, j.carbon.2008.09.051.

[44] Machino, T., Takeuchi, W., Kano, H., Hiramatsu, M., & Hori, M. (2009). Synthesis of platinum nanoparticles on two-dimensional carbon nanostructures with an ultrahigh aspect ratio employing supercritical fluid chemical vapor deposition process. *Applied Physics Express*, 2, APEX.2.025001.

[45] Hiramatsu, M., & Hori, M. (2010). Preparation of Dispersed Platinum Nanoparticles on a Carbon Nanostructured Surface Using Supercritical Fluid Chemical Deposition. *Materials*, 3(3), 1559-1572, ma3031559.

[46] Watkins, J. J., & Mc Carthy, T. J. (1995). Polymer/Metal Nanocomposite Synthesis in Supercritical CO_2. *Chemistry of Materials*, 7(11), 1991-1994, cm00059a001.

[47] Saquing, C. D., Kang, D., Aindow, M., & Erkey, C. (2005). Investigation of the supercritical deposition of platinum nanoparticles into carbon aerogels. *Microporous and Mesoporous Materials*, 80(1-3), 11-23, j.micromeso.2004.11.019.

[48] Zhang, Y., Kang, D., Saquing, C. D., Aindow, M., & Erkey, C. (2005). Supported Platinum Nanoparticles by Supercritical Deposition. *Industrial & Engineering Chemistry Research*, 44(11), 4161-4164, ie050345w.

[49] Zhang, Y., & Erkey, C. (2006). Preparation of supported metallic nanoparticles using supercritical fluids: A review. *The Journal of Supercritical Fluids*, 38(2), 252-267, j.supflu.2006.03.021.

[50] Bayrakceken, A., Kitkamthorn, U., Aindow, M., & Erkey, C. (2007). Decoration of multi-wall carbon nanotubes with platinum nanoparticles using supercritical deposition with thermodynamic control of metal loading. *Scripta Materialia*, 56(2), 101-103, 10.1016/j.scriptamat.2006.09.019.

[51] Erkey, C. (2009). Preparation of metallic supported nanoparticles and films using supercritical fluid deposition. *Journal of Supercritical Fluids*, 47(3), 517-522, j.supflu.2008.10.019.

[52] Hiramatsu, M., Machino, T., Mase, K., Hori, M., & Kano, H. (2010). Preparation of Platinum Nanoparticles on Carbon Nanostructures Using Metal-Organic Chemical Fluid Deposition Employing Supercritical Carbon Dioxide. *Journal of Nanoscience and Nanotechnology*, 10(6), 4023-4029, jnn.2010.1996.

[53] Pitchon, V., & Fritz, A. (1999). The relation between surface state and reactivity in the $DeNO_X$ mechanism on platinum-based catalysts. *Journal of catalysis*, 186(1), 64-74, jcat.1999.2543.

[54] Kobayashi, K., Tanimura, M., Nakai, H., Yoshimura, A., Yoshimura, H., Kojima, K., & Tachibana, M. (2007). Nanographite domains in carbon nanowalls. *Journal of Applied Physics*, 101(9).

[55] Mase, K., Kondo, H., Kondo, S., Hori, M., Hiramatsu, M., & Kano, H. (2011). Formation and mechanism of ultrahigh density platinum nanoparticles on vertically grown graphene sheets by metal-organic chemical supercritical fluid deposition. *Applied Physics Letters*, 98(19).

[56] Zhang, X., & Shen, Z. (2002). Carbon fiber paper for fuel cell electrode. *Fuel*, 81(17), 2199-2201, S0016-2361(02)00166-7.

[57] Wang, L., Husar, A., Zhou, T., & Liu, H. (2003, International Journal of Hydrogen Energy). A parametric study of PEM fuel cell performances. 28(11), 1263-1272, DOI:S0360-3199(02)00284-7.

[58] Ge, J., Higier, A., & Liu, H. (2006). Effect of gas diffusion layer compression on PEM fuel cell performance. *Journal of Power Sources*, 159(2), 922-927, j.jpowsour.2005.11.069.

[59] Waje, M. M., Wang, X., Li, W., & Yan, Y. (2005). Deposition of platinum nanoparticles on organic functionalized carbon nanotubes grown in situ on carbon paper for fuel cells. *Nanotechnology*, 16, S395-S400.

[60] Lisi, N., Giorgi, R., Re, M., Dikonimos, T., Giorgi, L., Salernitano, E., Gagliardi, S., & Tatti, F. (2011). Carbon nanowall growth on carbon paper by hot filament chemical

vapour deposition and its microstructure. *Carbon,* 49(6), 2134-2140, j.carbon. 2011.01.056.

[61] Shin, S. C., Yoshimura, A., Matsuo, T., Mori, M., Tanimura, M., Ishihara, A., Ota, K., & Tachibana, M. (2011). Carbon nanowalls as platinum support for fuel cells. *Journal of Applied Physics,* 110(10).

[62] Zhou, M., Zhai, , & Dong, S. (2009). Electrochemical Sensing and Biosensing Platform Based on Chemically Reduced Graphene Oxide. *Analytical Chemistry,* 81(14), 5603-5613, ac900136z.

[63] Pumera, M., Ambrosi, A., Bonanni, A., Chng, E. L. K., & Poh, H. L. (2010). Graphene for electrochemical sensing and biosensing. *Trends in Analytical Chemistry,* 29(9), 954-965, j.trac.2010.05.011.

[64] Hill, E. W. (2011). Graphene Sensors. *IEEE Sensors Journal,* 11(12), 3161-3170, JSEN. 2011.2167608.

[65] Fujishima, A., Einaga, Y., Rao, T. N., & Tryn, D. A. (2004). *Diamond electrochemistry,* BKC Inc./ Elsevier BV, Tokyo/Amsterdam, 28.